CLIMATE CHANGE

What the Science Tells Us

CLIMATE CHANGE

What the Science Tells Us

Charles Fletcher

University of Hawai'i

WILEY

The climax of every tragedy lies in the deafness of its heroes.
— Albert Camus, The Rebel

VP & Executive Publisher:	Jay O'Callaghan
Executive Editor:	Ryan Flahive
Editorial Assistant:	Julia Nollen
Marketing Manager:	Margaret Barrett
Designer:	Jasmine Lee
Associate Production Manager:	Joyce Poh
Cover Photograph:	Jim Kruger/E+/Getty Images, Inc.

Cover: Sunrise over the Pacific Ocean. This photo reveals several major elements of the climate system: the rising Sun, exerting the ultimate control on Earth's climate; clouds, somewhat enigmatic agents of both cooling and warming; the vast heat reservoir of the oceans, responsible for storing more than 90% of the excess heat now in the climate system; the troposphere, the lowest layer of air that has warmed 0.8°C (1.4°F) over the past 130 years; and early morning footprints, reminders of human impact on the planet.

This book was set by MPS Ltd, Macmillan Company. Cover and text printed and bound by RR Donnelley.

This book is printed on acid free paper.

Founded in 1807, John Wiley & Sons, Inc. has been a valued source of knowledge and understanding for more than 200 years, helping people around the world meet their needs and fulfill their aspirations. Our company is built on a foundation of principles that include responsibility to the communities we serve and where we live and work. In 2008, we launched a Corporate Citizenship Initiative, a global effort to address the environmental, social, economic, and ethical challenges we face in our business. Among the issues we are addressing are carbon impact, paper specifications and procurement, ethical conduct within our business and among our vendors, and community and charitable support. For more information, please visit our website: www.wiley.com/go/citizenship.

Library of Congress Cataloging-in-Publication Data

Fletcher, Charles.
  Climate change : what the science tells us / Charles Fletcher. — 1st ed.
      p. cm.
  Includes index.
  ISBN 978-1-118-05753-7 (pbk.)
  1. Climatic changes.   2. Global warming. I.   Title.
  QC903.F64 2013
  551.6—dc23
                              2012028592

Printed in the United States of America

10 9 8 7 6 5 4 3 2 1

To Ruth and our children and grandchildren.

BRIEF CONTENTS

TABLE OF CONTENTS

WHAT WE KNOW

Earth's climate has always changed. Modern climate change does not, however, fit geologic history: In the past half century, the rate and extent of climate change has been extraordinary. Despite extensive searching, no known natural processes can account for the present climate trend of extremely rapid warming of the temperature of the lower atmosphere. Furthermore, industrial exhaust, deforestation, and large-scale agribusiness are known producers of heat-trapping gas in the atmosphere. It is only logical to conclude that there is a strong likelihood that these human activities are causing the extraordinary warming. Modern climate change is a consequence of human-caused global warming; in fact, among scientists, this has been known for decades.[1]

Every professional scientific organization in the United States and globally has arrived at this same conclusion. For instance, the Union of Concerned Scientists,[2] representing more than 250,000 U.S. citizens and scientists, states:

> *The Earth is warming and human activity is the primary cause.*
> *Climate disruptions put our food and water supply at risk,*
> *endanger our health, jeopardize our national security, and threaten*
> *other basic human needs. Some impacts—such as record high*
> *temperatures, melting glaciers, and severe flooding and droughts—*
> *are already becoming increasingly common across the country*
> *and around the world. So far, our national leaders are failing to act*
> *quickly to reduce heat-trapping emissions.*[3]

However, this understanding of climate change is not popular among the American public, and there are many skeptics of global warming who do not form their opinions using critical thinking. In response, I have written *Climate Change: What the Science Tells Us,* a concise, comprehendible presentation of the most recent research that focuses on the causes and effects of climate change. The book is produced in the hope that it will help learners understand why and how scientists have come to this conclusion.

WHAT WE WANT TO KNOW

Sometimes we stop listening. Social scientists have revealed that when a person is confronted with a seemingly unstoppable tide of bad news, they might simply stop listening. Thus, there is a possibility that when reading the overwhelming information about climate change presented here, one might just turn away. Don't. This is a chance to build your expertise, to learn, and thereby to change your world.

The classroom is a fertile environment in which to pursue this goal. I expect this text to be put to effective use in the classroom, especially where it might accompany

---

[1] See the film clip at the end of this foreword, "Global Warming: What We Knew in 82."

[2] See the Union of Concerned Scientists website, http://www.ucsusa.org/global_warming/ (accessed July 15, 2012).

[3] Quote from the Union of Concerned Scientists Web page on global warming, http://www.ucsusa.org/global_ warming/ (accessed July 15, 2012).

the delivery of more-traditional content such as biological or physical science, humanities, law, architecture, and a plethora of other disciplines that all touch on the relevance of climate change and human society. I also hope this book finds a home among the public, startled at the heat waves and storms sweeping the Northern Hemisphere, and wanting to learn more about how their planet is changing.

Like most scientists, I found attributing modern climate changes to human activities was not a stretch; all forms of life change their environment, and humans are no exception. In fact, the human population is now of such size (numbering 7 billion as of October 27, 2011, Halloween Eve) and technological sophistication that change is evident in many global systems. For instance, humans affect the extent and health of the global forest, global fish stocks, global water quality and availability, global sedimentation, global river discharge (an estimated 36,000 dams interrupt the flow of nearly all of Earth's river systems), and others. Accepting that climate is affected by human activities actually makes sense because it accounts most elegantly for the global phenomena scientists have observed over the past half century (and, it has been suggested, much longer).

Because the cause of modern climate change is largely industrial, when we talk about the kinds of measures we could take to protect ourselves and our children from its worst effects, the discussion inevitably turns to jobs, taxes, government policies, and human livelihoods. Unfortunately, when the discourse veers down these paths, the bright line around the science of climate change is blurred by political opinion, personal world view, and individual beliefs. In fact, climate has even become political dogma[4]: at present, some assume that if you vote Democratic you accept climate theory, if you vote Republican you do not. Climate change can also have religious connotations: "It is the height of human arrogance to think that we could control God's creation" is an opinion I have heard more than once.

Climate change enters the discussion of what to teach our children, what is polite conversation, what kind of car to buy, the design of our buildings and cities, our source of electricity, and more. There are many examples of what is now known as the "climate debate." The irony? That among mainstream scientists, there is no climate debate. To paraphrase the National Research Council in their 2011 report *America's Climate Choices*,[5] it is "settled fact" that the climate system is warming and much of this warming is very likely due to human activities. Why use the phrase "very likely"? Because volcanic eruptions, the El Niño southern oscillation, and variations in the Sun's energy also affect global temperature. But these have been intensely studied, and research indicates their influence on global climate has been to cool down Earth but the global warming trend has been strong enough to overpower them.[6]

WHERE DOES THE SCIENCE END AND WHERE DO THE OPINIONS BEGIN?

As a life-long science educator, I wanted to produce a book that helps my students, the public, and elected officials understand the scientific thinking on the topic of climate change. There are many climate science books, and they mostly do a good job of summarizing the state of knowledge. This book, however, gives you direct access to the science. The science behind statements made in this book is referenced at the bottom of the page, not at the back of the book (or in the case of many books, not

[4] "No Green Tea: What Americans Think about Climate Change, by Political Allegiance," *The Economist,* 2011, http://www.economist.com/blogs/dailychart/2011/09/american-public-opinion-and-climate-change (accessed July 15, 2012).

[5] National Research Council, *America's Climate Choices: Panel on Advancing the Science of Climate Change* (Washington, D.C., National Academies Press, 2010), pp. 21–22, http://www.nap.edu/catalog.php?record_id=12782 (accessed July 15, 2012).

[6] G. Foster and S. Rahmstorf, "Global Temperature Evolution 1979–2010," *Environmental Research Letters* 6 (2011): 044022, doi:10.1088/1748-9326/6/4/044022.

at all). This simple difference, I believe, makes the text easier to read, the material more accessible, the content more credible, and the learning process ultimately more effective.

More than other texts, *Climate Change* exposes the general public, decision makers, and students to the processes of peer-reviewed scientific publishing, and connects published science papers to current events. This shows that even the boldest statements of climate scientists are backed up by the scientific system of skeptical peer review. Skeptical peer review is the process scientists use to filter strongly developed research from weak research. The process of peer review invokes critical thinking by competitive, judgmental scientists to gauge the appropriateness of research results to be published for widespread reading.

In the classroom, this text can stand alone as the backbone of a semester-long class, or it can accompany any curriculum that touches on Earth processes where the instructor wants students to delve deeper into climate change. Its content will augment many classes, including geomorphology, climatology, historical and physical geology, meteorology, earth science, oceanography and marine science, environmental science, planning, civil engineering, environmental law, American studies, political science, sociology, and many others.

Today's scientists know that if strong action to counteract climate change is not successfully achieved, within one generation the world will be a place characterized by intense heat waves, widespread disease, drought, food shortages, and deadly super storms. The beginning signs of these disasters are already evident.

Unfortunately, because scientists have done an inadequate job of sharing this knowledge, non-critical thinking rooted in politics, religion, and other sources has given rise to a loud climate-denier voice (one who denies the existence of global warming). In my opinion, this voice was created and is perpetuated by some of the media in its need to sell controversy. When the media permit people with questionable credentials to challenge scientific findings, or when media personalities are allowed to question published research that has gone through the system of peer review, they are not offering two sides of a debate, they are creating a false controversy contrived to sell headlines.

ENGAGING LEARNERS

I believe that climate education should be a purposeful effort among scientists. In addition to contributing to published, peer-reviewed scientific literature and accepting invitations to speak to audiences, we researchers must actively create opportunities to pass on our knowledge. A readable, easy-to-understand discussion of global warming and its impacts, *Climate Change* is an attempt to communicate that knowledge. To make the material user-friendly, the text includes the following features:

- **Each chapter title poses a question** designed to parallel the kinds of questions scientists have asked and to provide the reader with a framework for the evidence.

- **Learning Objectives** that open each chapter orient the learner toward fundamental core concepts that I hope will result in lasting knowledge[7].

- **Chapter Summary** provides a brief answer to the chapter title question, followed by a bulleted list that summarizes the content of the chapter.

- **Footnotes** at the bottom of each page allow readers to easily trace statements back to their source. These footnotes are also a handy source for further research and list dozens of websites for additional learning.

- **Illustrations** are designed to show observations and model results of the impacts of global warming and as such are a key learning feature.

---

[7] G. Wiggins and J. McTighe, *Understanding by Design* (Alexandria, Va., Association for Supervision and Curriculum Development, 2005).

- **Links to Animations and Videos** are listed at the back of each chapter. These offer access to online resources and videos of climate discussions as well as the work of research organizations and prominent scientists.
- **Comprehension Check** is a list of 10 questions at the back of each chapter that allow students to verify their understanding of key terms and concepts.
- **Thinking Critically** is a list of 10 thought-provoking questions that encourage readers to go a step beyond the chapter material, make connections to previously presented material, and apply the content to real-world situations and their own lives.
- **Activities** at the back of each chapter use videos, Web links, and other visually stimulating resources from credible sources. These activities extend content beyond the text and bring it to life, motivating students to gain a deeper understanding of the topics presented.

OUTLINE

The book is organized to build the reader's understanding of climate change with every turn of the page. Knowing that some readers will keep the book close at hand as a reference, I've designed Chapter 1 to be an introduction to the basic concepts of the atmosphere and ocean, the greenhouse effect, the concept of radiative forcing, and the carbon cycle. Chapter 2 is a short but powerful summary of the evidence that global warming is changing Earth's climate and humans are the primary cause. Chapter 3 is a detailed review of geologic changes in climate and answers some basic questions about the cause of global warming. Chapter 4 introduces climate modeling and reviews the critical natural processes (e.g., volcanism, El Niño Southern Oscillation, solar variability, clouds) that need to be accurately depicted in models. Chapter 5 discusses sea-level rise, and Chapter 6 presents a review of climate impacts in North America. The text ends with Chapter 7, which touches on recent topics of climate research such as Arctic amplification, severe weather, drought, ecosystem impacts, and others.

Chapter 1: *What is the greenhouse effect and how is it being altered by human activities?* The term greenhouse effect describes the role of certain atmospheric gases (such as carbon dioxide, water vapor, methane, and others) in trapping heat that radiates from Earth's surface after it has been heated by the Sun. The term greenhouse effect compares these atmospheric gases to the glass panels of a greenhouse, which lets sunlight in, isolates warm air, and impedes the loss of heat. Although the greenhouse effect is a natural and beneficial process, it has gotten a bad name because greenhouse gases, especially carbon dioxide, are increasing as a result of human activities, such as fossil fuel burning, deforestation, and industrialization, which are responsible for global warming.

Chapter 2: *What is the evidence for climate change?* Climate change is a result of global warming, a genuine phenomenon about which there is little debate within the scientific community. Rather, scientists debate the questions "How sensitive is climate to greenhouse gas buildup?" and "What will climate change look like regionally and locally?" There is abundant, convincing, and reproducible scientific evidence that the increase in Earth's surface temperature is having measurable impacts on human communities and natural environments: Glaciers are melting, spring is coming earlier, the tropics are expanding, sea level is rising, the global water cycle is amplified, ecosystems are shifting, global wind speed has increased, drought and extreme weather are more common. These and many other observations document that the Earth system is rapidly changing in response to global warming.

Chapter 3: *How do we know that humans are the primary cause of climate change?* Climate change has been a natural process throughout geologic history. But modern global warming is not the product of the Sun, natural cycles, or bad data. Every imaginable test has been applied to the hypothesis that humans are causing global warming. The simplest, most objective explanation for the many independent lines of clear, factual evidence is that humans are the primary drivers of climate change.

Chapter 4: *How do scientists project future climate?* Climate models successfully reproduce the past 100 years of climate change, but only when greenhouse gases, produced by human activities, are included. Models published by the International Panel on Climate Change use a range of potential future scenarios of greenhouse gas emissions to predict that surface air warming in the 21st century will likely (better than 66% probability) range from a low of 1.1°C to a high of 6.4°C (2.0°F to 11.5°F). Climate models provide important results for understanding future global climate, but their ability to project regional and localized climate is still limited.

Chapter 5: *What is the reality of sea-level rise?* Today, rising seas threaten coastal wetlands, estuaries, islands, beaches, reefs, and all types of coastal environments. Human communities living on the coast are subject to flooding by rainstorms that are coincident with high tides, accelerated coastal erosion, and saltwater intrusion into streams and aquifers. Sea-level rise threatens cities, ports, and other areas with passive flooding due to rising waters and with damaging inundation that will increase in magnitude when hurricanes and tsunamis strike. Because sea-level rise has enormous economic and environmental consequences, it is important to understand how global warming is creating this threat.

Chapter 6: *How does global warming affect our community?* Climate change impacts to human communities include: stresses to water resources, threats to human health, shifting demand on energy supply, disruptions to transportation and agriculture, and increased vulnerability of society and ecosystems to future climate change. In the United States, extreme weather events have increased in number and magnitude and are likely to do so in the future. Severe heat waves and record-setting temperatures are occurring with greater frequency. Among other impacts are the spread of diseases not historically prevalent in North America, retreat of tundra and northern and arctic ecosystems, increased occurrence of drought and flooding, sea-level rise, decreased snow pack and retreating glaciers, changes in the timing of seasons, and ecological impacts, among others.

Chapter 7: *What is the latest word on climate change?* It is useful to review the latest evidence from the scientific realm confirming that global warming and climate change are still actively changing the planet we call home. This last chapter provides a review of some of the important climate issues we have touched on: climate change confirmed, a new record in global emissions, warming the high latitudes (Arctic and Antarctic), extreme weather, drought, dangerous climate, ecosystem impacts, and climate sensitivity.

(NOT THE) FINAL WORD

National polls[8] have revealed that the number of Americans who said that they were "extremely sure" that global warming was happening slid from 35% in November of 2008, to only 22% in November of 2011. Simultaneously, among American global

---

[8] D.L. Wheeler, "Inside the Clash over Climate Change," *The Chronicle of Higher Education*, May 11, 2012, citing data from the Yale Project on Climate Change Communication: http://environment.yale.edu/climate/ (accessed July 15, 2012).

warming deniers, those who said they were extremely or very certain of their views rose from 35% in 2010 to 53% in 2011. Also alarming is the statistic that 65% of Americans say they have never heard of the United Nations Intergovernmental Panel on Climate Change, a key source of climate change information for scientists and media alike.

Climate Change embodies a hope that rather than leaving readers feeling paralyzed by the magnitude of the problem, increased knowledge will provide you with the confidence to ask politicians and other decision makers for action to address the impacts. And if this climate knowledge is applied in readers' personal decision making, such as voting, then the book has achieved an important purpose.

VIDEO

"Global Warming: What We Knew in 82"
http://www.youtube.com/watch?v=OmpiuuBy-4s&list=UU-KTrAqt2784gL_I4JisF1w&index=1&feature=plcp (accessed July 15, 2012)

WHAT IS THE GREENHOUSE EFFECT AND HOW IS IT BEING ALTERED BY HUMAN ACTIVITIES?

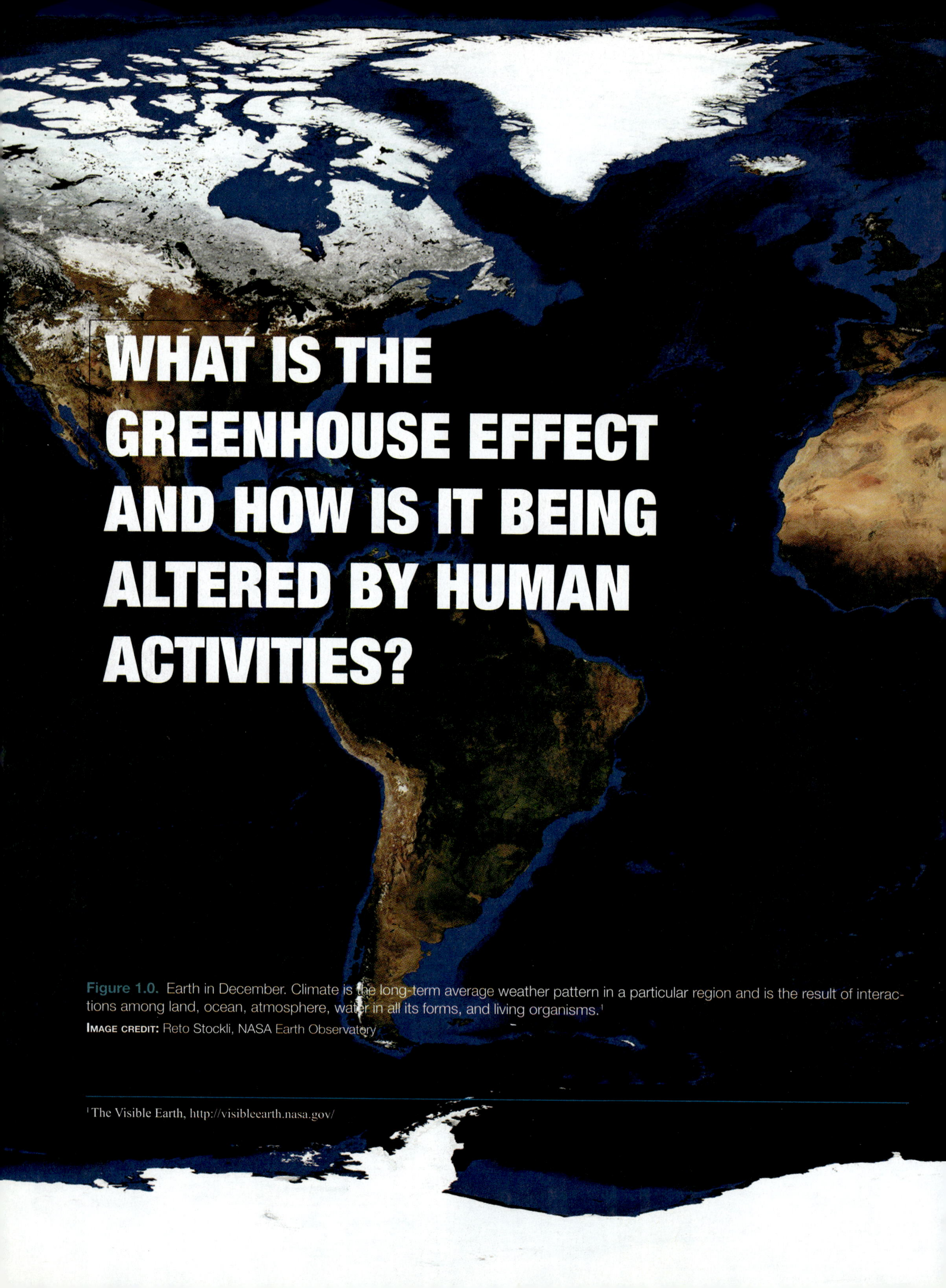

Figure 1.0. Earth in December. Climate is the long-term average weather pattern in a particular region and is the result of interactions among land, ocean, atmosphere, water in all its forms, and living organisms.[1]

IMAGE CREDIT: Reto Stockli, NASA Earth Observatory

[1] The Visible Earth, http://visibleearth.nasa.gov/

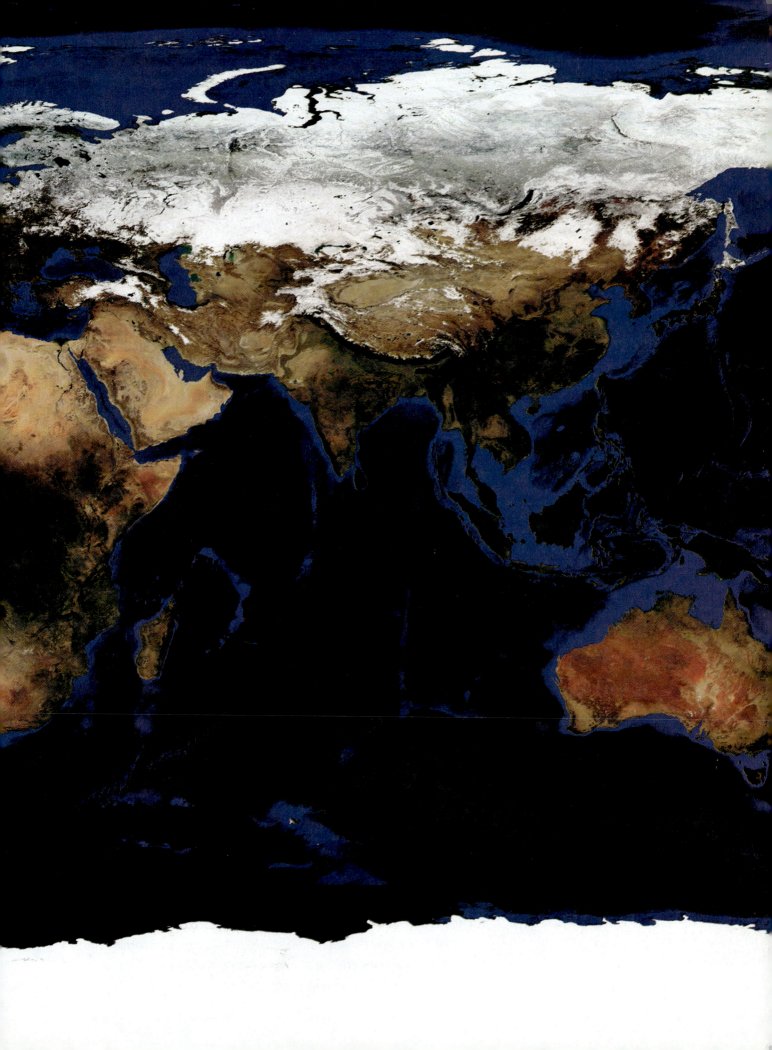

CHAPTER

1

CHAPTER SUMMARY

The term greenhouse effect describes the role of certain atmospheric gases (such as carbon dioxide, water vapor, methane, and others) in trapping heat that radiates from Earth's surface after it has been heated by the Sun. The term greenhouse effect compares these atmospheric gases to the glass panels of a greenhouse, which lets sunlight in, isolates warm air, and impedes the loss of heat. Although the greenhouse effect is a natural and beneficial process, it has gotten a bad name because greenhouse gases, especially carbon dioxide and methane, are increasing as a result of human activities and causing global warming. Human activities that magnify the greenhouse effect and cause global warming include burning petroleum (gasoline and diesel fuel) for transportation, industrialized agriculture (a major source of methane), burning household biofuels (wood and dung), and deforestation (to clear land for agriculture), all of which release heat-trapping gases into the atmosphere.

In this chapter you will learn that:

- Weather is the short-term state of the atmosphere at a given location. It affects the well-being of humans, plants, and animals and the quality of our food and water supply.

- Climate is the long-term average weather pattern in a particular region and is the result of interactions among land, ocean, atmosphere, water in its many forms and living organisms, together known as the climate system.

- The general circulation of the atmosphere is a system of winds that transport heat from the equator, where solar heating is greatest, toward the cooler poles. This pattern gives rise to Earth's climate zones.

- The oceans influence the weather and climate. Ocean water moderates air temperatures by absorbing heat from the Sun and transporting that heat toward the poles as well as down toward the seafloor.

- The overall outlook for the global ocean is not healthy. Warming, acidification, and anoxia have been identified as the "deadly trio" that threatens mass extinctions in the marine ecosystem.

- The greenhouse effect is a natural process by which heat radiated from Earth's surface is trapped by gases (called greenhouse gases), such as water vapor (H_2O), carbon dioxide (CO_2), nitrous oxide (N_2O), ozone (O_3), methane (CH_4), and others. When stable, this process maintains Earth's average surface temperature at a life-sustaining 14°C (57.2°F).

Learning Objective

The greenhouse effect is a natural phenomenon that regulates the temperature of the lowest layer of the atmosphere, known as the troposphere. Human activities have enhanced the greenhouse effect, leading to global warming and climate change.

- Directly or indirectly over the past 200 years, human activities involving fossil fuel consumption and land-use changes have increased all of the greenhouse gases, leading to an increase in Earth's average surface temperature of approximately 0.8°C (1.4°F). As greenhouse gases accumulate in the atmosphere, the amount of heat they trap also increases.

- Once in the atmosphere, carbon dioxide causes climate change that is essentially irreversible for the next 1,000 years.

- A 1°C (1.8°F) change in atmospheric temperature caused by CO_2 will stimulate a water vapor increase causing the temperature to go up another 1°C (1.8°F). This is an example of a climate process called positive feedback.

- Some aerosols (particles and droplets) in the atmosphere and some clouds that scatter sunlight offset warming to some degree, although not all aerosols and not all clouds scatter sunlight.

- Today, and for the next decade or so, cars, trucks, and buses are the greatest contributors to atmospheric warming.

- Burning coal, oil, and natural gas instantly releases carbon that took millions of years to accumulate in Earth's crust. Over 34 billion tons of carbon dioxide are released into the atmosphere annually as a result of industrialization and deforestation[2] and has resulted in a disruption of the carbon cycle.

CLIMATE LITERACY

Established in 1989 under the Executive Office of the President, the U.S. Global Change Research Program (USGCRP[3]) coordinates and integrates the climate change activities of 13 federal departments and agencies. The program is a ready source of peer-reviewed summaries on the subject of climate change and its impacts in the United States and the world.

The USGCRP has produced a short guide for educators to promote climate literacy among individuals and communities: The Climate Literacy Guide.[4] This guide provides a summary of essential principles underlying how Earth's climate system works and how climate change is occurring. The guide lists seven principles:

---

[2] A.P. Ballantyne et al., "Increase in Observed Net Carbon Dioxide Uptake by Land and Oceans During the Past 50 years," *Nature*, 488, no. 7409 (2012); see also the CO2 Now website that tracks carbon emissions http://co2now.org/ (accessed July, 13, 2012).

[3] See the USGCRP home page at http://www.globalchange.gov/ (accessed July 9, 2012).

[4] See USGCRP, "Climate Literacy: The Essential Principles of Climate Science," http://www.globalchange.gov/resources/educators/climate-literacy (accessed July 9, 2012).

1. The Sun is the primary source of energy for Earth's climate system.
2. Climate is regulated by complex interactions among components of the Earth system.
3. Life on Earth depends on, is shaped by, and affects climate.
4. Climate varies over space and time through both natural and human-made processes.
5. Our understanding of the climate system is improved through observations, theoretical studies, and modeling.
6. Human activities are affecting the climate system.
7. Climate change will have consequences for the Earth system and human lives.

This and following chapters expand on these principles.

WEATHER AND CLIMATE

Weather[5] is the short-term state of the atmosphere at a given location. It affects the well-being of humans, plants, and animals and the quality of our food and water supply. Weather is somewhat predictable because of our understanding of Earth's global climate patterns. For instance, in certain seasons we can expect precipitation events of either rain or snow, and these can be predicted a few days in advance by a combination of computer modeling and the modern technology of satellites and radar.[6]

Climate is the long-term average weather pattern in a particular region and is the result of interactions among land, ocean, atmosphere, water in all of its forms, and living organisms. Climate[7] is described by many weather elements, such as temperature, precipitation, humidity, sunshine, and wind. Both climate and weather result from processes that accumulate and move heat within and between the atmosphere and the oceans.

Heat

The key to understanding climate change is to follow the heat, because changes in the accumulation and movement of heat in the oceans and atmosphere result in changes to climate. To understand both natural and human influences on global climate, we must explore the physical processes that govern heat movement in the atmosphere and oceans.

Heat in Earth's climate system originates with sunlight that warms the land, oceans, and atmosphere. When Earth emits to space the same amount of energy as it absorbs, its energy budget is in balance, and its average temperature remains stable. Changes in the amount of heat coming from the Sun cause Earth to warm or cool. Satellite measurements taken over the past 30 years show that the Sun's output has changed only slightly and in both directions. Thus changes in the Sun's energy are thought to be too small to be the cause of the recent warming observed on Earth.[8]

---

[5] *Wikipedia* has a great "Weather" entry: http://en.wikipedia.org/wiki/Weather (accessed July 9, 2012).

[6] Radar (the use of pulses of radio waves to measure remote objects) is used in weather forecasting to identify various types of precipitation (rain, snow, hail, etc.). Weather radars can detect the motion of rain droplets in addition to the intensity of precipitation. This is used to characterize storms and their potential to cause severe weather.

[7] See "Climate," *Wikipedia*, http://en.wikipedia.org/wiki/Climate (accessed July 9, 2012).

[8] M. Lockwood, "Recent changes in solar outputs and the global mean surface temperature. III. Analysis of contributions to global mean air surface temperature rise," *Proceedings of the Royal Society* A 464 (2008): 1387–1404, doi:10.1098/rspa.2007.0348.

Energy Budget

Dr. James E. Hanson and co-researchers at NASA's Goddard Institute for Space Studies examined Earth's energy (heat) budget from 2005 to 2010.[9] Over this period the Sun entered a prolonged solar minimum that reduced the amount of energy reaching Earth's surface, yet the planet continued to absorb more energy than it returned to space. This energy imbalance underscores the fact that greenhouse gases generated by human fossil fuel burning—not changes in solar activity—are the primary force driving global warming. Hansen's team concluded that Earth has absorbed more than half a watt (W) more solar energy per square meter of Earth's surface than it let off throughout the study period. The calculated value of the imbalance (0.58 W of excess energy per square meter) is more than twice as much as the reduction in the amount of solar energy supplied to the planet between maximum and minimum solar activity (0.25 W per square meter). As a result of this energy imbalance, the researchers concluded that global warming has continued over the period and that as the Sun returns to normal levels of activity, sea-level rise and other environmental changes resulting from global warming will accelerate in the next decade.

LAYERS OF THE ATMOSPHERE ·

The atmosphere[10] is the envelope of gases that surround Earth, extending from its surface to an altitude of about 145 km (90 mi; Figure 1.1). Around the world, the composition of the atmosphere is similar, but when looked at in cross section, the atmosphere is not a uniform blanket of air. It can best be described as having four layers, each with distinct properties, such as temperature and chemical composition. The red line in Figure 1.1 shows how atmospheric temperature changes with altitude.

Thermosphere. The highest layer of the atmosphere, the thermosphere (also called the ionosphere), gradually merges with space. Temperatures increase with altitude in the thermosphere because it is heated by cosmic radiation from space.

Mesosphere. Below the thermosphere is the mesosphere, which extends to an altitude of about 80 km (50 miles). This layer grows cooler with increasing altitude.

Stratosphere. Below the mesosphere is the stratosphere, where the protective "ozone layer" absorbs much of the Sun's harmful ultraviolet radiation. This layer extends to an altitude of about 50 km (31 mi), it becomes hotter with increasing altitude, and it is vital to the survival of plants and animals on Earth because it blocks the intense solar radiation that damages living tissue.

Troposphere. In the layer nearest Earth, the troposphere (or weather zone), the air becomes colder with increasing altitude; you might have noticed this if you have ever hiked in the mountains. This layer extends to an altitude of about 8 km (5 mi) in the Polar Regions and up to nearly 17 km (10.5 mi) above the equator. It is also the layer where greenhouse gases are trapped and global warming occurs.

The boundaries between these layers are called pauses. For example, the tropopause is the boundary between the troposphere and the stratosphere.

---

[9] J. Hansen, M. Sato, P. Kharecha, and K. von Schuckmann, "Earth's Energy Imbalance and Implications," *Atmospheric Chemistry and Physics* 11 (2011), 13421–13449, doi: 10.5194/acp-11-13421-2011.

[10] See Earth's Atmosphere, *Wikipedia*, http://en.wikipedia.org/wiki/Earth%27s_atmosphere (accessed July 9, 2012).

Figure 1.1. Around the world, the composition of the atmosphere is similar, but when looked at in cross section, the atmosphere is not a uniform blanket of air. It has several layers, each with distinct properties, such as temperature and chemical composition. The red line shows how atmospheric temperature changes with altitude.

Source: Fletcher, *Physical Geology: The Science of Earth,* 2012.

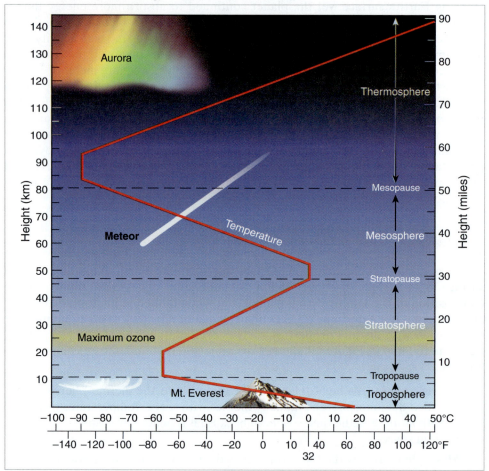

There is little vertical mixing of gases between layers of the atmosphere, and one layer can be warming while at the same time another is cooling. For example, global warming in the troposphere, the layer closest to Earth's surface, causes cooling in the stratosphere[11] because as more heat is trapped in the lower atmosphere, less heat reaches the upper atmosphere.[12] To an observer in space, Earth would appear to be cooling, but that is only true of the upper atmosphere.

GLOBAL CIRCULATION OF THE ATMOSPHERE

An essential component of climate, the atmosphere is the most rapidly changing and dynamic of Earth's physical systems, and it constantly interacts with Earth's other systems: the hydrosphere (water in all its forms), biosphere (living organisms), and lithosphere (rock, soil, and Earth's geology). In the troposphere, global winds circulate the air and interact with the ocean surface, mixing water vapor and

[11] Q. Fu, C. M. Johanson, S. G. Warren, and D. J. Seidel, "Contribution of Stratospheric Cooling to Satellite-Inferred Tropospheric Temperature Trends," *Nature* 429 (2004): 55–58.

[12] J. Laštovička, R. Akmaev, G. Beig, J. Bremer, and J. Emmert, "Global Change in the Upper Atmosphere," *Science* 314, no. 5803 (2006): 1253–1254, doi: 10.1126/science.1135134; see also B. D. Santer, T. M. L. Wigley, and K. E. Taylor, "The Reproducibility of Observational Estimates of Surface and Atmospheric Temperature Change," *Science* 334, no. 6060 (2011): 1232–1233.

Figure 1.2. The general circulation of the atmosphere is driven by heat from the Sun and rotation of the planet.

Source: Fletcher, *Physical Geology: The Science of Earth,* 2012.

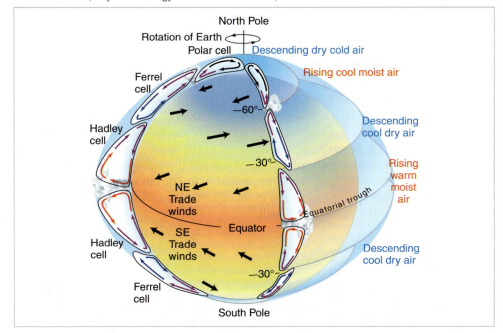

heat.[13] Close to Earth's surface, atmospheric circulation is so vigorous that air can travel around the world in less than a month.

Global circulation (Figure 1.2) is essentially driven by heat from the Sun and by the rotation of Earth. The worldwide system of winds that transport warm air from the equator, where solar heating is greatest, toward the cooler high latitudes is called the general circulation of the atmosphere. This pattern gives rise to Earth's climate zones.

Global atmospheric circulation is a primary factor determining variations in temperature, precipitation, surface winds, storminess and, hence, the weather and climate. The basic components of global atmospheric circulation are the Hadley cell, the Ferrel cell, and the Polar cell (see Figure 1.2). There is one of each cell type in the Northern Hemisphere and one of each in the Southern Hemisphere.

Atmospheric circulation starts with the basic principle that hot air rises and cool air sinks. Therefore, air heated by the Sun rises at the equator, where solar heating is greatest. As the air moves toward the poles, it cools and eventually sinks. Rising air causes low air pressure (at the equator), and sinking air causes high air pressure (at the poles). If Earth were perfectly still and smooth, we might have a single cell in each hemisphere where hot air rises at the equator, moves north or south toward the poles, and then sinks to ground level as it cools at the poles. This air would then flow back to the equator along the ground surface. We would see this pattern expressed in the Northern Hemisphere as a constant north wind and in the Southern Hemisphere as a constant south wind. Fortunately, however, Earth is neither still nor smooth. Earth spins on its axis, causing the changes of day and night, and large mountain ranges deflect the direction of surface winds. Life on Earth is much more interesting this way.

The Hadley Cell

By the time an air mass that has risen at the equator has traveled to about 30° latitude, it has cooled sufficiently to sink back to Earth's surface (forming an area of

---

[13] See the animation "Global Circulation of the Atmosphere" at the end of the chapter.

high pressure). When this air reaches the surface, it must flow away, and it moves back either toward the equator or toward the pole. The air that flows back to the equator is reheated and rises again to repeat the process. This completes the Hadley cell.

The Polar Cell and the Ferrel Cell

At the poles, cold, dense air descends, causing a high-pressure area. Air flows away from the high pressure and toward the equator. By the time this air nears 60° latitude, it begins to meet the air flowing poleward from the Hadley cell. When these two air masses meet, they have nowhere to go but up. As they rise, they cool and lose moisture, causing high precipitation. Once high in the atmosphere, they must head poleward, where they cool and sink again, or toward the equator, where they meet the flow heading poleward from the equator and sink. The circulatory cell sinking at the poles and rising at 60° latitude is the Polar cell, and the cell sinking at 30° latitude and rising at 60° latitude is the Ferrel cell.

The Coriolis Effect

In 1856, William Ferrel demonstrated that owing to the rotation of Earth, air and water currents moving distances of tens to hundreds of kilometers tend to be deflected to the right in the Northern Hemisphere and to the left in the Southern Hemisphere. This phenomenon is known as the Coriolis effect, named after the French scientist Gaspard-Gustave Coriolis (1792–1843), who described the transfer of energy in rotating systems. Because surface winds in a Hadley cell are moving south (in the Northern Hemisphere) when they are deflected to the right, they turn westward and are called the northeast trade winds. In the Southern Hemisphere, they turn left to become the southeast trade winds. The surface winds in the Northern Hemisphere's Ferrel cell are moving north, and when deflected right they become the mid-latitude westerlies. The surface winds in the northern Polar cell are heading south, and when deflected right they become the polar easterlies. Check a globe to convince yourself of these patterns and figure out what part of the global atmospheric circulation system you live in.

How Global Circulation Affects Climate

As air rises, it cools and expands. This is due to the increased distance from the warming effects of Earth's surface and the lower air pressure found at higher altitude. As a rising air mass cools and expands, so does the water vapor contained in it. As the water vapor cools and expands, more water condenses than evaporates, causing water droplets and then clouds to form. Continued condensation produces precipitation, which falls as rain or snow. Therefore, in areas where relatively warm moist air is rising, such as near the equator and around 60° latitude, there is ample precipitation in all seasons.

The opposite is also true: Air warms and contracts as it sinks closer to Earth's surface. This causes evaporation to exceed condensation. No clouds form in locations with lots of sinking air. These areas, such as at the poles and around 30° latitude, have few clouds and little precipitation, thus forming a great belt of arid climate (and deserts) that girdles the globe. Many of the world's deserts are clustered around 30°N and 30°S latitudes for this reason (Figure 1.3).

Atmospheric processes distribute heat, water vapor, and winds across the face of the planet, which in turn determines the level of precipitation, the character of the seasons, how cold or warm it is at various times of the year—in short, the climate. Oceans, because they carry heat from the tropics toward the poles, also play a significant role in regulating the climate.

Figure 1.3. Global climate is governed by the atmospheric circulation; the rising and falling air of circulation cells govern the movement of surface winds, water vapor, and aspects of the temperature.

SOURCE: Fletcher, *Physical Geology: The Science of Earth,* 2012.

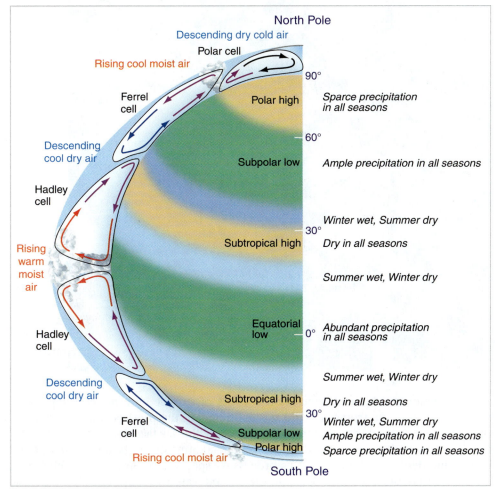

OCEAN CURRENTS CARRY HEAT

Climate change is the product of changes in the accumulation and movement of heat in the atmosphere and oceans. The oceans[14] influence many other natural systems on the planet, and they especially affect the weather and climate. Ocean water moderates surface temperatures by absorbing heat from the Sun and transporting that heat toward the poles as well as down toward the seafloor. Restless ocean currents[15] distribute this heat around the globe, warming the land and air during winter and cooling it in summer. The fact that cold water is denser and so is heavier than warm water also plays into the way ocean water circulates.

Ocean Circulation

There are basically two types of large-scale oceanic circulation: surface circulation, which is stimulated by winds and the Coriolis effect, and deep circulation, which is the result of cool water at the poles sinking and moving through the lower ocean. Both are driven by the exchange of heat.

[14] There are five oceans. They are the Atlantic, Pacific, Indian, Arctic, and Southern oceans. Smaller bodies, known as "seas," include the Mediterranean and China seas.

[15] Explore the science of oceanography at NASA, NASA Science Earth: NASA Oceanography, http://nasascience.nasa.gov/earth-science/oceanography/ (accessed July 9, 2012).

Figure 1.4. There are five major basin-wide gyres, each controlled by the interaction of winds and the Coriolis effect. Currents within the gyres carry heat from the equator toward the poles and thus strongly influence climate.

Source: Fletcher, *Physical Geology: The Science of Earth,* 2012.

The general pattern of circulation consists of surface currents carrying warm water away from the tropics toward the poles and, in the process, releasing heat to the atmosphere. Winter at the poles further cools this surface water. Once the surface water is cooler, it sinks to the deep ocean, creating currents along the seafloor and at mid depths in the ocean. This process is especially pronounced in the North Atlantic and in the Southern Ocean in the coastal waters of Antarctica, where cooling is the strongest. Deep ocean water gradually returns to the surface nearly everywhere in the ocean. Once at the surface, it is carried back to the tropics by surface currents, where it is warmed again and the cycle begins anew. The more efficient the cycle, the more heat is transferred from the tropics to the poles, and the more this heat warms the climate.[16]

SURFACE CURRENTS Owing to Earth's rotation (Coriolis effect), ocean currents are deflected to the right in the Northern Hemisphere and to the left in the Southern Hemisphere. In surface circulation, this process creates large-scale circulation systems called gyres that sweep the major ocean basins.

There are five major basin-wide gyres (Figure 1.4): the North Atlantic, South Atlantic, North Pacific, South Pacific, and Indian Ocean gyres. Each gyre is composed of a strong and narrow western boundary current and a weak and broad eastern boundary current. Each of the five major gyres in the oceans has parallel systems of currents, and these currents each carry heat and govern climate where they flow. The surface circulation of the North Pacific Gyre is a typical example of how winds and the Coriolis effect combine to create surface circulation.

In the North Pacific atmosphere, a descending column of dry air that originated at the equator (the northern end of the Hadley cell) blows toward the equator but is deflected to the west (right) by the Coriolis effect. This southwest-flowing wind is

---

[16] See the animation "Ocean Currents" at the end of the chapter.

known as the Pacific trade wind. The Pacific trade wind drives the North Equatorial Current to the west just north of the equator at about 15°N latitude. This current is deflected north near the Philippines to create the warm western boundary current known as the Japan or Kuroshio Current. The Kuroshio Current carries warm water away from the tropics until it turns to the east at approximately 45°N latitude and becomes the North Pacific Current, which moves across the basin toward North America.

As it approaches the North American continent, the North Pacific Current splits, sending one arm north to circulate through the Gulf of Alaska and the Bering Sea as the Alaska Current. The southern arm becomes the cool, slow-moving eastern boundary current called the California Current. The California Current moves from about 60°N to 15°N latitude and merges with the North Equatorial Current. From there it once again travels thousands of miles across the basin to Asia. Each of the five major gyres in the oceans has similar systems of currents, and these each carry heat and govern climate where they flow.

DEEP CIRCULATION   In the North Atlantic basin, the western boundary current is known as the Gulf Stream. The Gulf Stream carries warm tropical water from the Caribbean to the cold waters of the North Atlantic. As it moves, the Gulf Stream cools and evaporates, thus greatly increasing its density. By the time it arrives in the North Atlantic as a cold, salty body of water, it can no longer stay afloat and begins a long descent toward the seafloor of 2 km to 4 km (1 to 2.5 mi), where it becomes a deep current known as the North Atlantic Deep Water. The North Atlantic Deep Water travels south through the Atlantic and eventually joins similar deep water that is forming in the Southern Ocean. These waters then become the Circumpolar Deep Water, which journeys throughout the Southern Ocean. An arm of the Circumpolar Deep Water migrates into the North Pacific and there, after a voyage of approximately 35,000 km (22,000 mi), water that originated in the North Atlantic Gulf Stream eventually surfaces into the sunshine.

It has been estimated that up to 1,300 years can pass before the cycle is completed and water returns to its place of origin.[17] This thermohaline circulation (Figure 1.5), also called the oceanic conveyor belt,[18] travels through all the world's oceans.[19] The process connects all of Earth's oceans in a truly global system that transports both energy (heat) and matter (solids, dissolved compounds, and gases), and in doing so influences global climate.

GLOBAL WARMING IS CHANGING THE OCEAN

The ocean covers 70% of Earth and is the largest single component of the planet's surface. It exerts a vast influence on the climate because it contains a huge amount of heat energy. The top 2 m (6.5 ft) of ocean water carries as much heat as the entire overlying atmosphere.[20] When ocean currents carry this heat around the globe, the warmth provides the driving force behind the weather and climate.

The ocean influences the atmosphere, but the atmosphere also influences the ocean, in that excess heat from global warming is largely stored in the waters of the ocean. This heat drives up the sea surface temperature (and the temperature

---

[17] S. Lozier, "Deconstructing the Conveyor Belt," *Science* 328, no. 5985 (2010), 1507–1511, doi: 10.1126/science.1189250.

[18] See "Thermohaline Circulation," *Wikipedia*, http://en.wikipedia.org/wiki/Thermohaline_circulation (accessed July 9, 2012).

[19] A. Mann, "Ocean-Conveyor Belt Model Stirred Up," *Nature News*, September 12, 2010, http://www.nature.com/news/2010/100912/full/news.2010.461.html, doi:10.1038/news.2010.461 (accessed July 9, 2012).

[20] See NOAA, "Modeling Sea Surface Temperature," *ClimateWatch*, http://www.climatewatch.noaa.gov/image/2009/modeling-sea-surface-temperature (accessed July 9, 2012).

Figure 1.5. The thermohaline circulation is a global pattern of currents that carries heat, dissolved gas, and other compounds on a round trip that can take up to 1,300 years to complete.

Source: Fletcher, *Physical Geology: The Science of Earth,* 2012.

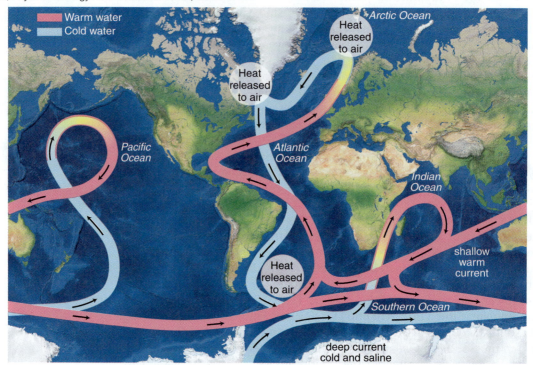

of the deep ocean), which further influences climate, and also influences critical aspects of the marine environment, such as the biology, chemistry, and physical attributes of the sea.

Warming Oceans

Because of global warming, the average sea surface temperature has increased by an average of 0.6°C (1.08°F) in the past 100 years[21] (Figure 1.6). The repercussions of a warmer ocean are far-reaching. Water expands as it warms, leading to sea-level rise.[22] Sea-level rise, a phenomenon that we study in Chapter 5, threatens many of the world's major cities, the global economy, coastal and marine ecosystems, and the livelihoods of thousands of communities and tens of millions of individuals. Warmer water also causes stress to corals, plankton, pelagic fish, and other marine plants and animals.

All living organisms are the product of evolution, a process that selects populations of species on the basis of their ability to successfully reproduce under given environmental conditions. When those conditions change, the natural framework that gave rise to a species is threatened. Marine ecosystems are faced with extinction[23] if they are not able to migrate away from warming waters and move into cooler regions. Marine ecosystems adapted to polar waters are left with no options whatsoever.

---

[21] D.S. Arndt, M. Baringer, and M. Johnson, eds, "State of the Climate in 2009," *Bulletin of the American Meteorological Society* 91, no 7 (2010), S1-S224.

[22] See "Current Sea Level Rise," *Wikipedia,* http://en.wikipedia.org/wiki/Current_sea_level_rise (accessed July 9, 2012).

[23] In 2011, an international panel of marine experts warned that the world's ocean is at high risk of entering a phase of extinction of marine species unprecedented in human history. See the report at International Programme on the State of the Ocean, home page, http://www.stateoftheocean.org/ (accessed July 9, 2012); see also "Multiple Ocean Stresses Threaten 'Globally Significant' Marine Extinction, Experts Warn," *ScienceDaily,* http://www.sciencedaily.com/releases/2011/06/110621101453.htm (accessed July 9, 2012).

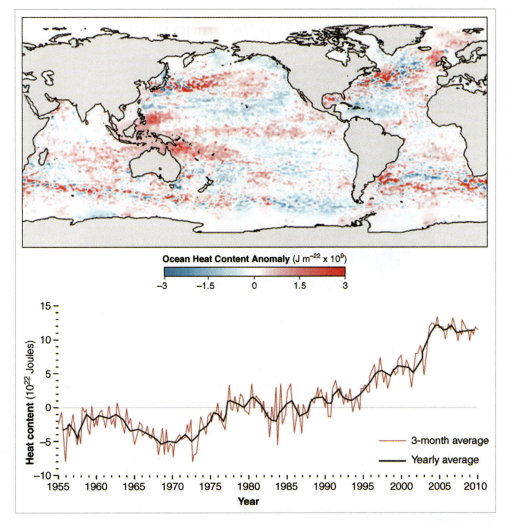

Ocean Heat Content Anomaly (J m^{-22} x 10^9)

-3   -1.5   0   1.5   3

Figure 1.6. Studies[24] show that the upper 750 m (2460 ft) of the oceans is absorbing heat owing to global warming. Map colors (top) show how the heat content of the oceans has changed compared to the average from 1993 to 2009. The graph (bottom) shows the average annual heat content of the oceans compared to the long-term baseline (gray line at zero).

Source: NOAA, "New Evidence on Warming Ocean," ClimateWatch, http://www.climatewatch.noaa.gov/image/2010/new-evidence-on-warming-ocean (accessed July 9, 2012).

One study[25] found that even though warming ocean water threatens many reef species, there may be a refuge adjacent to equatorial islands where deep, cool, nutrient-rich water rises to the surface. The Pacific nation of Kiribati was found to have 33 atoll islets that are bathed in cooling waters that rise from below along the westward flanks of each atoll. These conditions result from currents 100 to 200 m (330–660 ft) deep that run counter (known as the Equatorial Undercurrent) to the surface flow driven by the (east-to-west blowing) trade winds. When the (west-to-east flowing) undercurrent encounters the submerged slope of an atoll, it is forced to the surface and envelops the reef there with cool water. Conditions such as these may be right for protecting coral species in the future as oceans continue to store heat.

The ocean is warming because it absorbs most of the extra heat being added to the climate system from the buildup of greenhouse gases in the atmosphere. The warmer atmosphere leads to a warmer ocean, and ocean circulation carries the warm water across Earth's surface as well as into the depths of the sea,[26] although most of the heat is accumulating in the ocean's near-surface layers. In fact, according to the United Nations

---

[24] See the discussion of ocean heating in NOAA, "New Evidence on Warming Ocean," *ClimateWatch*, http://www.climatewatch.noaa.gov/image/2010/new-evidence-on-warming-ocean (accessed July 9, 2012).

[25] K. Karnauskas and A. Cohen, "Equatorial Refuge Amid Tropical Warming," *Nature Climate Change* 2 (2012): 530–534, doi: 10.1038/nclimate1499.

[26] The deep ocean is also warming. See Y. T. Song and F. Colberg, "Deep Ocean Warming Assessed from Altimeters, Gravity Recovery and Climate Experiment, in situ Measurements, and a Non-Boussinesq Ocean General Circulation Model," *Journal of Geophysical Research* 116 (2011): C02020, doi:10.1029/2010JC006601.

Figure 1.7. The vast majority of excess heat in the Earth system has been stored in the ocean.

Source: Figure by Skeptical Science Graphics, www.skepticalscience.com; calculated from IPCC AR4 5.2.2.3.

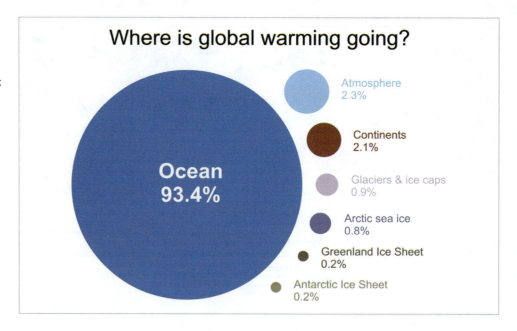

Intergovernmental Panel on Climate Change (IPCC) assessment report number 4 (AR4),[27] during the past 50 years the ocean has stored more than 90% of the increase in heat content of the Earth system.[28] A popular climate change website, Skeptical Science,[29] has provided a graphic that dramatically illustrates this point (Figure 1.7).

Phytoplankton

Marine life across the planet depends on tiny ocean plants called phytoplankton. The surface temperature of the ocean influences where and when phytoplankton grow. Because they cannot survive in water that is excessively warm, global warming is having a negative impact on them, which in turn is affecting the entire web of organisms in the ocean.

According to researchers,[30] phytoplankton account for half of all the production of organic matter on Earth, a colossal characteristic of a microscopic organism; however, worldwide phytoplankton levels are down 40% since the 1950s. In fact, their numbers have been declining in eight out of 10 ocean regions at a global rate of about 1% per year. Scientists identify rising sea surface temperatures as the cause of this decline, because warmer water makes it hard for phytoplankton to get vital nutrients. As these tiny components of the food chain decline, the entire marine ecosystem will be affected.

---

[27] Under the joint auspices of the United Nations Environmental Program (UNEP) and the World Meteorological Organization (WMO), the Intergovernmental Panel on Climate Change (IPCC) produces global assessments of climate change every five to seven years, representing the state of understanding. Past IPCC reports have been published in 1990, 1995, 2001, and 2007; the next report is dated 2014. The IPCC does not carry out original research, nor does it do the work of monitoring climate or related phenomena itself. Its primary role is publishing special reports on topics relevant to the implementation of the United Nations Framework Convention on Climate Change, which is an international treaty that acknowledges the possibility of harmful climate change. See IPCC, "Special Report on Managing the Risks of Extreme Events and Disasters to Advance Climate Change Adaptation," http://www.ipcc.ch/ (accessed July 9, 2012).

[28] See IPCC, "Climate Change 2007: Working Group I: The Physical Science Basis," http://www.ipcc.ch/publications_and_data/ar4/wg1/en/ch5s5-2-2-3.htm (accessed July 9, 2012).

[29] See SkepticalScience.com "Explaining Climate Change Science and Rebutting Global Warming Misinformation," http://www.skepticalscience.com/ (accessed July 9, 2012).

[30] S. Boyce, M. Lewis, B. Worm, "Global Phytoplankton Decline over the Past Century," *Nature* 466 (2010): 591–596, doi:10.1038/nature09268. See B. Borenstein, "Climate Change: Plankton in Big Decline, Foundation of Ocean's Food Web," *Huffington Post*, http://www.huffingtonpost.com/2010/07/29/climate-change-plankton-i_n_663488.html (accessed July 9, 2012).

Acidifying Oceans

In addition to absorbing heat, the oceans have absorbed about half of the carbon dioxide[31] emitted by humans over the past two centuries. This is a great environmental service[32] that has slowed warming of the atmosphere; unfortunately, the chemistry of the ocean is changing as a result. Increasing ocean acidity, brought on by dissolved carbon dioxide (CO_2) that mixes with seawater (H_2O) to form carbonic acid (H_2CO_3), makes it difficult for calcifying organisms (corals, mollusks, and many types of plankton) to secrete the calcium carbonate ($CaCO_3$) they need for their skeletal components. Calcium carbonate—the stuff of which shells, corals, and many types of plankton are made—is impeded from forming in low levels of carbonic acid and openly dissolves in the presence of high levels of carbonic acid. This ocean acidification[33] is one of the consequences of carbon dioxide buildup that could have a great impact on the world's ocean ecology, which depends on the secretion of calcium carbonate by thousands of different species.

For instance, one study[34] predicts that rising carbon emissions and acidification of seawater might kill off the ocean's coral reefs by 2050. But the reality may be more complicated than this. An analysis[35] of coral along the entire length of Australia's Great Barrier Reef found that as ocean temperatures and acidity rise, some species of corals are likely to succeed, some species might not, and the mix of species making up any single reef will change. The loss of healthy coral reefs affects all the species that dwell there (such as turtles, seals, mollusks, crabs, and fish), as well as the animals that depend on reef habitats as a food source (including seabirds, mammals, and humans). One quarter of all sea animals spends time in coral reef environments during their life cycle. Acidification has already been seen to damage the ability of oyster larvae on the Oregon coast to successfully develop their shells and grow at a pace that would allow them to be commercially harvested.[36]

Ocean acidification is measured using the pH scale. pH is a number scale[37] that ranges from 1 (acidic) to 14 (basic); a pH of 7 is considered neutral. Field studies[38] of locations where carbon dioxide seeps out of the ocean floor on the submerged slopes of a volcano have been used by scientists to calibrate coral response

---

[31] C. L. Sabine, R. A. Feely, N. Gruber, et al., "The Oceanic Sink for Anthropogenic CO_2," *Science* 305, no. 5682 (2004): 367–371, doi: 10.1126/science.1097403.

[32] See John Pickrell, "Oceans Found to Absorb Half of All Man-Made Carbon Dioxide," *National Geographic News*, July 15, 2004, http://news.nationalgeographic.com/news/2004/07/0715_040715_oceancarbon.html (accessed July 9, 2012).

[33] R. E. Zeebe, J. C. Zachos, K. Caldeira, T. Tyrrell, "Carbon Emissions and Acidification," *Science* 321, no. 5885 (2008): 51–52, doi: 10.1126/science.1159124.

[34] O. Hoegh-Guldberg, P. J. Mumby, A. J. Hooten, et al., "Coral Reefs Under Rapid Climate Change and Ocean Acidification," *Science* 318, no. 5857 (2007): 1737–1742.

[35] T. P. Hughes, A. H. Baird, E. A. Dinsdale, et al., "Assembly Rules of Reef Corals are Flexible along a Steep Climatic Gradient," *Current Biology* 22, no. 8 (2010): 736–741. See the video "Coral Winners" at the end of this chapter.

[36] A. Barton, B. Hales, G. Waldbusser, C. Langdon, R. Feely, "The Pacific Oyster, *Crassostrea gigas*, Shows Negative Correlation to Naturally Elevated Carbon Dioxide Levels: Implications for Near-Term Ocean Acidification Effects," *Limnology and Oceanography* 57, no. 3 (2012): 698, doi: 10.4319/lo.2012.57.3.0698

[37] Acidic and basic are two extremes that describe chemicals, just as hot and cold are two extremes that describe temperature. Mixing acids and bases can cancel out their extreme effects, much as mixing hot and cold water can even out the water temperature. A substance that is neither acidic nor basic is neutral. See EPA, "What is pH?" http://www.epa.gov/acidrain/measure/ph.html (accessed July 9, 2012).

[38] K. Fabricius, C. Langdon, S. Uthicke, C. Humphrey, et al., "Losers and Winners in Coral Reefs Acclimatized to Elevated Carbon Dioxide Concentrations," *Nature Climate Change*, 1 (2011): 165–169, http://www.reefrelieffounders.com/science/2011/06/07/nature-com-climate-change-losers-and-winners-in-coral-reefs-acclimatized-to-elevated-carbon-dioxide-concentrations-by-katharina-e-fabricius-et-al/ (accessed July 9, 2012).

to acidification.[39] It was found that as pH declines from 8.1 to 7.8 (equivalent to the seawater change expected if atmospheric carbon dioxide concentration increases from 396 ppm[40] [present day] to 750 ppm [possible by the end of this century]), reefs show a reduction in coral diversity, recruitment (new populations of coral on barren substrate), and abundances of reef-building corals. Reef development ceased below a pH of 7.7. Researchers concluded that these responses are consistent with previous model results and that together with temperature stress of warming seawater will probably lead to severely reduced resiliency of Indo-Pacific coral reefs this century.

There are economic impacts as well. Tourism tied to coral reefs and commercial fisheries generate billions of dollars in revenue annually. Biodiversity, food supplies, and economics thus could all be affected by ocean impacts. Reef loss is a complex issue, however. Reefs can suffer from coastal pollution,[41] overfishing, and other types of human stresses as well as at the hands of warming and acidification. Exactly what roles warming temperatures, ocean acidification, and other anthropogenic impacts play in global marine health have yet to be fully defined by researchers, but they are all negative factors.[42]

Deadly Trio

The overall picture for the global ocean is not healthy. Warming, acidification, and spreading anoxia[43] have been identified as the deadly trio that threatens mass extinctions in the marine ecosystem. Anoxia (also referred to as hypoxia) produces oceanic dead zones, where excess nutrients (particularly nitrogen and phosphorous) from fertilizers used in agriculture and human sewage collect in coastal waters. These nutrients fuel massive, short-lived blooms of phytoplankton. The algae produce oxygen during the day (through the process of photosynthesis), but at night they take oxygen out of the water column (through the process of respiration), and when they die, the decay process takes additional oxygen out of the water. The net result produces anoxia, regions where marine life cannot be supported owing to oxygen deficiency. Oceanographers first began noticing dead zones in the 1970s. In 2004, 146 dead zones[44] were reported, and by 2008 the number had increased to 405.[45]

Scientists have concluded[46] that the combination of stressors (warming, acidification, and anoxia) on the ocean today is creating the conditions associated with every previous major extinction of species in Earth's history.[47] The rate of degeneration in

---

[39] Carbon dioxide dissolved in the ocean reacts with seawater to form carbonic acid. Carbonic acid produces positively charged hydrogen ions (that lower pH) and negatively charged bicarbonate ions. Bicarbonate ions may lose a hydrogen ion (further lowering pH) to produce the carbonate ion (used by marine organisms to build calcium carbonate exoskeletons). Ocean acidification decreases the bicarbonate to carbonate reaction, and leads to some carbonate recombining with hydrogen to form bicarbonate. The result is that acidification reduces carbonate available for corals, some plankton, oysters, clams and other organisms

[40] Ppm means "parts per million." It is a measurement of abundance (or concentration) the same way that "per cent" means parts per hundred. In this case, ppm means molecules of CO_2 per million molecules of air.

[41] See "Mass extinctions and 'Rise of Slime' Predicted for Oceans," *Science Daily* http://www.sciencedaily.com/releases/2008/08/080813144405.htm (accessed July 9, 2012).

[42] See E. Wiese, "Scientists: Global Warming Could Kill Coral Reefs by 2050," *USA Today* http://www.usatoday.com/weather/climate/globalwarming/2007-12-13-coral-reefs_N.htm (accessed July 9, 2012).

[43] See D. Biello, "Oceanic Dead Zones Continue to Spread," *Scientific American* http://www.scientificamerican.com/article.cfm?id=oceanic-dead-zones-spread (accessed July 9, 2012).

[44] See "Dead Zone (Ecology)," *Wikipedia* http://en.wikipedia.org/wiki/Dead_zone_%28ecology%29#cite_note-sfgate-1 (accessed July 9, 2012).

[45] R. J. Diaz and R. Rosenberg, "Spreading Dead Zones and Consequences for Marine Ecosystems," *Science* 321, no. 5891 (2008): 926–929.

[46] A. D. Rogers and D. d'A. Laffoley, "International Earth System Expert Workshop on Ocean Stresses and Impacts." Summary Report. IPSO Oxford, 18 pp. http://www.stateoftheocean.org/ipso-2011-workshop-summary.cfm (accessed July 9, 2012).

[47] See several animations, "Ocean Threats," at the end of the chapter.

the ocean is faster than anyone has predicted. Many of the negative impacts we have already discussed are greater than the worst predictions, and although difficult to assess because of the unprecedented rate of change, the first steps to globally significant extinction may have already begun with a rise in the threat to marine species, such as reef-forming corals, open-ocean fishing stocks, and phytoplankton.

For example, experts[48] have determined that the rate at which carbon is being absorbed by the ocean is already far greater now than at the time of the last globally significant extinction of marine species: some 55 million years ago, when up to 50% of some groups of deep-sea animals were wiped out. Researchers point to these events as possible signals that extinction is under way:

- A single mass coral bleaching event in 1998 that killed 16% of all the world's tropical coral reefs[49]
- Overfishing, which has reduced some commercial fish stocks and populations of by-catch species by more than 90%[50]
- The widespread release of pollutants, including flame-retardant chemicals and synthetic compounds found in detergents. The presence of these pollutants has been traced to the polar seas, and throughout all the oceans, these pollutants are being absorbed by tiny plastic particles that are in turn ingested by marine creatures.

Are the oceans in trouble? The weight of scientific evidence indicates that the combined effects of warming, acidification, and human pollution of various types are assembling a deadly framework for marine ecosystems.

THE GLOBAL ENERGY BALANCE

With the oceans and atmosphere modulating the climate, Earth's average surface temperature has been until recently a moderate and comfortable 14°C (57.2°F), a temperature, changing with the seasons, that allows water to exist in all three physical states (solid, liquid, and gas). Life depends on the presence of liquid water. Without the greenhouse effect, the average temperature of Earth's atmosphere would be 18°C (0°F)—well below the freezing point of water—and life would not exist as we know it.[51]

The Greenhouse Effect

The term greenhouse effect compares the atmosphere to the glass panels of a greenhouse, which lets sunlight in, isolates warm air, and partially prevents heat from radiating away. In truth, atmospheric greenhouse gases operate somewhat differently, in that they actually absorb (trap) and emit heat, thus partially preventing it from escaping the atmosphere.

Life sustaining, the greenhouse effect is a natural process of trapping in the atmosphere the heat that originates from Earth's surface after it has been warmed by the Sun. Sunlight warms the oceans and the land. The ocean and land, in turn, warm the air by giving off infrared radiation, which we feel as heat. Carbon dioxide (CO_2), water vapor (H_2O), methane (CH_4), and other heat-trapping gases are responsible for the greenhouse effect by absorbing some of this infrared radiation and re-emitting it in all directions. If the concentration of these gases is too slight, the resulting cold conditions will not allow life to exist. If the concentration is too high, the atmosphere can overheat. Venus, for instance, has an atmosphere that is 98% carbon dioxide. A runaway greenhouse effect is responsible for raising the surface

[48] Rogers and Laffoley, "International Earth System Expert Workshop on Ocean Stresses and Impacts."

[49] See C .Wilkinson, *The 1997–1998 Mass Bleaching Event Around the World* http://www.oceandocs.net/bitstream/1834/545/1/BleachWilkin1998.pdf (accessed July 9, 2012).

[50] Rogers and Laffoley, "International Earth System Expert Workshop on Ocean Stresses and Impacts."

[51] See the animation "The Greenhouse Effect" at the end of the chapter.

Figure 1.8. Earth's heat budget governs the climate. (See text for discussion).

Source: Adapted from Australian Government, Bureau of Meteorology, "The Greenhouse Effect and Climate Change," http://www.bom.gov.au/info/climate/change/gallery/7.shtml.

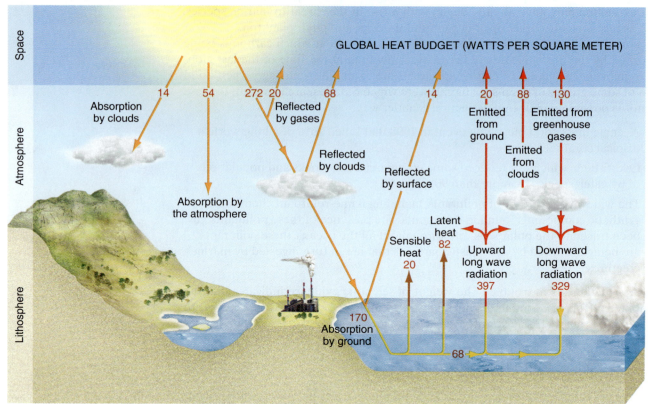

temperature to a deadly 477°C (858°F), one reason life is unlikely to exist on that planet.

Among all the known planets, these moderate conditions exist only on Earth. They exist because of Earth's heat budget (Figure 1.8), a complex balance of heat distribution and exchange among the air, water, rock, and living organisms on the planet.

In a natural state, the Sun's radiation is balanced at the top of the atmosphere, so that the amount of energy entering the atmosphere equals the amount leaving it.[52] The total incoming solar energy is about 340 W/m² (watts of energy per square meter [10.8 ft²] on Earth's surface). Part of this energy is absorbed by clouds and gases, and part is reflected by clouds, gases, and Earth's land and water surfaces. Approximately half (170 W/m²) is absorbed by Earth's surface. Some of the energy absorbed by the surface is reradiated upward, some is transferred to the atmosphere as sensible heat (heat that can be measured by a thermometer), and some is transferred to the atmosphere as latent heat (heat that is released by processes that change the physical state of matter, such as evaporation, freezing, melting, condensation, or sublimation). The atmosphere radiates this energy in all directions. When balance is achieved in the atmosphere, the total radiation leaving the top of the atmosphere equals the 340 W/m² received from the Sun.

Most of the light energy that penetrates Earth's atmosphere is short-wave ultraviolet (UV) radiation, which is mostly absorbed by the protective ozone (O₃) layer in the upper atmosphere. Of the radiation that reaches the lower atmosphere (mostly visible radiation), approximately half is reflected back into space. The remainder reaches the surface and is absorbed by the oceans and land, then reradiated back into the atmosphere in the form of long-wave infrared radiation. This long-wave radiation is absorbed (and reradiated) by greenhouse gases in the

---

[52]See the animation "Earth's Energy Balance" at the end of the chapter.

atmosphere. This process of warming the surface and the lower atmosphere, and maintaining a temperature that sustains life, constitutes the greenhouse effect.[53]

Greenhouse Gases

Earth's atmosphere is composed mostly (99%) of oxygen and nitrogen, but neither of these gases absorbs infrared energy, so they do not play a role in warming Earth. There are six principal greenhouse gases in Earth's atmosphere that absorb long-wave radiation and keep Earth warm:

- carbon dioxide (CO_2)
- methane (CH_4)
- ozone (O_3)
- nitrous oxide (N_2O)
- chlorofluorocarbons (CFCs)
- water vapor (H_2O)

Combined, these gases make up less than 1 percent of the atmosphere, but their heat-trapping ability is strong. Because greenhouse gases are efficient at trapping long-wave radiation from Earth's surface (Figure 1.9), even their small percentage is enough to keep temperatures in the ideal range for liquid water (and life) to exist on Earth. If the abundance of these gases increases, more heat is trapped. If their abundance decreases, less heat is trapped. Theoretically, as greenhouse gases increase, sensors in space should detect a cooling Earth, while on the surface it should be getting warmer. Indeed, this is exactly what is observed.[54]

Each greenhouse gas contributes differently to warming the atmosphere. Its role is affected by both the characteristics of the gas and its abundance. For example, methane is stronger at trapping heat than carbon dioxide, but it is not as abundant, so its total contribution is smaller. However, as its abundance increases, its role in global warming increases. Also, different gases have different residence times in the atmosphere: Water recycles within a few hours to a few days, methane resides only a decade or so, and carbon dioxide may stay in the atmosphere a few decades to over 1,000 years. These gases can be described by their net contribution to the greenhouse effect:[55] Water vapor contributes about 50%, clouds contribute approximately 25%, carbon dioxide contributes about 20%,[56] and the other gases such as methane and ozone contribute minor amounts.[57]

Although this accounting can identify water vapor as the dominate greenhouse gas, in reality water vapor only resides in the atmosphere for a few days. But as the temperature of the atmosphere rises, more water is evaporated and the amount of water vapor increases. Hence, it accumulates in the atmosphere as a positive feedback to the warming caused by other gases, principally carbon dioxide. As a result, the most powerful greenhouse gas is carbon dioxide, because once it is in the atmosphere it is only removed in any significant abundance by dissolving in water (such

---

[53]See USGCRP, "Climate Literacy: The Essential Principles of Climate Sciences."

[54]J. Harries, H. Brindley, P. Sagoo, and R. Bantges, "Increases in Greenhouse Forcing Inferred from the Outgoing Longwave Radiation Spectra of the Earth in 1970 and 1997," *Nature* 410 (2001): 355–357.

[55]G. A. Schmidt, R. A. Ruedy, R. L. Miller, and A. A. Lacis, "Attribution of the Present Day Total Greenhouse Effect," *Journal of Geophysical Research* 115 (2010): D20106, doi:10.1029/2010JD014287.

[56]A. Lacis, G. A. Schmidt, D. Rind, and R. A. Ruedy, "Atmospheric CO_2: Principal Control Knob Governing Earth's Temperature," *Science* 330, no. 6002 (2010): 356. doi 10.1126/science.1190653. A discussion is available at NASA, "Carbon Dioxide Control's Earth's Temperature," http://www.nasa.gov/topics/earth/features/co2-temperature.html (accessed July 9, 2012).

[57]E. E. Trenberth, J. T. Fasullo, and J. Kiehl, "Earth's Global Energy Budget," *Bulletin of the American Meteorological Society* 90, no. 3 (2009): 311–324, doi: 10.1175/2008BAMS2634.1. See also K. E. Trenberth, "An Imperative for Climate Change Planning: Tracking Earth's Global Energy." *Current Opinion in Environmental Sustainability* 1, no. 1 (2009): 19–27, http://www.sciencedirect.com/science/article/pii/S1877343509000025 (accessed July 9, 2012).

Figure 1.9. Energy from the Sun powers the climate system. Some solar radiation is reflected off the atmosphere, clouds, and Earth's surface; about half is absorbed by Earth's surface and warms it. Heat that is emitted from Earth's surface can be absorbed and re-emitted in all directions by greenhouse gases. As more greenhouse gases build up in the atmosphere, more heat accumulates.

Source: Based on figure from S. Solomon, D. Qin, M. Manning, et al., eds., *Contribution of Working Group I to the Fourth Assessment Report (AR4) of the Intergovernmental Panel on Climate Change.* Cambridge, U.K., Cambridge University Press, 2007.

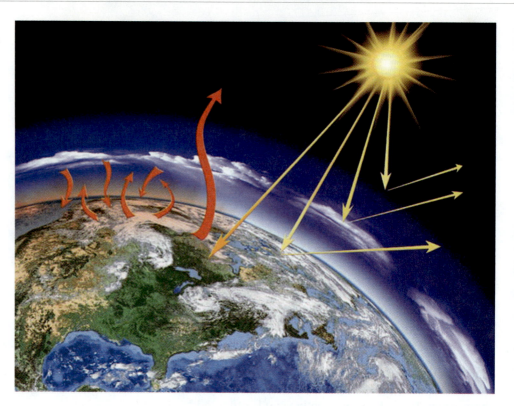

as seawater), or by geologic processes (such as the formation of limestone, $CaCO_3$), requiring long time scales to complete. Additionally, its ability to trap heat is magnified because it drives the positive feedback of other gases, such as water vapor.

Clouds are an important though still poorly understood contributor to the greenhouse effect. This is because they absorb and emit infrared radiation similar to greenhouse gases, but they also reflect sunlight, a cooling effect. Different types of clouds do more of one than the other depending on whether they are high or low altitude, more or less abundant, thick or thin, and depending on how these factors change in a warming world. The net effect of clouds on climate change depends on so many variables that cloud research is a growing and major area of study. For example[58] as climate warms, more water evaporates; thus it should be cloudier, with thicker and denser clouds. However, warmer air requires more water molecules to reach saturation and condense into clouds, thus limiting cloud development. Similarly, although summer is warmer and more humid than winter, the sky is not noticeably cloudier. Despite the complexities of these and other factors, research is indicating that clouds are in fact strongly amplifying global warming.[59]

Carbon Dioxide (CO_2)

Carbon dioxide and the other greenhouse gases have increased in abundance because of human activity, but the reasons differ for each gas. The amount of carbon dioxide in the atmosphere has varied significantly during Earth's history, and it began doing so long before modern humans inhabited the planet. Natural sources of carbon dioxide include volcanic outgassing (volcanoes exhaust CO_2 released from molten rock during and between eruptions), animal respiration, and decay of organic matter (decaying tissue is made of carbon, C, which combines with O_2 in the atmosphere to

---

[58]See NASA, "Clouds and Climate Change: The Thick and Thin of It," http://www.giss.nasa.gov/research/briefs/delgenio_03/ (accessed July 9, 2012).

[59]A. C. Clement, R. Burgman, and J. R. Norris, "Observational and Model Evidence for Positive Low-Level Cloud Feedback," *Science* 325, no. 5939 (2009): 460–464.

make CO_2). The air we inhale is roughly 78% by volume nitrogen (N_2), 21% oxygen (O_2), 0.96% argon (Ar) and 0.04% carbon dioxide (CO_2), helium (He), water (H_2O), and other gases. The permanent gases we exhale are roughly 4% to 5% more carbon dioxide and 4% to 5% less oxygen than was inhaled (the difference in O_2 is used to fuel our metabolism, and CO_2 is a waste product that we expel in our breath).

Scientists can measure the concentration of past atmospheric carbon dioxide and other gases by analyzing air bubbles trapped in ice cores (Figure 1.10) and the chemistry of ancient sediments[60] and by employing other techniques using geologic proxies of climate such as the chemical composition of fossils (e.g., corals) and various plankton from freshwater and marine ecosystems. These methods have helped scientists understand long-term trends in carbon dioxide variability and global climate change caused by natural factors.

Excess carbon dioxide is added to the atmosphere by human activities, in particular, the burning of fossil fuels (e.g., oil, natural gas, and coal), the burning of solid waste for fuel (e.g., dung, peat, wood, etc., mostly in Asia), deforestation (e.g., logging

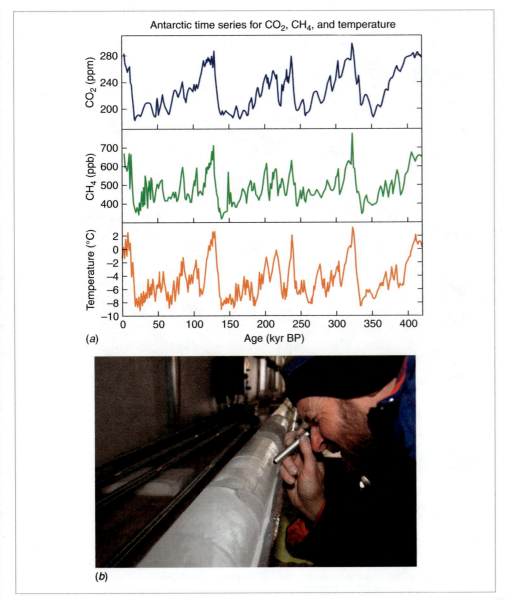

(a)

(b)

Figure 1.10. **a,** Global carbon dioxide content (CO_2 in parts per million [ppm]), methane content (CH_4 in parts per billion [ppb]), and temperature (in degrees Celsius) over the past 400,000 years have been measured using fossil air trapped in ice in Antarctica. **b,** Scientists drill ice cores on mountain glaciers, as well as in Greenland and Antarctica, to obtain evidence of past atmospheric composition.

SOURCE: (a) Fletcher, *Physical Geology: The Science of Earth,* 2012; (b) Karim Agabi/PhotoResearchers

[60] A. K. Tripati, D. R. Roberts, and R. A. Eagle, "Coupling of CO_2 and Ice Sheet Stability over Major Climate Transitions of the Last 20 Million Years." *Science* 326, no. 5958 (2009): 1394–1397, http://www.sciencemag.org/cgi/content/abstract/1178296 (accessed July 9, 2012).

and clearing land for farming and development), industrial agricultural (a large CO_2 source and a source of other greenhouse gases such as methane), and cement production (cement plants account for 5% of global emissions of CO_2[61]).

These anthropogenic emissions are at the center of research on global warming. The sources and heat-trapping properties of greenhouse gases are undisputed, but there is uncertainty about the details of how Earth's climate will respond to increasing concentrations of the various gases. There is wide consensus among scientists around the world (including leading climate researchers in the U.S., like those at the National Climate Data Center[62] in Asheville, North Carolina, and at the National Center for Atmospheric Research[63] in Boulder, Colorado) that if anthropogenic emissions of carbon dioxide continue to rise (Figure 1.11), by the end of this century average global

Figure 1.11. The level of carbon dioxide in Earth's atmosphere has been on the rise since the late 19th century. Originally collected in flasks by hand, and then by instrument, CO_2 is now measured by a global network of monitoring stations as well as one satellite, the Atmospheric Infrared Sounder (AIRS).[64] The satellite has been able to pinpoint the influence of specific carbon dioxide sources. For instance, it identified a large amount of carbon dioxide cycling around 40°S to 50°S latitude—the Roaring 40s—fed by two huge anthropogenic sources: a coal liquefaction plant in South Africa that is the largest single source of carbon dioxide on Earth, and a cluster of power generation plants in eastern Australia.

Source: PIA11194: Global Carbon Dioxide Transport from AIRS Data, July 2008 (http://photojournal.jpl.nasa.gov/catalog/PIA11194)

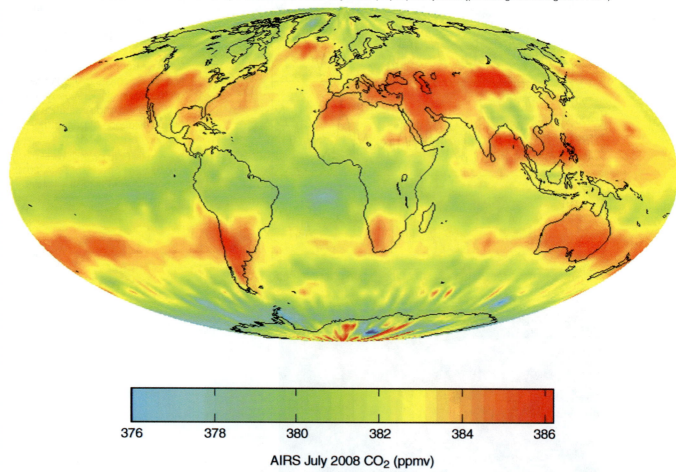

AIRS July 2008 CO_2 (ppmv)

[61]E. Rosenthal, "Cement Industry Is at Center of Climate Change Debate," *New York Times*, October 26, 2007, http://www.nytimes.com/2007/10/26/business/worldbusiness/26cement.html (accessed July 9, 2012).

[62]See NOAA Satellite and Information Service, "NCDC Frequently Asked Questions," http://www.ncdc.noaa.gov/faqs/index.html (accessed July 9, 2012).

[63]See discussion of "Climate of the Future" at NCAR, "Learn More about Climate," http://ncar.ucar.edu/learn-more-about/climate (accessed July 9, 2012).

[64]See NASA Jet Propulsion Lab, "AIRS and Carbon Dioxide: From Measurement to Science," http://airs.jpl.nasa.gov/story_archive/Measuring_CO2_from_Space/Measurement_to_Science/ (accessed July 9, 2012).

temperatures will increase, perhaps by as much as 2.4°C to 6.4°C (4.3°F to 11.5°F).[65] Scientists also agree that industrial emissions have been the dominant influence on climate change for the past 50 years, overwhelming natural causes.[66]

In the geologic past, climate changes occurred naturally. In the past half million years, carbon dioxide concentration remained between about 180 ppm already aligned earlier during ice ages (also called glacial periods), and 280 to 300 ppm during warm periods (also called interglacial periods), such as today (more on glacial and interglacial events in Chapter 3). But carbon dioxide content has been much greater in other periods of Earth's history.

Estimates of carbon dioxide content during geologic history are based on the chemistry of fossilized soils, fossil plants, and fossil shells of plankton. These indicate that concentrations as high as 1,000 to 4,000 ppm[67] may have occurred for sustained periods, and even reached twice this level. The cause of such high levels is controversial among researchers who study this history: Episodes of extreme global volcanism, changes in land surface area as a result of plate tectonics, reorganization of ocean circulation, absence of polar ice, mountain building, and other mechanisms have all been suggested. However, it is clear that the level of only 180 ppm during glaciations is not far from the lowest that has ever occurred since the rise of macroscopic life on Earth in the past half billion years.

In the 200 years since the Industrial Revolution in the early 1800s, humans have altered Earth's environment through agricultural and industrial practices. The growth of the human population and activities such as deforestation and burning of fossil fuels have affected the mixture of gases in the atmosphere. We know from ice cores that the amount of carbon dioxide in the atmosphere prior to the Industrial Revolution was about 280 ppm. Today (July 2012) the concentration of carbon dioxide is 396 ppm and rising, higher than at any other time in the past 15 million years.[68]

Although carbon dioxide is not the most effective absorber of heat compared to other greenhouse gases, it is one of the most abundant, and once in the atmosphere it can stay there for a very long time. Carbon dioxide can reside in the atmosphere for more than 1,000 years, resulting in essentially irreversible climate change.[69] Monthly records of atmospheric CO_2 concentration collected at the Mauna Loa Observatory in Hawaii[70] show seasonal oscillations superimposed on a long-term increase in CO_2 in the atmosphere (Figure 1.12). This increase is attributed to the activities of humans associated with the rise of modern industry, primarily the burning of fossil fuels and deforestation.

Of all the greenhouse gases released by human activities, carbon dioxide is the largest individual contributor to the enhanced greenhouse effect, accounting for about 60% compared to the other human sources. Unfortunately, the increase in global emissions of carbon dioxide from fossil fuels over the five years of 2005 to 2009 was four times greater than the increase over the preceding 10 years. Rather

[65]This is the IPCC-AR4 best estimate for a "high scenario." IPCC, *Climate Change 2007: Synthesis Report. Contribution of Working Groups I, II and III to the Fourth Assessment Report of the Intergovernmental Panel on Climate Change.* (Geneva, IPCC, 2007). http://www.ipcc.ch/publications_and_data/ar4/syr/en/contents.html (accessed July 9, 2012).

[66]A. Lacis et al., "Atmospheric CO_2: Principal Control Knob Governing Earth's Temperature."

[67]R. A. Berner, "The Rise of Plants and Their Effect on Weathering and Atmospheric CO_2." *Science* 276 (1997): 544–546. M. Pagani, M. Arthur, and K. Freeman, "Miocene Evolution of Atmospheric Carbon Dioxide," *Paleoceanography* 14 (1999): 273–292. See "Carbon Dioxide in Earth's Atmosphere," *Wikipedia*, http://en.wikipedia.org/wiki/Carbon_dioxide_in_Earth's_atmosphere (accessed July 9, 2012).

[68]A. K. Tripati et al., "Coupling of CO_2 and Ice Sheet Stability over Major Climate Transitions of the Last 20 Million Years."

[69]S. Solomon, G.-K. Platter, R. Knutti, and P. Friedlingstein, "Irreversible Climate Change Due to Carbon Dioxide Emissions," *Proceedings of the National Academy of Science* 106 (2009): 1704–1709, doi: 10.1073/pnas.-9128211-6.

[70]The NOAA Earth System Research Laboratory on Mauna Loa, Hawaii, measures carbon dioxide daily: http://www.esrl.noaa.gov/gmd/ccgg/trends/ (accessed July 9, 2012).

Figure 1.12. Concentration of the most important greenhouse gases in Earth's atmosphere. Clockwise from upper left: Levels of carbon dioxide and nitrous oxide continue to climb. Levels of CFCs have declined since the Montreal Protocol (see text for discussion) was implemented in 1987. The concentration of methane stabilized early in this century as a result of droughts and a temporary decline in industrial emissions, but it has since returned to its previous pattern of steady increases.

Source: Figure 2, The National Oceanic and Atmospheric Administration Global Monitoring Division, http://www.esrl.noaa.gov/gmd/aggi/.

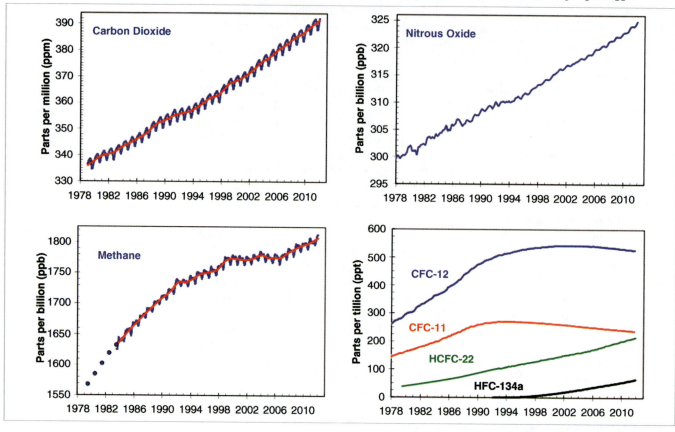

than recognizing that the enhanced greenhouse effect is a potentially dangerous trend that can be curtailed by decreasing emissions of heat-trapping gases, humans are accelerating carbon-burning activities.

How has the global economic decline of 2008 to 2012 affected carbon dioxide emissions? After a 1% decline in 2009, global CO_2 emissions increased[71] by more than 6% in 2010, producing the highest annual net increase in carbon pollution ever. The world pumped about 564 million more tons of carbon into the air in 2010 than it did in 2009, a rate that exceeds the worst-case scenario (A1FI) of the Intergovernmental Panel on Climate Change. Carbon dioxide production went up in most of the major economies, led by increases in China (10% increased emissions) and India (9% increased emissions).[72] The average annual growth rate in CO_2 emissions over the three years of the global recession, including a 1% increase in 2008 when the first impacts became visible, is 1.7%, almost equal to the long-term annual average of 1.9% for the preceding two decades back to 1990. However, as of this writing, most industrialized countries have not recovered fully from their decreases in emissions of 7% to 12%, and carbon dioxide emissions will continue to climb.

[71]Seth Bornstein, "Biggest Jump Ever Seen in Global Warming Gases," Associated Press, November 4, 2011, http://news.yahoo.com/biggest-jump-ever-seen-global-warming-gases-183955211.html (accessed July 9, 2012).

[72]PBL Netherlands Environmental Assessment Agency, "Long-Term Trend in Global CO_2 Emissions, 2011 Report," September 21, 2011, http://www.pbl.nl/en/publications/2011/long-term-trend-in-global-co2-emissions-2011-report (accessed July 9, 2012).

Methane (CH₄)

Natural sources of methane include the activity of microbes and insects in wetlands, seawater, and soils; wildfires; and the release of gases stored in ocean sediments. The present global atmospheric concentration of methane is more than 1,800 parts per billion (ppb), more than double what it was before the Industrial Revolution. Methane levels increased steadily in the 1980s, but the rate of increase slowed in the 1990s and was close to zero from 2000 to 2007. Researchers attribute this lull to a temporary decrease in emissions during the 1990s related to the decline of industry and farming when the former Soviet Union collapsed, along with a slowdown in wetland emissions during prolonged droughts. Scientists warn that with methane levels on the rise again, a more typical rate of increase will have a significant impact on climate.[73]

Methane is more potent as a greenhouse gas than carbon dioxide, but there is far less of it in the atmosphere and it is measured in parts per billion. When related climate effects are taken into account, methane's overall climate impact is less than half that of carbon dioxide; thus, methane is second only to carbon dioxide as a cause of global warming.

About 60% of annual methane emissions come from anthropogenic sources. Human activities that release methane into the atmosphere include deforestation (burning logged tracts of forest), mining and burning fossil fuels, processing human waste, and cultivating rice in paddies (industrial wetlands). Methane has increased owing to manure production on farms and ranches, landfill emissions, and industrial activities.

Methane is also trapped in ice, glaciers, frozen seafloor sediment, and the permafrost in tundra and under the rapidly disappearing sea ice of the Arctic Ocean;[74] as melting of all these frozen sources occurs, the gas is released to the atmosphere. There is fear that as frozen regions thaw, methane released from the ice will add to atmospheric concentrations and constitute a positive feedback.[75, 76]

Climate feedbacks are processes that can amplify (positive feedback) or suppress (negative feedback) the effects of a temperature change. For example, as climate warms, snow and ice melt, and the formerly white surface is replaced by dark land and water. The darker surfaces absorb more of the Sun's heat, causing more warming, which causes more melting, and so on, in a self-reinforcing cycle. In the case of methane, climate change melts permafrost (Figure 1.13), releasing more methane, causing more warming, melting more permafrost, releasing more methane, and so on. Unlike CO_2, methane is destroyed by reactions with other chemicals in the atmosphere and soil, so its atmospheric lifetime is about a decade. But if it is released rapidly and in large quantities it could drive a potent positive feedback process.

In March 2010, the National Science Foundation (NSF) issued a remarkable press release warning that methane escaping from the Arctic continental shelf of Siberia has been observed to be much larger and faster than anticipated.[77] Researchers identified a section of the Arctic Ocean seafloor with vast stores of frozen methane showing signs of instability and widespread venting of the gas. A paper published in Science[78] showed that permafrost under the East Siberian Arctic Shelf,

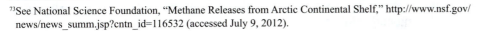

[73]See National Science Foundation, "Methane Releases from Arctic Continental Shelf," http://www.nsf.gov/news/news_summ.jsp?cntn_id=116532 (accessed July 9, 2012).

[74]E. Kort, S. Wofsy, B. Daube, et al., "Atmospheric Observations of Arctic Ocean Methane Emissions up to 82° North," Nature Geoscience 2012, doi: 10.1038/ngeo1452.

[75]McGuire, A., Anderson, L., Christensen, T., Dallimore, S., Guo, L., Hayes, D., Heimann, M., Lorenson, T., Macdonald, R., Roulet, N. (2009) "Sensitivity of the carbon cycle in the Arctic to climate change." Ecological Monographs 79, no. 4:523–555, doi: 10.1890/08-2025.1. See: http://www.esajournals.org/doi/abs/10.1890/08-2025.1 (accessed July 9, 2012).

[76]See the animation "Thawing Permafrost-Changing Planet" at the end of the chapter.

[77]See: http://www.nsf.gov/news/news_summ.jsp?cntn_id=116532&org=NSF&from=news (accessed July 9, 2012).

[78]N. Shakova, I. Semiletov, A. Salyuk, et al., "Extensive Methane Venting to the Atmosphere from Sediments of the East Siberian Arctic Shelf," Science 327, no. 5970 (2010): 1246–1250.

Figure 1.13. Permafrost bluffs from Barter Island in northeastern Alaska. As permafrost thaws, quantities of methane are liberated; as the soil warms, microbes digest vegetation contained in the frozen ground, releasing methane as a byproduct. Potentially, this process could set a feedback cycle into motion, amplifying atmospheric warming, increasing permafrost thaw, and promoting the release of more methane.[79]

SOURCE: Photograph by Ben Jones of the USGS Alaska Science Center.

long thought to be an impermeable barrier sealing in methane, is instead perforated and starting to leak gas into the atmosphere. The amount of methane being released from just this one area is comparable to the amount coming out of all of the world's oceans combined. High volumes of methane released by decaying permafrost in the oceans and on the land have been identified by some researchers as a potential climate tipping point. A tipping point[80] occurs when a positive feedback cannot be recovered, implying that human efforts to decrease greenhouse gas emissions might not be sufficient to stop the warming.

Tropospheric Ozone (O_3)

The role of ozone[81] is complicated. There is "good ozone" in the upper atmosphere that blocks the Sun's harmful UV radiation and does not play a role in climate change. There is "bad ozone" at ground level that damages people's lungs and contributes to smog. There is also mid-altitude ozone that acts as a greenhouse gas. Greenhouse ozone, also known as tropospheric ozone, is on the rise because cars and coal-fired power plants release air pollutants that react with oxygen to produce more tropospheric ozone. Natural sources of ozone include chemical reactions that occur among carbon monoxide (CO), hydrocarbons, and nitrous oxides (N_2O), as well as lightning and wildfires.

Human activities increase ozone concentrations indirectly by emitting pollutants that are precursors of ozone. These include CO, N_2O, sulfur dioxide (SO_2), and

---

[79] See NOAA, "Hot on Methane's Trail" http://researchmatters.noaa.gov/news/Pages/NOAAHotonMethane%27sTrail. aspx (accessed July 9, 2012).

[80] See http://www.sciencedaily.com/releases/2007/05/070531073748.htm (accessed July 9, 2012).

[81] See the NASA page on ozone: http://www.nasa.gov/missions/earth/f-ozone.html (accessed July 9, 2012); the U.S. Environmental Protection Agency also has an ozone report that explains the role of ozone in climate; see http://www.epa.gov/oar/oaqps/gooduphigh/ (accessed July 9, 2012).

hydrocarbons that result from the burning of biomass and fossil fuels. Ozone is a strong absorber of heat but it does not stay in the atmosphere for long, only a few weeks to a few months. Nonetheless, its concentration is increasing at a rapid rate. Concentrations of tropospheric ozone have risen by around 30% since the preindustrial era, and ozone is now considered by the IPCC to be the third most important greenhouse gas after carbon dioxide and methane.[82]

Nitrous Oxide (N_2O)

Nitrous oxide[83] is a clear, colorless gas with powerful greenhouse properties. Because it has a long atmospheric lifetime (approximately 120 years) and is about 310 times more powerful than carbon dioxide at trapping heat, it is important to track where N_2O comes from and where it is stored in nature. The main natural source of nitrous oxide is the activity of microbes in swamps, soil, rainforests, and the ocean surface. Human sources of this greenhouse gas include fertilizers, industrial production of nylon and nitric acid, the burning of fossil fuels, and solid waste. The present concentration of N_2O is 323 ppb. Human activities have caused it to increase by 16% since the beginning of the Industrial Revolution.

Nitrous oxide is also produced by permafrost thaw.[84] About 25% of the land surface in the Northern Hemisphere is underlain by permafrost. Global warming's thawing of these soils does not initially stimulate nitrous oxide production. However, as meltwater from the frozen soils flows back into the thawed sediment it stimulates increased nitrous oxide production by more than 20 times. Nearly a third of the nitrous oxide produced in this process escapes into the atmosphere, adding to the positive feedback aspects of permafrost thaw.

Fluorocarbons

A number of very powerful heat-absorbing greenhouse gases in the atmosphere do not occur naturally. They include chlorofluorocarbons (CFCs), hydrofluorocarbons (HFCs), perfluorocarbons (PFCs), and sulfur hexafluoride (SF_6), all of which are produced by industrial processes. CFCs are used as coolants in air conditioning (Freon is a CFC), aerosol sprays, and the manufacture of plastics and polystyrene. CFCs did not exist on Earth before humans created them in the 1920s. They are very stable compounds, have long atmospheric lifetimes, and are now abundant enough to cause global changes in air chemistry and climate.

Fluorocarbons contribute to warming by enhancing the greenhouse effect in the lower atmosphere. CFCs also chemically react with and destroy ozone (O_3) in the upper atmosphere, creating the "ozone hole" over the Southern Hemisphere (Figure 1.14). Depending on where ozone resides, it can protect or harm life on Earth. Most ozone resides in the stratosphere, where it shields Earth's surface from harmful ultraviolet (UV) radiation emitted by the Sun. However, because chlorofluorocarbons destroy ozone, stratospheric ozone has been declining at a rate of about 4% per decade. At the same time, a much stronger, but seasonal decrease in ozone over Earth's poles has opened an "ozone hole" over the Antarctic and the Arctic. Without ozone, humans are more liable to develop skin cancer,[85] cataracts, and impaired immune systems. Closer to Earth in the troposphere, ozone is a harmful pollutant that causes damage to lung tissue and plants.

---

[82]See the discussion of greenhouse gases by the National Climatic Data Center at http://www.ncdc.noaa.gov/oa/climate/gases.html#introduction (accessed July 9, 2012).

[83]See Environmental Protection Agency, "Nitrous Oxide: Sources and Emissions," http://www.epa.gov/nitrousoxide/sources.html (accessed July 9, 2012).

[84]B. Elberling, H. Christiansen, and B. Hansen, "High Nitrous Oxide Production from Thawing Permafrost," *Nature Geoscience* 3 (2010): 332–335, doi:10.1038/ngeo803.

[85]Overexposure to UV radiation is believed to be contributing to the increase in melanoma, the most fatal of all skin cancers. Since 1990, the risk of developing melanoma has more than doubled. See http://www.epa.gov/oar/oaqps/gooduphigh/good.html#1 (accessed July 9, 2012).

Figure 1.14. The ozone hole over Antarctica. Blue and purple are where there is the least ozone; green, yellow, and red are where there is more ozone. Ozone abundance is measured in Dobson units; one Dobson unit is the number of molecules of ozone that would be required to create a layer of pure ozone 0.01 mm thick at a temperature of 0°C and a pressure of 1 atmosphere (the air pressure at sea level).

IMAGE CREDIT: Courtesy of the TOMS Science Team & the Scientific Visualization Studio, NASA GSFC.

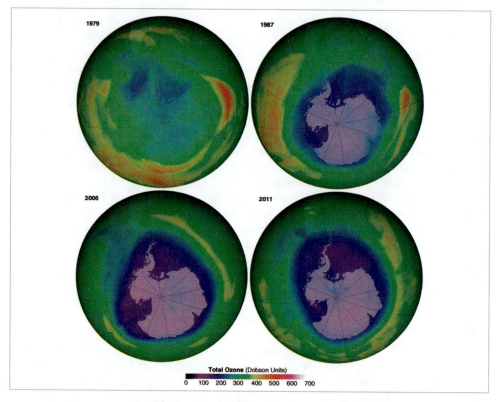

The good news is that the effects of many CFCs are reversible. Thanks to the Montreal Protocol, signed by 27 nations in 1987,[86] CFCs were recognized as dangerous pollutants and their production and use was significantly reduced. The United States, one of the signers of the Protocol, banned the use of CFCs in aerosols and ceased their production by 1995. CFCs already in the atmosphere have lifetimes of 75 to 150 years, so ozone depletion could continue for decades. However, the first signs that the ozone hole in the Southern Hemisphere is beginning to heal have surfaced,[87] and scientists are hopeful that the trend of ozone depletion in the stratosphere over the Antarctic may be reversing.

Unfortunately, an ozone hole has opened in the Arctic that is nearly as large as the hole over the Antarctic. Depletion of Arctic ozone is mainly due to unusually cold temperatures in the stratosphere that drive reactions involving CFCs and which destroy ozone. Researchers[88] have calculated that over the past thirty years the stratosphere in cold Arctic winters has cooled down by about 1°C (1.8°F) per decade. What is driving the cooling trend? The likely culprit is that when heat is trapped by greenhouse gases in the troposphere it produces cooling in the overlying stratosphere. If this trend to colder stratospheric temperatures continues, the Arctic

[86]See the Wiki entry on "Montreal Protocol," *Wikipedia*, http://en.wikipedia.org/wiki/Montreal_Protocol (accessed July 9, 2012). The same treaty is being considered by governments today as a way to reach agreement on reducing greenhouse gas production.

[87]M. Salby, E. Titova, and L. Deschamps, "Rebound of Antarctic Ozone," *Geophysical Research Letters* 38 (2011): L09702, doi:10.1029/2011GL047266.

[88]B.-M. Sinnhuber, G. Stiller, R. Ruhnke, T. von Clarmann, S. Kellmann, and J. Aschmann. "Arctic Winter 2010/2011 at the Brink of an Ozone Hole," *Geophysical Research Letters* 38 (2011): doi: 10.1029/2011GL049784.

ozone hole can be expected to persist and widen. Further decrease in temperature by just 1°C (1.8°F) would be sufficient to cause a nearly complete destruction of the Arctic ozone layer in certain areas including densely populated areas in northern Russia, Greenland and Norway.

Water Vapor (H_2O)

Earth's climate is able to support life because of the greenhouse effect and the availability of water. Water vapor (a gas) is a key component of both of these processes. It is the most abundant and powerful greenhouse gas and an important link between Earth's surface and its atmosphere. The concentration of water in the atmosphere is constantly changing, controlled by the balance between evaporation and precipitation (rain and snowfall). In fact, the average water molecule spends only about nine days in the air before precipitating back to Earth's surface (Figure 1.15).

Water vapor constitutes as much as 2% of the atmosphere and accounts for the largest percentage of the natural greenhouse effect. The abundance of water vapor varies from one spot to another based on natural evaporation and precipitation. Normally, human activity does not significantly affect water vapor concentrations except in local circumstances (such as irrigating fields or building reservoirs in arid areas). However, as global warming increases the average temperature of the troposphere, the rate of evaporation increases; hence, the amount of water vapor increases in a warmer atmosphere—a powerful positive feedback effect.[89] Increases in other heat-trapping gases, such as carbon dioxide, lead to more heating and thus more water vapor (increased water vapor in the atmosphere has already been observed[90]). This increase in atmospheric water vapor in turn produces increased heating, more water vapor, and so on. Like methane being released from melting permafrost, water vapor can drive a positive feedback in the global warming system.

Basic theory, observations, and climate models all show the increase in water vapor is around 6% to 7.5% per degree Celsius (or per 1.8°F) warming of the lower atmosphere. Notably, a study by NASA[91] confirmed that the heat-amplifying effect of water vapor is potent enough to double the climate warming caused by increased levels of carbon dioxide in the atmosphere. So if there is a 1°C (1.8°F) change caused by CO_2, the resulting water vapor increase will cause the temperature to go up another 1°C (1.8°F).

A general rule has developed among climate scientists who study the water cycle: "In a warmer world wet places will get wetter and dry places will get drier." This is based on the simple observation that places that already experience abundant rainfall will see more moisture as air temperature rises and humidity increases, and dry places, such as around 30°N and 30°S latitudes, where the dry limb of the Hadley Cell descends to the surface, will continue to see the same dry air.

This was unambiguously confirmed by a study[92] of the 50-year salinity history (1950 to 2000) of the ocean surface. Researchers tracked the changing salinity of ocean water using shipboard data and the 3,500-float armada of the Argo array.[93] Gauging the oceans' changing salinity reveals the movement of water between the

[89]See "Water Vapor," *Wikipedia*, http://en.wikipedia.org/wiki/Water_vapor (accessed July 9, 2012).

[90]B. D. Santer, C. Mears, F. J. Wentz, et al., "Identification of Human-Induced Changes in Atmospheric Moisture Content," *Proceedings of the National Academy of Sciences* 104, no. 39 (2007): 15248–15253, http://www.pnas.org/content/104/39/15248.full.pdf (accessed July 9, 2012).

[91]NASA/Goddard Space Flight Center, "Water Vapor Confirmed as Major Player in Climate Change," *ScienceDaily,* November 18, 2008, http://www.sciencedaily.com /releases/2008/11/081117193013.htm (accessed July 9, 2012).

[92]P. Durack, S. Wijffels, and R. Matear, "Ocean Salinities Reveal Strong Global Water Cycle Intensification during 1950 to 2000," *Science* 336, no. 6080 (2012): 455–458, doi: 10.1126/science.1212222.

[93]Argo is an observation system for the Earth's oceans that provides real-time data for use in climate, weather, oceanographic, and fisheries research. Argo consists of a large collection of small drifting oceanic robotic probes deployed worldwide. The probes float as deep as 2 km. Once every 10 days, the probes surface, measuring conductivity and temperature profiles to the surface. From these, salinity and density can be calculated. The data are transmitted to scientists on shore via satellite.

Figure 1.15. Nearly 577,000 km³ (138,500 mi³) of water circulates through the water cycle every year. The cycle consists of five major processes: condensation (cloud formation), precipitation (rain and snowfall), infiltration (water soaking into the ground), runoff (water draining off the land in streams), and evapotranspiration (evaporation plus transpiration; transpiration is a process wherein plants take water in through the roots and release it through the leaves). These processes keep water continuously moving through Earth's environments.[94]

atmosphere and ocean; salinity drops where there is more rain and salinity rises where there is more evaporation. Researchers discovered that areas of high rainfall, such as the high-latitude and equatorial parts of the oceans, became even less salty during the period of study. In the middle latitudes, where evaporation dominates, ocean salinity increased. The scientists calculated that over 50 years the water cycle had sped up roughly 4% while the surface warmed 0.5°C (0.9°F), an 8% increase per degree Celsius of warming. Because the water cycle over land behaves the same way

---

[94] See the animation "Water Cycle Animation" at the end of the chapter.

as over the oceans, and because much of the rain over land comes from the ocean, these results likely apply to rainfall on the continents as well.

If (as predicted) the world warms 2°C to 3°C (3.6°F to 5.4°F) by the end of the century, the water cycle could accelerate 16% to 24%. This would be an ominous development for several reasons: Evaporation carries heat from the surface to the atmosphere that can fuel violent storms, from tornadoes to tropical cyclones, and increased evaporation would enhance this relationship. Increasing rainfall in wet places can lead to more-severe and more-frequent flooding. Decreasing rainfall in arid and semiarid regions would mean longer and more-intense droughts.

Aerosols

Burning fossil fuels not only produces heat-trapping gases, it also produces aerosols, fine solid particles or liquid droplets suspended in the atmosphere that scatter (reflect) or absorb sunlight. Scattering behavior increases Earth's albedo, the tendency to reflect sunlight, and thus has a cooling effect. On the other hand, heat absorption, such as by black soot produced by biomass burning, has a warming effect. Most anthropogenic aerosols are sulfates (SO_4) that are released with the pollution from burning coal, wood, dung, and petroleum. So much aerosol production accompanied industrial growth in the middle of the 20th century that global cooling occurred in the decades of the 1950s to 1970s. Today we track atmospheric particles with sensors aboard NASA's Terra satellite (Figure 1.16).

Volcanic eruptions can have the same effect. They blast huge clouds of particles and gases (including sulfur dioxide, SO_2) into the atmosphere. Most of these particles stay in the troposphere and fall out within a few days to weeks. But if a volcanic eruption is especially large, particles may be blasted into the stratosphere and can remain in the air for years. In the upper atmosphere, sulfur dioxide converts to tiny, persistent, sulfuric acid (called sulfate) particles that reflect sunlight. Particularly

Figure 1.16. The image (from NASA's Terra satellite) shows the concentration of particles in the atmosphere (aerosols) during March 2010. A dark brown plume extends west from Africa, where thick dust blew over the Atlantic Ocean. Dark brown patches also cover parts of China and Southeast Asia where aerosols clouded the sky. Dust contributed to the aerosols in northern Asia, but smoke is the likely culprit for high aerosols in southern Asia. Fires burned extensively in Southeast Asia through March, veiling the region in a pall of smoke. See the NASA Earth Observatory website (http://earthobservatory.nasa.gov) for more satellite imagery.

IMAGE CREDIT: NASA Earth Observatory Image by Kevin Ward, based on data provided by the NASA Earth Observations (NEO) Project.

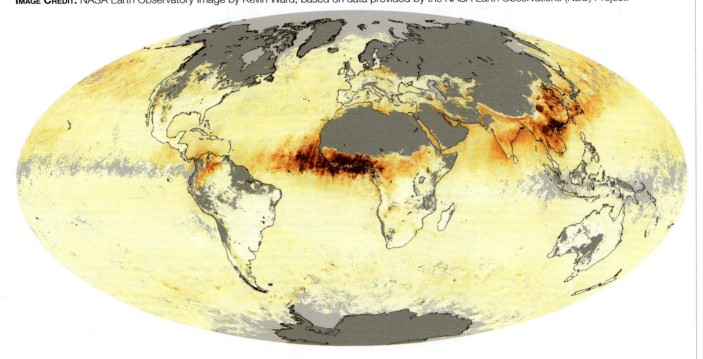

large eruptions can produce global cooling. For example, Mount Pinatubo in the Philippines erupted in June 1991 and cooled the planet nearly 1°C (1.8°F), temporarily offsetting the greenhouse effect for more than one year. Other major recent volcanic eruptions that produced temporary global cooling were from Mount Agung (Indonesia, 1963) and El Chichon (Mexico, 1982).

Scientists continue to investigate the role that stratospheric aerosols play in the climate system. One study[95] concluded that global climate models used to project future patterns in Earth's climate miss an important cooling factor if they do not account for the influence of stratospheric aerosol or do not include recent changes in stratospheric aerosol levels. Researchers found that a previously unmeasured increase in the abundance of particles high in the atmosphere has offset about a third of the warming influence of carbon dioxide change during the first decade of the 21st century. Since 2000, stratospheric aerosols have caused a slower rate of climate warming than would have occurred without them. The reasons for this increase are not known, but because there were no large-scale volcanic eruptions over the period, the particles could have come from several sources: smaller volcanic eruptions, sulfur compounds from Earth's surface such as biomass burning and industrial emissions, and even meteoric dust arriving from space.

When we burn coal, animal dung, diesel fuel, wood, vegetable oil, and other fuels made of biomass, part of the exhaust is black soot. Soot consists of microscopic particles of carbon that are carried into the atmosphere and that contribute to global warming. Soot has been found to cause climate changes in areas of higher latitude where ice and snow are more common.[96]

Typically, ice and snow reflect sunlight rather than absorb it, owing to their white background. When black soot collects on the snow, the particles absorb heat, accelerating the melting of snow and ice, and replace part of the reflective white surface with heat-absorbing black particles. As the snow and ice disappear, the water and barren earth that are revealed also absorb heat; hence, the formerly reflective surface is replaced by heat-absorbing water and rock. According to computer simulations, soot may be responsible for 25% of observed global warming over the past century. One study[97] found that soot may be contributing to the trend of early spring in the Northern Hemisphere. Earlier springs are a contributing factor to the thinning of Arctic sea ice and the melting of glaciers and permafrost.

In summary, aerosols can have a cooling effect (e.g., sulfate particles from volcanic eruptions and industrialization) or a warming effect (e.g., black soot from biomass burning) on climate.

RADIATIVE FORCING

When discussion turns to global warming[98] or the greenhouse effect, it is the concept of radiative forcing that provides the underlying scientific principle. Radiative forcing describes the fact that energy is constantly flowing into Earth's atmosphere in the form of sunlight, and if there is a difference between the amount of energy going back out into space and the amount coming in, the planet has to be either heating or cooling. As we saw in our discussion of the greenhouse effect, about half of incoming sunlight

---

[95]S. Solomon, J. Daniel, R. Neely, J. Vernier, and E. Dutton, "The Persistently Variable 'Background' Stratospheric Aerosol Layer and Global Climate Change," *Science* 333, no. 6044 (2011): 866–870, doi: 10.1126/science.1206027.

[96]B. Xu, J. Cao, J. Hansen, et al., "Black Soot and the Survival of Tibetan Glaciers," *Proceedings of the National Academies of Sciences* 106, no 52 (2009): 22114–22118, http://www.pnas.org/content/early/2009/12/07/0910444106 (accessed July 10, 2012). See also J. Hansen, "Science Briefs: Survival of Tibetan Glaciers," 2009 http://www.giss.nasa.gov/research/briefs/hansen_14/ (accessed July 9, 2012).

[97]R. Gutro, "NASA Study Finds Soot May be Changing the Arctic Environment," *NASA News Archive*, March 2005 (accessed July 9, 2012).

[98]See the animation "NASA: A Warming World" at the end of the chapter.

is reflected back to space and the rest is absorbed by the planet's surface. The planet releases this heat back into the atmosphere in the form of invisible infrared light, some of which is trapped by greenhouse gases. The growth of greenhouse gases implies that more heat is being trapped and less is being lost to space (indeed, measurements confirm this effect[99]). This equilibrium is described by the concept of radiative forcing, which uses watts of energy per square meter of Earth's surface (W/m^2) to describe the balance.[100]

The Sun

Radiative forcing allows scientists to identify imbalances in the energy budget of the atmosphere. An energy imbalance is the difference between the amount of solar energy absorbed by Earth's surface and the amount returned to space as heat. Researchers have calculated that despite unusually low solar activity between 2005 and 2010, Earth continued to absorb more energy than it returned to space; this is a result of heat-trapping greenhouse gases. The Sun undergoes a regular cyclical oscillation in energy output called the sunspot cycle. Solar irradiance, the amount of energy produced by the Sun that reaches the top of Earth's atmosphere, typically declines by about a 0.1% during low periods in the sunspot cycle. These solar minimums occur about every 11 years and last a year or so. However, the most recent minimum persisted more than two years longer than normal, making it the longest minimum recorded during the satellite era; NASA scientists saw that this was an opportunity to assess the impact of the Sun's energy on the temperature of Earth's atmosphere.

As we discussed earlier in the chapter, NASA scientists studied[101] the solar lull from 2005 to 2010 and calculated that despite the decrease in sunlight, Earth nonetheless accumulated 0.58 W of excess heat per square meter than it released back to space. This extra heat was more than twice as much as the reduction in solar energy between the maximum and minimum points of the sunspot cycle ($0.25\ W/m^2$). Lead researcher Dr. James Hansen stated "The fact that we still see a positive imbalance despite the prolonged solar minimum isn't a surprise given what we've learned about the climate system, but it's worth noting because this provides unequivocal evidence that the Sun is not the dominant driver of global warming."

Global Warming Potential

As we have learned, molecule for molecule some greenhouse gases are stronger than others. Each differs in its ability to absorb heat and in the length of time it resides in the atmosphere.[102] The ability to absorb heat and warm the atmosphere is expressed by its global warming potential (GWP), usually compared to CO_2 over some given time period. Methane traps 21 times more heat per molecule than carbon dioxide. Nitrous oxide absorbs 270 times more heat per molecule than CO_2. Fluorocarbons are the most heat-absorbent, with GWPs that are up to 30,000 times those of CO_2. The GWPs of various gases are very useful for understanding the impact of human emissions and determining what changes in emissions can accomplish the most positive effect in mitigating global warming; they also allow the attribution of warming to various types of human activities.[103]

[99]J. Hansen, R. Ruedy, M. Sato, K. Lo, "Global Surface Temperature Change," *Reviews of Geophysics* 48 (2010): RG4004, doi:10.1029/2010RG000345. See http://www.giss.nasa.gov/research/news/20110113/ (accessed July 9, 2012).

[100]For a simple description see "Explained: Radiative Forcing," *MIT News*, http://web.mit.edu/newsoffice/2010/explained-radforce-0309.html (accessed July 9, 2012).

[101]Hansen, M. Sato, P. Kharecha, K. von Schuckmann, "Earth's Energy Imbalance and Implications."

[102]G. Schmidt, R. Ruedy, R. Miller, and A. Lacis, "The Attribution of the Present-Day Total Greenhouse Effect." *Journal of Geophysical Research* 115 (2010): D20106, doi:10.1029/2010JD014287. See also A. Lacis, G. Schmidt, D. Rind, and R. Ruedy, "Atmospheric CO_2: Principal Control Knob Governing Earth's Temperature," *Science* 330 (2010): 356–359, doi:10.1126/science.1190653.

[103]N. Unger, T. Bond, J. Wang, et al., "Attribution of Climate Forcing to Economic Sectors," *Proceedings of the National Academy of Science* 107, no. 8 (2010): 3382–3387, www.pnas.org/cgi/doi/10.1073/pnas.0906548107.

Figure 1.17. Radiative forcing (RF) is the net effect of various factors that cool (blue bars) or warm (red bars) the atmosphere. RF is measured in watts/m² (bottom axis). The top three quarters of the box reports on human-induced RF, and the lower portion reports on the Sun before 2006, the only persistent natural factor (volcanic and ENSO effects are short-lived). The total net effect of human activities is strong warming (bottom), and CO_2 is the most important human factor.

Source: Figure from IPCC Fourth Assessment Report, Climate Change 2007, Synthesis Report; see http://www.ipcc.ch/publications_and_data/ar4/syr/en/figure-2-4.html. See also U.S. Global Change Research Program: Global Climate Change image gallery: http://www.globalchange.gov/resources/gallery.

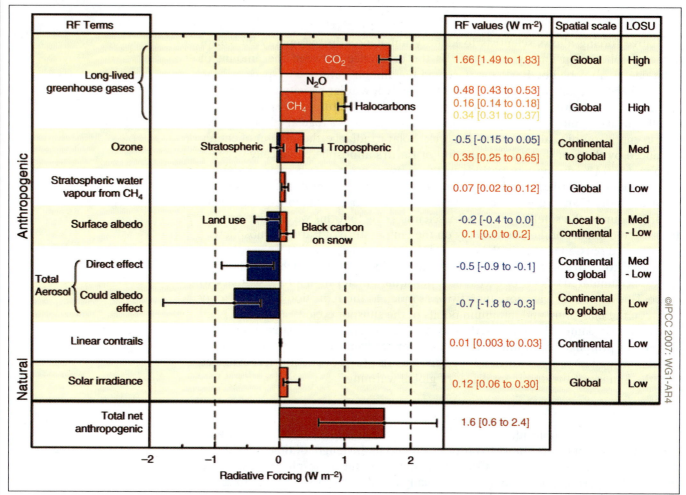

As greenhouse gases accumulate in the atmosphere, the amount of heat they trap also increases. Some factors offset this process, including the reflection of sunlight by aerosols and increases in the reflectivity of land cover and perhaps clouds (though it is becoming more apparent[104] that clouds provide a positive feedback to warming). The overall impact of compounds that alter the balance between radiation entering and exiting the atmosphere can be assessed by calculating the amount that they change the overall radiative forcing in watts per square meter (Figure 1.17).

Assigning Radiation Values to Human Behavior

Another approach to understanding the factors driving Earth's radiation balance is to assign radiation values to specific kinds of human behavior. Each segment of the economy, such as operating automobiles, doing agricultural work, generating power, or burning dung to boil water in Southeast Asia, emits a specific combination of gases and aerosols that influence the greenhouse effect in different ways and at different

---

[104]A. Lauer, K. Hamilton, Y. Wang, V. T. J. Phillips, and R. Bennartz, "The Impact of Global Warming on Marine Boundary Layer Clouds over the Eastern Pacific—A Regional Model Study," *Journal of Climate* 23, no. 21 (2010): 5844, doi: 10.1175/2010JCLI3666.1. See "Cloud Study Predicts More Global Warming," *ScienceDaily*, http://www.sciencedaily.com/releases/2010/11/101122172010.htm (accessed July 9, 2012).

Figure 1.18. By 2020 (left), transportation, household biofuels, and animal husbandry have the greatest warming impact on the climate, and the shipping, biomass burning, and industrial sectors have a cooling impact. By 2100 (right), the power and industrial sectors become strongly warming as the impacts of long-lived carbon dioxide accumulate.[105]

Source: NASA's Earth Science News Team at http://www.nasa.gov/topics/earth/features/road-transportation.html (accessed July 9, 2012).

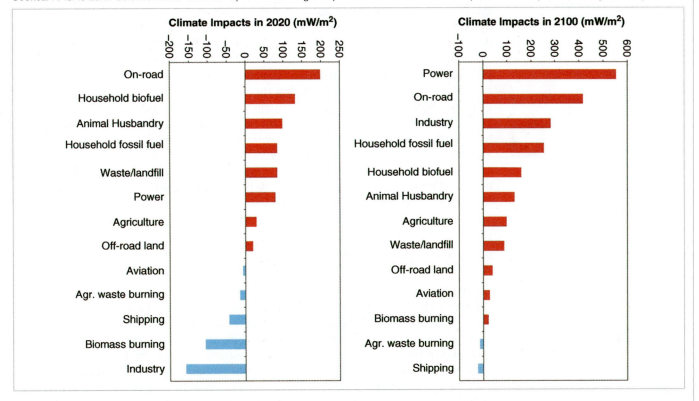

times. By reorganizing Figure 1.17 according to economic sectors, a profile emerges of how humans are affecting climate.

In this version[106] of radiative forcing, the impacts of various human activities can be calculated for the near future (2020) and the end of the century (2100). Today, and for the next decade or so, cars, trucks, and buses emerge as the greatest contributor to atmospheric warming (Figure 1.18). Motor vehicles release greenhouse gases that promote warming, and they emit few aerosols that counteract it. The next most important contributor is burning household biofuels, primarily wood and animal dung for heating and cooking. Third in line is the methane produced by livestock, particularly methane-producing cattle (whose numbers have grown enormously above natural levels due to industrial agriculture operations and whose methane emissions are amplified by a diet dedicated to rapid growth).

But the picture changes somewhat by the end of this century. Assuming that greenhouse gas emissions today remain relatively constant in the near-term future, electric power generation will overtake road transportation as the biggest promoter of warming, and the industrial sector will shift from the smallest contribution in 2020 to the third largest by 2100. These changes would occur because the aerosols produced by household biofuels have short lifetimes in the atmosphere and eventually rain out. But power generation and industrialization generate long-lived CO_2, and their impacts would accumulate and intensify over time.

Are There Activities that Promote Cooling?

Industrialization releases a high number of sulfates and other aerosols, leading to a significant amount of cooling. Biomass burning (such as tropical forest fires, deforestation,

[105]See NASA's Earth Science News Team at http://www.nasa.gov/topics/earth/features/road-transportation.html (accessed July 9, 2012).

[106]N. Unger, T. Bond, J. Wang, et al., "Attribution of Climate Forcing to Economic Sectors."

and savannah and shrub fires) produces black soot and greenhouse gases, but it also emits particles that block solar radiation. Poor air quality can produce health problems, however, and many developed countries have been reducing aerosol emissions through technology improvements driven by policies promoting public health (e.g., the Clean Air Act in the United States passed in 1963 and amended in 1970, 1977, and 1990 eliminated many cooling aerosols in the U.S.). By reducing air pollution, such efforts also decrease the cooling effect of aerosol production, likely leading to accelerated warming.

These results indicate that to reduce radiative forcing caused by human activities, policy makers can focus on decreasing emissions from transportation, household biofuel, and animal husbandry. Targeting the transportation sector may be particularly effective, because it would yield both short-term and longer-term climate benefits. Public health research indicates that traffic-related particulate matter is more toxic than particulates from the power sector,[107] and by reducing industrial particles there are benefits for human health. To protect Earth's climate in the longer term[108] and tackle concerns about climate change toward the end of this century, emphasis can be placed on reducing emissions from the power and industry sectors,[109] a conclusion that is consistent with findings of other research.[110]

MITIGATING GLOBAL WARMING REQUIRES MANAGING CARBON

Many of the chemical compounds found on Earth's surface move between the air, the water cycle, Earth's crust, and living organisms. Along the way, they go through biologic, geologic, and chemical exchanges and reactions, as does everything they come in contact with. The worldwide movement of these chemical compounds as they pass through, interact with, change, and are changed by Earth's atmosphere, crust, water supply, and life forms is known as global biogeochemical cycling.[111]

The key elements required for life move through biogeochemical cycles; they include oxygen, carbon, phosphorus, sulfur, and nitrogen. The rates at which elements and compounds move between places where they are temporarily stored (reservoirs) and where they are exchanged (processes) can be measured directly and modeled using computer programs.

The Carbon Cycle

One of the most important cycles that affects global climate is that of the element carbon (Figure 1.19). Most carbon stored on Earth is in the form of geologic (long-term) reservoirs (e.g., coal, oil, limestone). For the past 200 years humans have been moving carbon out of these reservoirs and into the atmosphere at greater rates than natural processes can move it back. This phenomenon is explained in the carbon cycle.

The global carbon cycle[112] describes the many forms that carbon takes in various reservoirs and processes. These include the following:

- Rocks in the crust, such as limestone ($CaCO_3$) and carbon-rich shale
- Gases in the atmosphere, such as carbon dioxide (CO_2) and methane (CH_4)
- Carbon dioxide dissolved in water (oceans and fresh water)

[107]T. Grahame, and R. Schlesinger, "Health Effects of Airborne Particulate Matter: Do We Know Enough to Consider Regulating Specific Particle Types or Sources?" *Inhalation Toxicology* 19 (2007): 457–481.

[108]See the animation "NASA Scientist James Hansen Talks about the Urgency of the Climate Crisis" at the end of the chapter.

[109]N. Unger, T. Bond, J. Wang, et al., "Attribution of Climate Forcing to Economic Sectors."

[110]M. Jacobson, "The Short-Term Cooling but Long-Term Global Warming Due to Biomass Burning." *Journal of Climate* 17 (2004): 2909–2926.

[111]See "Biogeochemical Cycle," *Wikipedia*, http://en.wikipedia.org/wiki/Biogeochemical_cycle (accessed July 9, 2012).

[112]See the carbon cycle explained by scientists at NASA: http://earthobservatory.nasa.gov/Features/CarbonCycle/ (accessed July 9, 2012).

Figure 1.19. Carbon is cycled through Earth's atmosphere, oceans, living organisms, and the crust. The values given here are global carbon reservoirs in gigatons (Gt; 1 Gt = 1 billion tons). Annual exchange and accumulation rates are in gigatons of carbon per year (Gt C/year).

SOURCE: Australian Government, Bureau of Meteorology, http://www.bom.gov.au/info/climate/change/gallery/9.shtml.

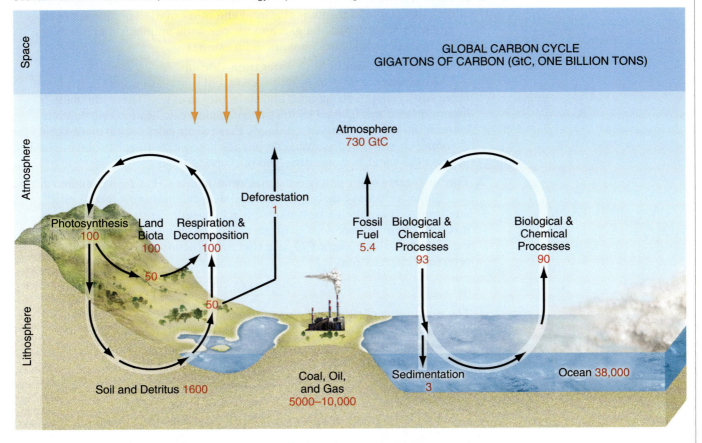

- Organic material in ecosystems, such as the simple carbohydrate glucose ($C_6H_{12}O_6$), found in plants and animals

Most of the carbon on Earth is contained in the rocks of the crust; it has been deposited slowly over tens of millions of years in the form of dead organisms (mostly plankton). Carbon is stored in the crust in two forms: (1) Oxidized carbon is buried as carbonate (CO_3) such as limestone, which is composed of calcium carbonate, $CaCO_3$, and (2) reduced carbon is buried as organic matter (such as dead plant and animal tissue).

Carbon moves through the carbon cycle via several processes:

1. **Limestone that forms under the ocean and surface waters traps carbon.**

    Most of Earth's carbon is contained in limestone, which provides effective long-term storage of carbon that has been taken from the atmosphere and transferred to the crust. The process of "storing" carbon in this way occurs in several steps:

    a. Carbon dioxide (CO_2), a gas made up of two oxygen molecules and one carbon molecule, is constantly moving from the atmosphere to the ocean and other surface waters, where it dissolves into the bicarbonate ion (HCO_3^-). In fact, it has been calculated that about half of the carbon dioxide released by human activities has been absorbed by the oceans. This (simplified) chemical reaction describes the process:

$$2CO_2 + 2H_2O \rightarrow 2HCO_3^- + 2H^+$$

    b. The bicarbonate ion combines with dissolved calcium (Ca^{2+}) in seawater to form calcium carbonate ($CaCO_3$, a mineral called calcite that is the primary component of limestone) in a reaction called calcification:

$$2HCO_3^- + Ca^{2+} \rightarrow CaCO_3 \text{ (limestone)} + CO_2 + H_2O$$

Note: In the first reaction, two molecules of CO_2 are taken from the atmosphere; in the second, one molecule of CO_2 is released. Thus, one molecule of carbon dioxide is stored in the limestone.

c. As limestone is formed, some atmospheric CO_2 is trapped and buried in the most stable of forms, rock. Coral and other marine organisms, such as mollusks, some types of algae, and the plankton animal foraminifera, are also excellent calcifiers (makers of limestone).

2. **Limestone that forms in fresh water traps calcium.**

   Most calcification occurs in the ocean, but some also occurs in fresh water. Have you ever seen stalagmites and stalactites in caves? These are made of limestone that was formed by the same chemical calcification reaction but without the help of plants and animals. Freshwater calcification often occurs by evaporation, wherein dissolved compounds precipitate (form a solid mineral) because the water they are dissolved in evaporates.

3. **The weathering of limestone consumes atmospheric CO_2, which contains carbon.**

   The movement of carbon doesn't end with calcification. Once formed, limestone can eventually be broken down by weathering, a natural process involving chemical reactions between rocks and atmospheric gases. Weathering consumes atmospheric CO_2 in a chemical reaction that is essentially the reverse of calcification:

$$CaCO_3 + CO_2 + H_2O \rightarrow Ca^{2+} + 2HCO_3^-$$

4. **The weathering of silica rocks, mostly in Earth's crust, also uses CO_2.**

   The weathering of types of rocks other than limestone also uses CO_2. These rocks are known as "silica" rocks, represented here as the mineral $CaSiO_3$, which symbolizes the rocks in Earth's crust. In this case a silica rock is changed by reaction with carbon dioxide and water into dissolved calcium, bicarbonate, and silica:

$$CaSiO_3 + 2CO_2 + H_2O \rightarrow Ca^{2+} + 2HCO_3^- + SiO_2$$

5. **Living organisms use CO_2 for photosynthesis and convert it into organic carbon.**

   Cycling of carbon also occurs among living organisms. Plants and some forms of bacteria can "use" inorganic CO_2 and convert it into organic carbon (such as carbohydrates and proteins), which is then consumed by all other forms of life, from zooplankton to humans, through the food chain. During photosynthesis, plants remove carbon dioxide from the atmosphere and convert it into organic carbon in plant tissues. Photosynthesis occurs on land in trees, grasses, and aquatic (freshwater) plants and in phytoplankton, algae, and kelp in ocean surface waters that are penetrated by sunlight. The reaction requires sunlight and chlorophyll, and in its simplest form it can be represented as follows:

$$6CO_2 + 6H_2O \rightarrow C_6H_{12}O_6 + O_2$$

(carbon dioxide + water → organic matter + oxygen)

Because CO_2 is a greenhouse gas, vegetation plays an important role in global climate. Through the process of photosynthesis, plants remove 200 billion tons of CO_2 from Earth's atmosphere each year. This is about 26% of the total amount of carbon in the atmosphere.

6. **Carbon returns to the atmosphere as gaseous CO_2, a byproduct of respiration and the decay of organic matter.**

   Some of the organic carbon created by plants during photosynthesis is consumed by animals and transferred through the food chain to higher forms of life. Eventually, the organic matter decays or is used in respiration, and the carbon is returned to the atmosphere as gaseous CO_2. Respiration is the reverse of photosynthesis, and it occurs when animals consume organic material to produce the energy they need to live. These organisms (from bacteria to humans) breathe, die, and decay, all processes that convert organic carbon into carbon dioxide, which is released back in the atmosphere. The basic chemical reaction for respiration is:

$$C_6H_{12}O_6 + O_2 \rightarrow 6CO_2 + 6H_2O$$

(organic matter + oxygen → carbon dioxide + water)

The cycling of carbon through photosynthesis and respiration is so rapid and efficient that all of the CO_2 in the atmosphere is estimated to pass through the global ecosystem every 4 to 5 years.

The Imbalance of the Carbon Cycle and Its Impact on Climate Change

Understanding the above steps of the carbon cycle, scientists are able to use computer software to measure, track, and model the movement of carbon dioxide and other forms of carbon throughout the carbon cycle. What they have learned is that many global events have changed the carbon cycle in the past: the coming and going of ice ages, changes in land surface and ocean currents related to plate tectonics, increased volcanism also related to plate tectonics, and others.

What scientists have also found is that as a result of human activities, more carbon is being released into the air than at any time in recent geologic history, resulting in the presence of more methane and carbon dioxide in the atmosphere, in turn resulting in a perturbation of the carbon cycle. For example, today more than 3.6 billion tons of carbon dioxide per year is released into the atmosphere by removing forests, which store carbon in tree trunks and leaves, and replacing them with crops or grasslands that store less carbon.

Burning fossil fuels (coal, oil, and natural gas) releases carbon that took millions of years to accumulate. This activity releases more than 36 billion tons of carbon dioxide into the atmosphere annually. These disruptions to the carbon cycle have caused the amount of carbon dioxide in the atmosphere to rise about 40% since the mid-1800s (Figure 1.20). As a consequence, extra carbon in the atmosphere is causing the planet to warm, threatening ecosystems worldwide,[113] and excess carbon dissolved in the oceans is causing the water to grow acidic, putting marine life in danger.

Now that you understand the basic components of Earth's climate system, in the next chapter we take a look at the evidence that climate has changed and that humans are the primary cause.

Figure 1.20. Emissions of carbon dioxide (shown in gigatons [billions of tons] of carbon) by human activities have been growing steadily since the onset of the Industrial Revolution. In 2010, global CO_2 emissions due to industrial activities grew by 5.9%, the largest annual increase on record.

SOURCE: NASA, http://earthobservatory.nasa.gov/Features/CarbonCycle/page1.php.

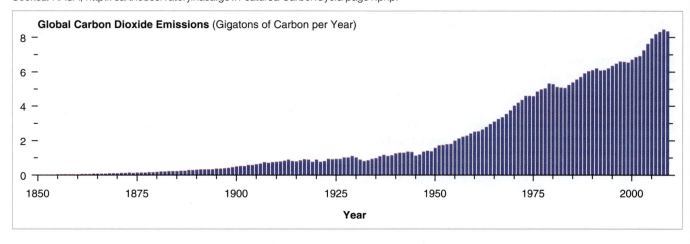

[113]C. Rosenzweig, D. Karoly, M. Vicarelli, et al., "Attributing Physical and Biological Impacts to Anthropogenic Climate Change," *Nature* 453, no. 7193 (2008): 353–357, doi:10.1038/nature06937.

ARBON

41

DEOS

mosphere,
tch?v=DHrapzHPCSA

ontereyinstitute.org/noaa/

hobservatory.nasa.gov/
ce/GlobalWarming/

Earth's Energy Balance, http://earthobservatory.nasa.gov/
Experiments/PlanetEarthScience/GlobalWarming/
GW_Movie2.php

Ocean Threats

Ocean Extinction, http://www.youtube.com/watch?v=vkNns0-
w79Q&feature=player_embedded

Ocean Pollution, http://www.youtube.com/watch?v=QxZs0B5
yqqo&feature=player_embedded

The Coral Story, http://www.youtube.com/watch?v=3WH_6Pg
NIQI&feature=player_embedded#at=62

Thawing Permafrost-Changing Planet, http://www.youtube.com/
watch?v=yN4OdKPy9rM

Water Cycle Animation, http://www.youtube.com/
watch?v=Az2xdNu0ZRk

NASA, "A Warming World," http://www.youtube.com/
watch?v=LjFz1FCKfT8

NASA Scientist James Hansen talks about the urgency of the
climate crisis, http://www.youtube.com/watch?v=f0hHlxaYN
b0&feature=related

Coral Winners, http://www.abc.net.au/catalyst/stories/3576802
.htm

Vital Role of the Oceans, http://www.youtube.com/watch?v=a
yb7zpXSs0g&feature=player_embedded

The Speed of Ocean Change, http://www.youtube.com/watch?
v=Giua4EmwPgw&feature=player_embedded#at=12

An Overview of Threats, http://www.youtube.com/
watch?feature=player_embedded&v=sup3XxHmBoo

COMPREHENSION QUESTIONS

1. What is climate literacy?
2. How is Earth's atmosphere organized?
3. Explain the global circulation of the atmosphere.
4. How do ocean currents transport heat and why is it important?
5. Describe the deadly trio and how they put the oceans at risk.
6. List the greenhouse gases in order of global warming potential.

7. For each greenhouse gas, describe at least one process that increases the amount of the gas in the atmosphere.
8. Why is carbon dioxide considered the most important greenhouse gas?
9. What is radiative forcing?
10. According to the paper by Unger et al. (2010), which five human activities have the greatest impacts on radiative forcing in the near future?

THINKING CRITICALLY

1. What could you and your friends do to decrease the impacts of global warming?
2. Describe some ways Earth's surface is becoming less reflective. What is decreased reflection's impact on climate change?
3. How are global warming and ocean acidification related?
4. Describe the role of ozone and nitrous oxide as greenhouse gases. What steps can policy makers take to reduce their future impact?
5. You are a politician running for President of the United States today. Describe what steps you would take to address future climate change likely to occur at the end of the century.

6. How does climate change affect the weather?
7. Why have carbon dioxide emissions increased so dramatically over the past 150 years?
8. How is the water cycle likely to change as the atmosphere gets warmer?
9. Study the human impacts on radiative forcing by the end of the century as modeled by Unger et al. (2010). Identify three activities that concern you; what can the United States do to mitigate their impacts?
10. Is climate change "dangerous," or is that too strong a word? Why?

CLASS ACTIVITIES (FACE TO FACE OR ONLINE)

ACTIVITIES

1. Visit the *Climate Literacy* website http://www.globalchange.gov/resources/educators/climate-literacy and answer the following questions.

   a. Why is it important for everyone to become informed on climate science?

   b. What are the essential principles of climate science?

   c. Describe the primary ways to improve understanding of the climate system.

   d. Describe how climate varies over space and time.

2. Explore this article at ABC News: http://abcnews.go.com/Technology/GlobalWarming/global-warming-common-misconceptions/story?id=9159877. What are the seven common misconceptions about global warming?

3. Visit the National Academy of Sciences *Global Warming Facts and Our Future* website http://www.koshland-science-museum.org/exhibitgcc/index.jsp and answer the following questions.

   a. What did you learn there about human impacts to the carbon cycle?

   b. What are the major causes of climate change?

   c. What does the term *amplified warming* mean?

   d. Describe some possible responses to global warming.

   e. How are carbon dioxide emissions expected to change between the years 2000 and 2025?

WHAT IS THE EVIDENCE FOR CLIMATE CHANGE?

Figure 2.0. Astronauts aboard the International Space Station captured this image of Earth's atmosphere and the Moon on July 31, 2011. Closest to Earth's surface, the orange-red glow reveals the troposphere—the lowest, densest layer of atmosphere. A brown transitional layer marks the upper edge of the troposphere, known as the tropopause. A milky white and gray layer rests above that, likely a slice of the stratosphere. The upper reaches of the atmosphere—the mesosphere, thermosphere, and exosphere—fade from shades of blue to the blackness of space.

IMAGE CREDIT: NASA Earth Observatory

CHAPTER SUMMARY

Climate change is a result of global warming, a genuine phenomenon about which there is little debate within the scientific community. Rather, scientists debate the questions "How sensitive is climate to greenhouse gas buildup?" and "What will climate change look like regionally and locally?" There is abundant, convincing, and reproducible scientific evidence that the increase in Earth's surface temperature is having measurable impacts on human communities and natural environments: Glaciers are melting, spring is coming earlier, the tropics are expanding, sea level is rising, the global water cycle is amplified, ecosystems are shifting, global wind speed has increased, drought and extreme weather are more common. These and many other observations document that the Earth system is rapidly changing in response to global warming.

In this chapter you will learn that:

- Excess heat in the atmosphere owing to rising levels of greenhouse gases is causing changes in ecosystems, weather patterns, and other climate-dependent aspects of Earth's surface.

- July 2012 marked the hottest month in U.S. history and the end of the hottest 12-month period in 117 years of record keeping. Drought expanded to cover 63% of the contiguous U.S. leading the U.S. Department of Agriculture to declare a national drought emergency.

- If global warming continues at its current rate, in the future Earth will be characterized by more abnormally hot days and nights; more heat waves; fewer cold days and nights; more frequent and severe droughts; greater storminess; a decrease in glaciers and ice sheets; erosion and inundation of coastal areas; and other effects.

- Leading research centers at universities, government offices, and institutions around the world conduct scientific investigations and publish their results in peer-reviewed, critically evaluated journals and reports. By and large, these are credible and abundant sources of information about climate change.

- The Intergovernmental Panel on Climate Change (IPCC) releases special reports on topics relevant to the implementation of the United Nations Framework Convention on Climate Change, an international treaty that

Learning Objective

Since 1880, the global mean annual air temperature has increased approximately 0.8°C (1.4°F). This is due to increased greenhouse gases in the atmosphere resulting from human activities. Because of global warming, scientists have observed widespread changes in climate, which in turn are causing significant changes in Earth's environments and ecosystems.

acknowledges the possibility of harmful climate change. The IPCC will publish its fifth assessment report in 2014.

- Certain "human fingerprints" on the climate system confirm that humans are the cause of global warming.

WHAT DO THE EXPERTS SAY?

There are many ways that humans affect Earth's environment and natural resources. In his book Earth in Mind,[1] David Orr writes that on a typical day we lose about 300 square kilometers (116 square miles) of rainforest to logging (one acre per second), 186 square kilometers (72 square miles) of land to encroaching deserts, and numerous species to extinction. Other sources tell us that in a day the world's human population increases by more than 200,000,[2] we add 100 million tons[3] of carbon dioxide to the atmosphere, and we burn an average of 84.4 million barrels of oil (1000 barrels per second[4]). By the end of the day, Earth's freshwater, soil, and ocean are more acidic,[5] its natural resources more depleted, and its temperature is a little hotter.[6]

These unrelenting impacts to Earths ecosystems and natural resources have led researchers to conclude[7] that our planet is perched on the edge of a tipping point, a planetary-scale critical transition as a result of human influence. Scientists are warning that human population growth, widespread destruction of natural ecosystems, and climate change are pushing Earth's ecosystems and resources toward irreversible change.[8]

[1] David W. Orr, *Earth in Mind* (Washington, D.C., Island Press, 2004).

[2] Answers.com, "How Much Does World Population Increase Each Day?" http://wiki.answers.com/Q/How_much_does_world_population_increase_each_day (accessed July 9, 2012).

[3] CO2Now.org, "What the World Needs to Watch," http://co2now.org/Current-CO2/CO2-Now/ (accessed July 9, 2012).

[4] Peter Tertzakian, *A Thousand Barrels a Second: The Coming Oil Break Point* (New York, McGraw-Hill, 2006).

[5] The U.S. Geological Survey has found that mining and burning coal, mining and smelting metal ores, and use of nitrogen fertilizer are the major causes of chemical oxidation processes that generate acid in the Earth-surface environment. These widespread activities have increased carbon dioxide in the atmosphere, increasing the acidity of oceans; produced acid rain that has increased the acidity of freshwater bodies and soils; produced drainage from mines that has increased the acidity of freshwater streams and groundwater; and added nitrogen to crop lands that has increased the acidity of soils. K. Rice and J. Herman, "Acidification of Earth: An Assessment across Mechanisms and Scales," *Applied Geochemistry* 27, no. 1 (2012): 1–14.

[6] Largest natural disaster in U.S. history declared today, see: http://www.examiner.com/article/largest-natural-disaster-u-s-declared-today. Additionally see also, USDA Announces Streamlined Disaster Designation Process with Lower Emergency Loan Rates and Greater CRP Flexibility in Disaster Areas; (accessed July 14, 2012).

[7] A. Barnosky, et al., "Approaching a State Shift in Earth's Biosphere," *Nature*, 486, no. 7401 (2012): 52–58, doi: 10.1038/nature11018.

[8] See the video "Conversation with climatologist Dr. James Hansen, Director of NASA's Goddard Institute for Space Studies, and Climate Crisis Coalition Coordinator Tom Stokes on May 10, 2008" at the end of the chapter.

National Academy of Sciences

Orr was not merely speculating. According to the U.S. National Academy of Sciences, it is "settled fact" that the Earth system is warming, and there is 90% to 99% probability that humans are the cause.[9]

> *Some scientific conclusions or theories have been so thoroughly examined and tested, and supported by so many independent observations and results, that their likelihood of subsequently being found to be wrong is vanishingly small. Such conclusions and theories are then regarded as <u>settled facts</u>. This is the case for the conclusions that the Earth system is warming and that much of this warming is very likely due to human activities.*[10]

This quotation, published in 2011 by a panel of scientists convened by the U.S. National Academy of Sciences, is included in a set of five volumes collectively called *America's Climate Choices*. The panel was compelled to conclude that "There is a strong, credible body of scientific evidence showing that climate change is occurring, is caused largely by human activities, and poses significant risks for a broad range of human and natural systems."[11]

Global warming causes climate change. As we learned in Chapter 1, warming is a consequence of deforestation, industrial agriculture, manufacturing, and other human activities that increase the concentration of heat-trapping greenhouse gases in the atmosphere. The greenhouse gases include carbon dioxide (CO_2), methane (CH_4), water vapor (H_2O), nitrous oxide (N_2O), ozone (O_3) and others (see Chapter 1 for discussion of these). Actions such as burning oil and coal[12] release these gases to the atmosphere in quantities that have increased with the rise of the industrial age (and, as one respected climatologist proposes, since humans first domesticated animals and cleared land for farms beginning 8,000 years ago[13]).

Climate Change Evidence

There is abundant, convincing, and reproducible scientific evidence that the resulting increase in Earth's surface temperature is having measurable impacts on human communities and natural ecosystems. In fact, within the scientific community, rather than debating whether climate change is happening, the debate centers on whether climate is changing faster than anticipated.[14]

- By 2015, the atmospheric carbon dioxide concentration will reach 400 parts per million (ppm[15]), growing at an average annual rate of about 2.1 ppm, almost three times the growth rate of the 1990s.[16] This concentration is the highest

---

[9] National Research Council, *Advancing the Science of Climate Change* (Washington, D.C., National Academies Press, 2011), 21–22, http://www.nap.edu/catalog.php?record_id=12782 (accessed July 9, 2012).

[10] National Research Council, *Advancing the Science of Climate Change*.

[11] National Research Council, *Advancing the Science of Climate Change*.

[12] Called "fossil fuels" because coal is made from fossil wetland plants, and oil is made of fossil marine algae.

[13] William Ruddiman has proposed the "anthropogenic hypothesis." It is supported by ice core data and calculations of the Earth-Sun orbital geometry, suggesting that the relatively warm climate of the past several thousand years is unnatural and should instead have been characterized by cooling. Ruddiman proposes that through the production of excess methane and carbon dioxide, human agricultural practices took control of Earth's climate as early as 5,000 to 8,000 years ago. See W. F. Ruddiman, "The Anthropogenic Greenhouse Era Began Thousands of Years Ago," *Climatic Change* 61 (2003): 261–293; and W. F. Ruddiman, "Cold Climate during the Closest Stage 11 Analog to Recent Millennia," *Quaternary Science Reviews* 24 (2005): 1111–1121; and W. F. Ruddiman, *Plows, Plagues, and Petroleum: How Humans Took Control of Climate* (Princeton, N.J., Princeton University Press, 2005).

[14] R. A. Kerr, "Amid Worrisome Signs of Warming, Climate Fatigue Sets In," *Science* 326 (2009): 926–928.

[15] Ppm means "parts per million." It is a measurement of abundance (or concentration) the same way that "per cent" means parts per hundred. In this case ppm means molecules of CO_2 per million molecules of air.

[16] R. A. Kerr, "Amid Worrisome Signs of Warming, Climate Fatigue Sets In."

since the Miocene Epoch (15 million years ago),[17] when sea level is estimated to have been 25 to 40 m (82–131 ft) higher and global temperature 3°C to 6°C (5°F to 10°F) warmer than present.

- In 2010 the average temperature on Earth's surface tied with 2005 as the warmest year since record keeping began in 1880[18] (Figure 2.1). The year 2011 was the ninth warmest[19] recorded (but it was the warmest year on record under the cooling influence of La Niña), and 9 of the 10 warmest years are in the 21st century, the only exception being 1998, which was warmed by the strongest El Niño of the past century.

- According to the National Aeronautics and Space Administration (NASA) Goddard Institute for Space Studies (GISS)[20] and the National Oceanic and Atmospheric Administration (NOAA),[21] 2009 was only a fraction of a degree cooler than 2005, and it tied with a cluster of other years—1998, 2002, 2003, 2006, and 2007—as the second warmest year since record keeping began.

- The first decade of the 21st century was the warmest decade since instrumental records began.

- During the past three decades, Earth's surface temperature has trended upward about 0.2°C (0.36°F) per decade.[22]

Figure 2.1. Nine of the 10 warmest years since 1880 have occurred since 2000, as Earth has experienced sustained higher temperatures than in any decade in recorded history. As greenhouse gas emissions and atmospheric carbon dioxide levels continue to rise, scientists expect the long-term temperature increase to continue as well.[23]

Source: NASA Earth Observatory, Robert Simmons.

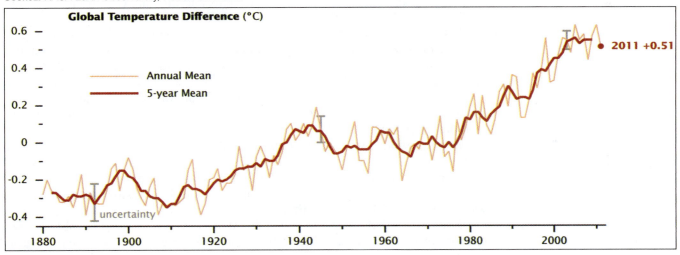

[17] A. K. Tripati, D. R. Roberts, R. A. Eagle, "Coupling of CO2 and Ice Sheet Stability over Major Climate Transitions of the Last 20 Million Years." *Science* 326, 5958 (2009): 1394–1397, http://www.sciencemag.org/cgi/content/abstract/1178296 (accessed July 9, 2012).

[18] National Oceanic and Atmospheric Administration (NOAA), "2010 Tied for Warmest Year on Record," http://www.noaanews.noaa.gov/stories2011/20110112_globalstats.html (accessed July 9, 2012). See also NASA, "NASA Research Finds 2010 Tied for Warmest Year on Record," http://www.nasa.gov/topics/earth/features/2010-warmest-year.html (accessed July 9, 2012).

[19] See NASA, "Global Temperature in 2011, Trends and Prospects," http://data.giss.nasa.gov/gistemp/2011/ (accessed July 9, 2012).

[20] See NASA, "Surface Temperature Analysis: Latest News," http://data.giss.nasa.gov/gistemp/ (accessed July 9, 2012).

[21] See NOAA Climate Services, *ClimateWatch*, http://www.climate.gov/#climateWatch (accessed July 9, 2012).

[22] See NASA, "2009: Second Warmest Year on Record; End of Warmest Decade," http://www.nasa.gov/topics/earth/features/temp-analysis-2009.html (accessed July 9, 2012).

[23] NASA, "NASA Research Finds 2010 Tied for Warmest Year on Record."

- Despite decades of discussion and scientific recognition of the problem of global warming, human communities have failed to reduce the production of greenhouse gases[24]—the known primary cause of warming—and instead have accelerated it. The respected journal Science reported: "Almost all climate scientists are of one mind about the threat of global warming: It's real, it's dangerous, and the world needs to take action immediately."[25]

The year 1998 set a record for warmth; 2005 did, too. In 2008, however, global mean temperature dropped, returning to temperatures not seen since the mid-1990s (although 2008 was, nonetheless, the ninth warmest year on record at the time). To the naked eye, a graph of annual temperatures from 1998 to 2008 looked as if global warming had stopped, when in fact average annual global temperature over the period still had a positive trend. Nevertheless, the drop in temperature from 2005 to 2008 influenced national attitudes, and a trend of "global cooling" was reported in some media. However, mistaking short-term variability (year-to-year changes in temperature) for long-term trends (climate change) is a fundamental error.

Scientists do not expect global warming to be expressed as a smooth annual rise in average atmospheric temperature from one year to the next. They understand that short-term climate processes (such as the El Niño Southern Oscillation, the sunspot cycle, and volcanic eruptions) dominate year-to-year temperatures[26] and that it does not get warmer everywhere at the same time. Like the rise in stock market value since the 1970s, climate is taking a bumpy ride of ups and downs as it undergoes a long-term increase in global temperature. In the same way, scientists do not take every snowstorm and cool day as evidence that global warming is not a looming issue or that it does not exist at all.

On the contrary, scientists recognize that global warming is a "noisy" process that requires analysis of both short-term events and the long-term trends. By 2009 studies emerged pointing this out.

In a blind test the Associated Press gave prominent statisticians global temperature data without identifying its source or what the numbers represented; the statisticians rejected global cooling.[27] Also, the U.S. National Climate Data Center published peer-reviewed research[28] reporting that climate history since the 1970s reveals many episodes when the average temperature of the atmosphere temporarily stopped rising, and even reversed its upward climb, but that strong net warming over the entire period is indisputable.

By the spring and summer of 2011 and 2012, all talk of global cooling ended as the United States experienced unprecedented heat waves and drought hit North America with dramatic intensity. July 2012 marked the hottest month in U.S. history[29] and drought expanded to cover 63% of the contiguous U.S. The average temperature was 25.33°C (77.6°F), 1.8°C (3.3°F) above the 20th century average, marking the hottest 12-month period the nation had endured in 117 years of record-keeping. NASA scientist James Hansen has written[30] that these record-breaking events are the logical result of global warming, "there is virtually no explanation other than climate change."

[24] The failure to act in response to the threat of global warming moved former Vice President Al Gore to speculate that future communities dealing with the worst consequences of global warming will be justified in looking back on us as a "criminal generation . . . the architects of humanity's destruction." A. Gore, *Our Choice: A Plan to Solve the Climate Crisis* (Emmaus, Penn., Rodale Press, 2009).

[25] R. A. Kerr, "Amid Worrisome Signs of Warming, Climate Fatigue Sets In."

[26] J. Hansen, R. Ruedy, M. Sato, K. Lo, "Global Surface Temperature Change," *Reviews of Geophysics* 48 (2010): RG4004, doi:10.1029/2010RG000345.

[27] S. Borenstein, "Statisticians Reject Global Cooling," Associated Press, October 26, 2009.

[28] D. Easterling and M. Wehner, "Is the Climate Warming or Cooling?" *Geophysical Research Letters* 36 (2009): L08706. See NASA, "The Ups and Downs of Global Warming," http://climate.nasa.gov/news/index.cfm?Fuse Action=ShowNews&NewsID=175 (accessed July 9, 2012).

[29] See NOAA State of the Climate, "July 2012: Hottest Month on Record for Contiguous United States," http://www.ncdc.noaa.gov/sotc/ (accessed August 12, 2012).

[30] See http://thinkprogress.org/climate/2012/08/09/666601/james-hansen-on-the-new-climate-dice-and-public-perception-of-climate-change/ (accessed August 12, 2012).

THE EARTH SYSTEM IS CHANGING

The circulation of heat through Earth's atmosphere and oceans links the planet's living organisms and environments, from soil at the equator to ice at the poles.[31] Even though Earth is 40,075 km (24,901 mi) in circumference and has a surface area of 509,600,000 square kilometers (196,757,000 square miles) the poles and tropics, deserts and forests, continents and oceans are all connected by certain global processes. These include mixing of the atmosphere and ocean, the water cycle, seasonal heating and cooling, and more.

Global warming causes changes to these processes on the scale of the whole Earth; the result is referred to as climate change. Detailed analysis of ice core records of climate over the past 20,000 years reveals that today's changes in climate are unique over that entire period; there is no "natural process" that can explain today's warming.[32]

Earth

Earth is not an unchanging ball of rock hurtling through space. Energy from within and without alter it. For example, heat diffuses upward from the core through the mantle, the thickest layer of Earth, causing rock in the mantle to flow and migrate. As heat moves through the crust, the outermost layer, it drives plate tectonics and causes volcanism.[33] Heat also arrives from the Sun. As this heat circulates through the atmosphere and oceans, and is carried by ocean and air currents around the planet, it too influences Earth's weather and climate.

These processes make Earth dynamic and cause it to constantly change, and it has been this way, in various forms, throughout its 4.6 billion-year history. For most of that history, those changes have been controlled by natural processes, and many of them have been enormous (such as the collision of continents and the increasing diversity of living forms). The natural processes that cause global climate change include plate tectonics, volcanic eruptions, solar cycles, extraterrestrial impacts, and variations in Earth's orbit (we will study these in following chapters); global climate change is also caused by human activities.

Irreversible Change

On modern Earth, human activities have indeed caused global changes in land use, air and water quality, and the abundance of natural resources,[34] particularly over the past two centuries. There is scientific consensus that human activities are also altering Earth's climate, largely owing to increasing levels of the heat-trapping gas carbon dioxide (CO_2) and other greenhouse gases.

Because it can reside in the atmosphere for more than 1,000 years,[35] carbon dioxide is the most powerful greenhouse gas. It is released when we burn fossil fuels, sources of energy provided by burning fossil carbon, such as petroleum (fossil marine algae) and coal (fossil continental wetland plants). A study by NOAA[36]

---

[31] See the video "General Circulation" the end of the chapter.

[32] S. Björck, "Current Global Warming Appears Anomalous in Relation to the Climate of the Last 20,000 Years," *Climate Research* 48, no. 1 (2011): 5, doi: 10.3354/cr00873.

[33] See the videos "Plate Tectonics" and "Heat Circulation within Earth" at the end of the chapter.

[34] According to the 2010 Edition of the National Footprint Accounts, humanity demanded the resources and services of 1.51 planets in 2007; such demand has increased 2.5 times since 1961. This situation, in which total demand for ecological goods and services exceeds the available supply for a given location, is known as *overshoot*. On the global scale, overshoot indicates that stocks of ecological capital may be depleting and/or that waste is accumulating. See the Global Footprint Network, http://www.footprintnetwork.org/en/index.php/GFN/page/at_a_glance/ (accessed July 9, 2012).

[35] S. Solomon, G.-K. Platter, R. Knutti, and P. Friedlingstein, "Irreversible Climate Change Due to Carbon Dioxide Emissions," *Proceedings of the National Academy of Science* 106 (2009): 1704–1709, doi: 10.1073/pnas.-9128211-6.

[36] S. Solomon, G.-K. Platter, R. Knutti, and P. Friedlingstein, "Irreversible Climate Change Due to Carbon Dioxide Emissions."

concluded that climate change is largely irreversible for the next 1,000 years because of the long lifetime of CO_2 in the atmosphere. As a result, at higher levels of carbon dioxide (450 to 600 ppm), sea-level rise, changes in rainfall, severe weather events, and other consequences of global warming will come to permanently (relative to human time scales) characterize the planet's surface.

Observed Impacts

Changes in precipitation (rain and snowfall), the source of our drinking water, the cause of flooding, and the crucial factor governing the health of critical ecosystems that provide us with natural resources are of special concern to humanity. Studies[37] document that global warming directly influences precipitation because the water-holding capacity of air increases by about 7% for each 1°C (1.8°F) of warming. Thus, storms that are provided with more moisture produce more-extreme precipitation events. Warmer air also results in greater evaporation that dries Earth's surface, increasing the intensity and duration of drought.

Global warming is producing a world that is drier, yet, ironically, prone to greater flooding. Because warming is producing only modest changes in winds, generalized precipitation patterns do not change much, and thus wet areas are becoming wetter and dry areas are becoming drier.[38] Notably, a warmer atmosphere produces more rainfall instead of snow, and winter snowpack melts earlier. This increases runoff in late winter and early spring, raising the risk of flooding and extending the duration and intensity of summer drought. Farmers, communities, and government agencies responsible for public safety and health all find it challenging to adapt to this new pattern.

Heat waves and drought are a consequence of warmer air temperatures that have been felt throughout the Northern Hemisphere. Researchers[39] have shown that Earth's land areas have become much more likely to experience an extreme summer heat wave than they were in the middle of the 20th century. Extremely hot temperatures covered about 0.1 to 0.2% of the globe from 1951 to 1980. But since then average temperatures have risen and extremely hot temperatures now cover about 10% of the globe.

Studies indicate that the climate change observed during the 20th and early 21st centuries is due to a combination of changes in solar radiation, volcanic activity, land use, and increases in atmospheric greenhouse gases. Of these, greenhouse gases are the dominant long-term influence, and they are causing the lower atmosphere, the air closest to Earth, to warm. This excess heat is causing dramatic changes in ecosystems, weather patterns, and other climate-dependent aspects of Earth's surface (Figure 2.2). These changes are listed in Box 2.1.

Future Change

How will all these changes play out in a future characterized by continued global warming? This question has been at the root of much of the research being conducted by climate scientists in recent years. For instance, in a report produced by the U.S. Global Change Research Program,[40] a combined effort of more than a dozen government science agencies, researchers found the following:

- Future abnormally hot days and nights and heat waves are very likely to become more common.

---

[37] K. Trenberth, "Changes in Precipitation with Climate Change," *Climate Research* (2011) doi: 10.3354/cr00953.

[38] P. Durack, S. Wijffels, and R. Matear, "Ocean Salinities Reveal Strong Global Water Cycle Intensification during 1950 to 2000," *Science* 336, no. 6080 (2012): 455–458, doi: 10.1126/science.1212222.

[39] J. Hansen, M. Sato, and R. Ruedy. PNAS Plus: "Perception of Climate Change." *Proceedings of the National Academy of Sciences*, 2012; doi: 10.1073/pnas.1205276109.

[40] T. R. Karl, G. A. Meehl, C. D. Miller, et al. (eds.), *Weather and Climate Extremes in a Changing Climate*. Report by U.S. Climate Change Science Program and Subcommittee on Global Change Research. (Washington, D.C., Department of Commerce, NOAA National Climatic Data Center, 2008); http://www.climatescience.gov/Library/sap/sap3-3/final-report/default.htm (accessed July 9, 2012).

Figure 2.2. Global warming is changing Earth's climate, leading to rising sea levels, changes in weather, and ecosystem impacts. These changes pose an extraordinary challenge to the natural environment.

SOURCE: After Skepticalscience.com and National Climatic Data Center.[41]

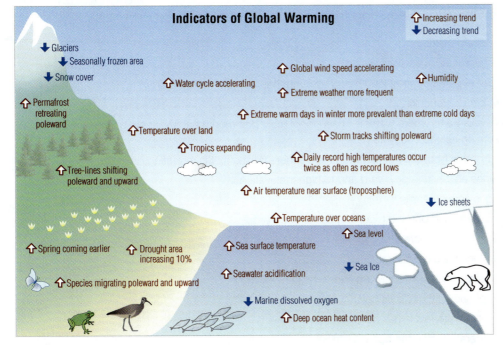

- Cold days and cold nights are very likely to become much less common. The number of days with frost is very likely to decrease.

- Future sea ice extent will continue to decrease and could even disappear entirely in the Arctic Ocean in summer in coming decades. Sea ice loss has increased coastal erosion in Arctic Alaska and Canada because of increased exposure of the coastline to wave action.

- Future precipitation is likely to be less frequent but more intense, and precipitation extremes are very likely to increase.

- Future droughts are likely to become more frequent and severe in some regions (e.g., U.S. Southwest, Mexico), leading to a greater need to respond to reduced water supplies, increased wildfires, and various ecological impacts.

- Future hurricanes in the North Atlantic and North Pacific are likely to have increased rainfall and wind speeds; for each 1°C (1.8°F) increase in tropical sea-surface temperatures, rainfall rates will increase by 6% to 18% and wind speeds of the strongest hurricanes will increase by 1% to 8%.

- Future strong cold-season storms in both the Atlantic and Pacific are likely to be more frequent, with stronger winds and more extreme wave heights.

Climate change has already transformed our planet. Air temperatures have risen, and as a result heat waves and drought are more common, storms have increased in frequency and intensity, seasons have shifted, the ranges of plant and animal life have moved, glaciers are melting, sea levels have risen, and the temperature of the oceans has increased. Climate change is rapidly altering the lands and waters we depend on for survival, and the cause is the buildup of greenhouse gases produced by human activities.

[41] See the detailed NCDC website, where you can view plots of datasets: http://www.ncdc.noaa.gov/bams-state-of-the-climate/2009-time-series/land; last viewed 1/12/12.

BOX 2.1

Climate Changes Resulting from Global Warming

Glaciers are melting.[1]

Air temperature over land is rising.[2]

The global percentage of land area in drought has increased about 10%.[3]

Snow cover is shrinking.[4]

Regions of Earth where water is frozen for at least one month each year are shrinking with impacts on related ecosystems.[5]

The southern boundary of Northern Hemisphere permafrost is retreating poleward.[6]

Tree lines are shifting poleward and to higher elevations.[7]

Spring is coming earlier.[8]

Plants are leafing out and blooming earlier each year.[9]

The lower atmosphere (troposphere) is warming.[10]

The tropics have expanded.[11]

Species are migrating poleward and to higher elevations.[12]

Atmospheric humidity is rising.[13]

The global water cycle has accelerated.[14]

Air temperature over the oceans is rising.[15]

[1] V. Radić and R. Hock, "Regionally Differentiated Contribution of Mountain Glaciers and Ice Caps to Future Sea-Level Rise," *Nature Geoscience* 4 (2011): 91–94.

[2] M. J. Menne, C. N. Williams, and M. A. Palecki, "On the Reliability of the U.S. Surface Temperature Record," *Journal of Geophysical Research* 115 (2010): D11108, doi:10.1029/2009JD013094.

[3] A. Dai, "Characteristics and Trends in Various Forms of the Palmer Drought Severity Index during 1900–2008," *Journal of Geophysical Research* 116 (2011): D12115, doi:10.1029/2010JD015541.

[4] S. J. Déry and R. D. Brown, "Recent Northern Hemisphere Snow Cover Extent Trends and Implications for the Snow Albedo-Feedback." *Geophysical Research Letters* 34 (2007): L22504.

[5] A. Fountain, J. Campbell, E. Schuur, S. et al, "The Disappearing Cryosphere: Impacts and Ecosystem Responses to Rapid Cryosphere Loss," *BioScience* 62, no. 4 (2012): 405–415, doi: 10.1525/bio.2012.62.4.11.

[6] S. Thibault and S. Payette, "Recent Permafrost Degradation in Bogs of the James Bay Area, Northern Quebec, Canada," *Permafrost and Periglacial Processes* 20, no. 4 (2009): 383, doi: 10.1002/ppp.660.

[7] P. S. A. Beck, G. P. Juday, C. Alix, et al, "Changes in Forest Productivity Across Alaska Consistent with Biome Shift," *Ecology Letters* 2011, doi: 10.1111/j.1461-0248.2011.01598.x.

[8] M. Kahru, V. Brotas, M. Manzano-Sarabia, and B. G. Mitchell, "Are Phytoplankton Blooms Occurring Earlier in the Arctic?" *Global Change Biology* 2010, doi: 10.1111/j.1365-2486.2010.02312.x.

[9] E. Wolkovich, B. Cook, J. Allen, et al, "Warming Experiments Underpredict Plant Phenological Responses to Climate Change," *Nature* 485, no. 7399 (2012): 494–497, doi: 10.1038/nature11014.

[10] P. W. Thorne, J. R. Lanzante, T. C. Peterson, D. J. Seidel, K. P. Shine, "Tropospheric Temperature Trends: History of an Ongoing Controversy," *Wiley Interdisciplinary Reviews: Climate Change* 2010, doi: 10.1002/wcc.80.

[11] D. J. Seidel, Q, Fu, W. J. Randel, and T. J. Reichler, "Widening of the Tropical Belt in a Changing Climate," *Nature Geoscience* 1 (2008): 21–24, doi:10.1038/ngeo.2007.38.

[12] S.R. Loarie, P.B. Duffy, H. Hamilton, et al, "The Velocity of Climate Change," Nature 462 (2009): 1052–1055.

[13] K. Willett, N. Gillett, P. Jones, and P. Thorne, "Attribution Of Observed Surface Humidity Changes To Human Influence," *Nature* 449 (2007): 710–712, doi:10.1038/nature06207.

[14] P. Durack, S. Wijffels, and R. Matear, "Ocean Salinities Reveal Strong Global Water Cycle Intensification during 1950 to 2000," *Science* 336, no. 6080 (2012): 455–458, DOI: 10.1126/science.1212222.

[15] NOAA National Climatic Data Center, "State of the Climate: Global Analysis for May 2011," published online June 2011, http://www.ncdc.noaa.gov/sotc/global/ (accessed July 9, 2012).

Sea surface temperature is rising.[16]

Ocean water is more acidic from dissolved CO_2, and this is negatively affecting marine organisms.[17]

Dissolved oxygen in the oceans is declining because of warmer water.[18]

Deep ocean temperature is rising.[19]

Continental ice sheets are shrinking.[20]

Arctic sea ice is shrinking as a result of global warming.[21]

Storm tracks are shifting poleward.[22]

Extreme weather events are more frequent.[23]

Daily record high temperatures occur twice as often as record lows.[24]

Sea level is rising and the rising has accelerated.[25]

Global wind speed has accelerated.[26]

Extreme warm events in winter are much more prevalent than cold events.[27]

Extreme weather is increasing.[28]

Global warming is changing life on Earth on a global scale.[29]

---

[16] S. Levitus, J. Antonov, T. Boyer, et al., "Global Ocean Heat Content in Light of Recently Revealed Instrumentation Problems," *Geophysical Research Letters* 36 (2008): L07608, doi:10.1029/2008GL037155.

[17] A. Barton, B. Hales, G. Waldbusser, C. Langdon, and R. Feely, "The Pacific Oyster, *Crassostrea gigas*, Shows Negative Correlation to Naturally Elevated Carbon Dioxide Levels: Implications for Near-Term Ocean Acidification Effects," *Limnology and Oceanography* 57, no. 3 (2012): 698, doi: 10.4319/lo.2012.57.3.0698.

[18] L. Stramma, E. Prince, S. Schmidtko, et al., "Expansion of Oxygen Minimum Zones May Reduce Available Habitat for Tropical Pelagic Fishes," *Nature Climate Change* 2 (2011): 33–37, doi:10.1038/nclimate1304.

[19] Y. T. Song and F. Colberg, "Deep Ocean Warming Assessed from Altimeters, Gravity Recovery and Climate Experiment, in situ Measurements, and a non-Boussinesq Ocean General Circulation Model," *Journal of Geophysical Research* 116 (2011): C02020, doi:10.1029/2010JC006601.

[20] E. Rignot, I. Velicogna, M. van den Broeke, A. Monaghan, and J. Lenaerts, "Acceleration of the Contribution of the Greenland and Antarctic Ice Sheets to Sea Level Rise," *Geophysical Research Letters* 38 (2011): L05503, doi:10.1029/2011GL046583.

[21] M. Serreze, M. Holland, and J. Stroeve, "Perspectives on the Arctic's Shrinking Sea-Ice Cover," *Science* 315 (2007): 1533–1536.

[22] F. A-M Bender, V. Ramanathan, and G. Tselioudis, "Changes in Extratropical Storm Track Cloudiness 1983–2008: Observational Support for a Poleward Shift," *Climate Dynamics* 38 (2012): 2037–2053, doi:10.1007/s0038-011-1065-6.

[23] D. Medvigy and C. Beaulieu, "Trends in Daily Solar Radiation and Precipitation Coefficients of Variation since 1984." *Journal of Climate* 25, no. 4 (2011): 1330–1339, doi: 10.1175/2011JCLI4115.1.

[24] G. Meehl, C. Tebaldi, G. Walton, D. Easterling, and L. McDaniel, "The Relative Increase of Record High Maximum Temperatures Compared to Record Low Minimum Temperatures in the U.S.," *Geophysical Research Letters* 36 (2009): L23701, doi:10.1029/2009GL040736.

[25] M.A. Merrifield, S.T. Merrifield, and G.T. Mitchum, "An Anomalous Recent Acceleration of Global Sea Level Rise," *Journal of Climate* 22 (2009): 5772–5781.

[26] I. R. Young, S. Zieger, and A. V. Babanin, "Global Trends in Wind Speed and Wave Height," *Science* 332, no. 6028 (2011): 451–455, doi: 10.1126/science.1197219.

[27] K. Guirguis, A. Gershunov, R. Schwartz, and S. Bennett, "Recent Warm and Cold Daily Winter Temperature Extremes in the Northern Hemisphere," *Geophysical Research Letters* 38 (2011): L17701, doi:10.1029/2011GL048762.

[28] P. O'Gorman, "Understanding the Varied Response of the Extratropical Storm Tracks to Climate Change," *Proceedings of the National Academy of Sciences* 107, no. 45 (2010): 19176–19180, doi: 10.1073/pnas.1011547107. See also T. R. Karl, G. A. Meehl, C. D. Miller, et al. (eds.), *Weather and Climate Extremes in a Changing Climate*. Report by U.S. Climate Change Science Program and Subcommittee on Global Change Research. (Washington, D.C., Department of Commerce, NOAA National Climatic Data Center, 2008); http://www.climatescience.gov/Library/sap/sap3-3/final-report/default.htm (accessed July 9, 2012).

[29] C. Rosenzweig, D. Karoly, Vicarelli, et al., "Attributing Physical and Biological Impacts to Anthropogenic Climate Change," *Nature* 453, no. 7193 (2008): 353–357.

During the summer of 2012, following the warmest spring on record, the average temperature of the continental U.S. was 2.5°C (4.5°F) above average. As a result, the 12 months ending July 31, 2012 were the warmest 12-month period on record for the U.S. This unusual heat wave, more intense than any on record, led to the declaration of the largest natural disaster in the history of the United States.[42] Sixty-three percent of the continental U.S. was enveloped in intense to extreme drought conditions and the U.S. Department of Agriculture declared a nationwide state of emergency. In the first six months of 2012, 27,042 new record highs were set in the United States, more than the 26,674 record highs set during the entire 12-month period of 2011. Researchers concluded[43] that extremely hot summers around the world are now 40 times more frequent than they were 10 years ago.

If we don't act to lessen the cause of global warming and adapt our societies to the new reality that has emerged, we may leave our children—and all living things— with a world characterized by the most dangerous consequences of climate change.[44]

RELIABLE SOURCES OF CLIMATE CHANGE INFORMATION

"Global warming" refers to an increase in the average temperature of Earth's surface, including the air, the land, and the oceans. Global warming is not a political position. It is a scientific certainty that has been verified by independent studies of literally thousands of scientists. A 2009 study[45] of scientific consensus on global warming published by the American Geophysical Union[46] concludes

> *The debate on the authenticity of global warming and the role played by human activity is largely nonexistent among those who understand the nuances and scientific basis of long-term climate processes. The challenge, rather, appears to be how to effectively communicate this fact to policy makers and to a public that continues to mistakenly perceive debate among scientists.*

Climate Data

Data on global temperature are collected by several groups. In the United States, climate data are collected, maintained, and analyzed by the NASA Goddard Institute of Space Studies (GISS)[47] and the National Oceanographic and Atmospheric Administration (NOAA).[48,49] Because climate knows no boundaries, both these organizations work closely with governments and researchers worldwide (Figure 2.3). In the United Kingdom, the Met Office Hadley Center (UKMET)[50] is the foremost climate change research

---

[42] See: http://www.examiner.com/article/largest-natural-disaster-u-s-declared-today.

[43] See ThinkProgress http://thinkprogress.org/climate/2012/08/01/622111/juiced-by-climate-change-extreme-weather-on-steroids/ (accessed August 12, 2012).

[44] USDA Announces Streamlined Disaster Designation Process with Lower Emergency Loan Rates and Greater CRP Flexibility in Disaster Areas: http://www.usda.gov/wps/portal/usda/usdahome?contentid=2012/07/0228. xml&navid=NEWS_RELEASE&navtype=RT&parentnav=LATEST_RELEASES& edeployment_action=retrievecontent.

[45] P. T. Doran and M. K. Zimmerman, "Examining the Scientific Consensus on Climate Change," *Eos* 90, no. 3 (2009), http://tigger.uic.edu/~pdoran/012009_Doran_final.pdf (accessed July 9, 2012).

[46] The American Geophysical Union is a prominent international scientific organization of 50,000 researchers, teachers, and students in 137 countries. You can read their position statement about human impacts on climate change here: http://www.agu.org/outreach/science_policy/positions/climate_change2008.shtml (accessed July 9, 2012).

[47] See their homepage at: http://www.giss.nasa.gov/ (accessed July 9, 2012).

[48] See their homepage at: http://www.noaa.gov/climate.html (accessed July 9, 2012).

[49] See the animation "NASA: Climate Change Visualization 1880–2010" at the end of the chapter.

[50] See their homepage at: http://www.metoffice.gov.uk/ (accessed July 9, 2012).

Figure 2.3. Numerous institutions monitor global surface temperatures. Temperature anomalies plotted here are deviations from normal values. Despite subtle differences in the ways scientists perform their analyses, these four widely referenced records show remarkable agreement.

SOURCE: NOAA, "Global Temperature Trends, http://www.arctic.noaa.gov/detect/global-temps.shtml.

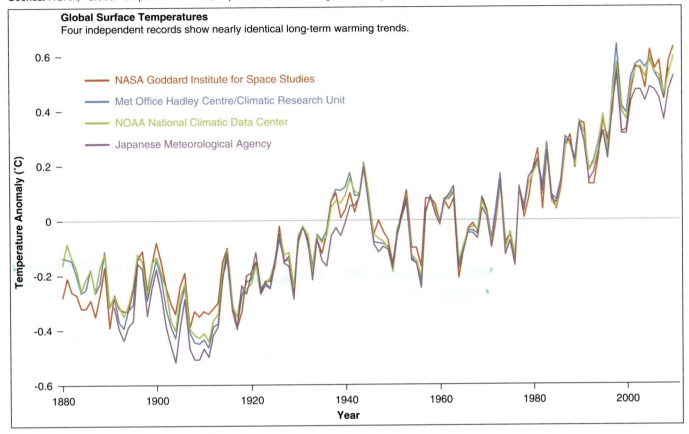

center, with responsibility to collect and analyze global climate information. The Japan Meteorological Agency[51] provides weather observation and forecasting and climate change and global environmental tracking services. These units along with researchers at universities, government offices, and institutions around the world conduct scientific investigations and publish their results in peer-reviewed journals and reports.

Media

One would expect that mainstream media accounts of science are generally reliable, but this is not always the case; headlines are conceived to sell controversy, not communicate fact. It is important to read past the headlines and filter personal opinion from scientific observation. Other sources of information include websites[52] and institutional reports and newsletters.[53]

Peer Review

The peer-review process, while not perfect, is the best available system for assessing the accuracy of scientific findings and ensuring that a rigorous standard is applied to

---

[51] See their homepage at: http://www.jma.go.jp/jma/indexe.html (accessed July 9, 2012).

[52] For instance, *Science Daily*: http://www.sciencedaily.com/ (accessed July 9, 2012). Also *Science News*: http://www.sciencenews.org/view/home (accessed July 9, 2012).

[53] For instance the Pew Center on Global Climate Change: http://www.pewclimate.org/ (accessed July 9, 2012). Also the Union of Concerned Scientists: http://www.ucsusa.org/ (accessed July 9, 2012). See as well the U.S. National Academy of Sciences: http://americasclimatechoices.org/ (accessed July 9, 2012).

the work of those who report on the results of their research. Typically, a scientist sends a manuscript describing the results of a research project to the editor of a scientific journal and requests publication. The editor reviews the work and sends it to other specialists in the field to get their opinions on its quality. On the basis of these reviews the editor makes a decision to publish the piece, reject it, or request revisions from the author subject to further review.

Peer-reviewed research forms the basis of improving our understanding of the details of climate change; what are the characteristics of changing air temperature? How rapidly is the ocean warming and how is this affecting marine ecosystems? Are there shifts in precipitation patterns, global winds, snow cover, and storminess? These and other questions drive the engine of climate research so that constantly advancing knowledge is the norm.

IPCC Assessments

Because key decision-makers may not keep up on the latest scientific research, it is important to provide summaries of our improving knowledge to policy makers and the public on a regular basis. This is a key role of the IPCC (Intergovernmental Panel on Climate Change, introduced in Chapter 1). The IPCC is an international organization under the joint auspices of the United Nations Environmental Program and the World Meteorological Organization.[54] The IPCC produces global assessments of climate change every five to seven years representing the state of understanding.

Past IPCC reports have been published in 1990, 1995, 2001, and 2007; the next report is dated 2014. The IPCC does not carry out original research, nor does it do the work of monitoring climate or related phenomena itself. Its primary role is publishing special reports on topics relevant to the implementation of the United Nations Framework Convention on Climate Change, which is an international treaty that acknowledges the possibility of harmful climate change.[55] The IPCC is organized in three working groups: Working Group I reports on the physical science basis of climate change; Working Group II reports on climate change impacts, adaptation, and vulnerability; and Working Group III reports on mitigation of climate change.

Knowledge about global warming can be acquired from IPCC assessment reports (for instance Assessment Report 4 [AR4] was published in 2007[56] and provides a detailed and thorough review of global, regional, and local climate patterns and processes) and from peer-reviewed scientific literature published between IPCC assessments in reputable journals such as Science,[57] Nature,[58] Nature Climate Change,[59] and others.

Global Change Research Program

The U.S. Global Change Research Program[60] published a report titled "Global Climate Change Impacts in the U.S.," which summarizes the science of climate

---

[54] See their homepage at: http://www.ipcc.ch/ (accessed July 9, 2012).

[55] Most countries are members of an international treaty, the United Nations Framework Convention on Climate Change, designed to consider what can be done to reduce global warming and to cope with whatever temperature increases are inevitable. See their homepage: http://unfccc.int/2860.php (accessed July 9, 2012).

[56] The 2007 fourth assessment report (AR4) consists of four elements, one from each of the three working groups and a Synthesis Report. All four reports, as well as past reports, can be found at: http://www.ipcc.ch/publications_and_data/publications_and_data_reports.htm (accessed July 9, 2012).

[57] See the Science journal homepage at: http://www.sciencemag.org/ (accessed July 9, 2012).

[58] See the Nature journal homepage at: http://www.nature.com/ (accessed July 9, 2012).

[59] See the Nature Climate Change journal homepage at: http://www.nature.com/nclimate/index.html (accessed July 9, 2012).

[60] See the U.S. Global Change Research Program: http://www.globalchange.gov/ (accessed July 9, 2012).

change and the impacts of climate change on the United States now and in the future. The report's key findings are as follows[61]:

- Global warming is unequivocal and primarily human induced.
- Climate changes are under way in the United States and are projected to grow.
- Widespread climate-related impacts are occurring now and are expected to increase.
- Climate change will stress water resources.
- Crop and livestock production will be increasingly challenged.
- Coastal areas are at increasing risk from sea-level rise and storm surge.
- Threats to human health will increase.
- Climate change will interact with many social and environmental stresses.
- Thresholds will be crossed, leading to large changes in climate and ecosystems.
- Future climate change and its impacts depend on choices made today.

This report is reviewed in detail in Chapter 6.

HOW UNUSUAL IS THE PRESENT WARMING?

To identify the difference between the present warming and natural climate changes it is useful to study climate in a longer geologic context. A number of studies have done this. One[62] approach is to search the geologic record of the past several thousand years for simultaneous changes in the Northern and Southern Hemispheres ("global" warming) such as has happened over the past century. Svante Björck, a climate researcher at Lund University in Sweden, has used this approach and shown that simultaneous warming of the two hemispheres has not occurred in the past 20,000 years. This is as far back as it is possible to analyze with sufficient precision to compare with modern climate changes occurring at a rapid pace. His study concludes that what is happening today is unique from a historical geological perspective. The field of paleoclimatology is reviewed in detail in Chapter 3.

Unprecedented Warming

Several independent studies confirm that recent warming is unprecedented in both magnitude (the amount of warming) and speed (the rate of warming). For instance, a study[63] of North Atlantic currents flowing into the Arctic highlights the fact that the Arctic is responding more rapidly to global warming than most other areas on our planet. This is called Arctic amplification. Researchers concluded that early 21st century temperatures of Atlantic water entering the Arctic Ocean are unprecedented over the past 2,000 years and are presumably linked to the Arctic amplification of global warming.

These facts have raised alarm among scientists, some of whom[64] have concluded that the Arctic Ocean is already suffering the effects of a dangerous climate change. Another study[65] concluded that 20th century warming of deep North Atlantic currents has had no equivalent during the last thousand years. Still another research

---

[61] Key findings of the report *Global Climate Change Impacts in the U.S.* may be accessed here: http://www.globalchange.gov/publications/reports/scientific-assessments/us-impacts/key-findings (accessed July 9, 2012).

[62] S. Björck S, "Current Global Warming Appears Anomalous in Relation to the Climate of the Last 20,000 years." *Climate Research* 48 (2011): 5–11.

[63] R. Spielhagen, K. Werner, S. Sørensen, et al, "Enhanced Modern Heat Transfer to the Arctic by Warm Atlantic Water," *Science* 331, no. 6016 (2011): 450–453, doi: 10.1126/science.1197397.

[64] C. Duarte, T. Lenton, P. Wadhams, and P. Wassman, "Abrupt Climate Change in the Arctic," *Nature Climate Change* 2 (2012): 60–62, doi:10.1038/nclimate1386.

[65] B. Thibodeau, A. de Vernal, C. Hillaire-Marcel, and A. Mucci, "Twentieth Century Warming in Deep Waters of the Gulf of St. Lawrence: A Unique Feature of the Last Millennium," *Geophysical Research Letters* 37 (2010): L17604, doi:10.1029/2010GL044771.

effort[66] concluded that the past few decades have been characterized by a global temperature rise that is unprecedented in the context of the last 1600 years. Research[67] by the National Center for Atmospheric Research concluded that Arctic temperatures in the 1990s and 2000s reached their warmest level of any decade of the past 2000 years. They found that the Arctic would be experiencing a long-term cooling trend (due to the nature of Earths orbital configuration with the Sun) were it not for greenhouse gases that are overpowering natural climate patterns.

The aggregate conclusion of these independent studies is unmistakable: Present warming is unprecedented in recent geologic history, no natural mechanism can be identified accounting for modern climate change, and human greenhouse gas emissions have the obvious potential to be the cause of the present warming.

THE AIR IS HEATING . . .

The distribution of heat on our planet, while sounding like an obscure subject, is important in every region and every environment on Earth. The total amount of heat and its variation across the planet surface drives global winds that circulate the atmosphere and control regional weather patterns, rainfall, growing seasons, and living conditions to which humans have adapted since civilization began. Earth is the right distance from the Sun (about 148 million kilometers; 92 million miles), has the right combination of gases in its atmosphere, and has water covering more than 70% of the planet's surface, which allow the origin and evolution of life and the resources necessary to sustain life. So far as we know, no other planet in our solar system has the thermal, physical, and chemical conditions that allow life to exist. This is what makes our blue planet unique and habitable.

However, global warming threatens severe changes to aspects of this system, including the temperature regime under which human civilization has developed, the location and distribution of agriculture and other protein sources that sustain us, the natural ecosystems that provide important services, and the supply of water around which we have built communities. By studying climate change, we gain critical knowledge that will support efforts to adapt to, and mitigate the negative impacts of, global warming (Figure 2.4).

In 2006, the U.S. Congress asked the National Research Council (NRC) to study Earth's climate and report on the levels of warming in recent history. The NRC concluded[68] that Earth's average surface temperature today is the highest of the past 1,300 years. Their report states that Earth's surface warmed 0.6°C (1°F) during the 20th century and is projected to warm by an additional (approximately) 2°C to 6°C (3.6°F–10.8°F) during the 21st century. Global average temperature measurements by instruments indicate a near-level trend from 1856 to about 1910, a rise to 1945, a slight decline to about 1975, and a rise to the present. Global warming is also verified by several independent sources including the National Climatic Data Center[69] (NCDC), NASA,[70] U.K. Met Office,[71] Japan Meteorological Agency,[72] the IPCC, and others.

---

[66] T. Kellerhals, S. Brütsch, M. Sigl, et al, "Ammonium Concentration in Ice Cores: A New Proxy for Regional Temperature Reconstruction?" *Journal of Geophysical Research* 115 (2010): D16123, doi:10.1029/2009JD012603.

[67] D. S. Kaufman, D.P. Schneider, N.P. McKay, et al, "Recent Warming Reverses Long-Term Arctic Cooling," *Science* 325 (2009): 1236–1239.

[68] National Research Council, *Surface Temperature Reconstructions for the Last 2,000 Years* (Washington, D.C., National Academies Press, 2006), p. 29, http://www.nap.edu/catalog.php?record_id=11676 (accessed July 9, 2012).

[69] See: http://www.ncdc.noaa.gov/oa/climate/globalwarming.html (accessed July 9, 2012).

[70] See: http://data.giss.nasa.gov/gistemp/ (accessed July 9, 2012).

[71] See: http://www.cru.uea.ac.uk/cru/info/warming/ (accessed July 9, 2012).

[72] See: http://www.jma.go.jp/jma/en/Activities/cc.html (accessed July 9, 2012).

Figure 2.4. (Bottom) Annual global mean observed temperatures (black dots) along with lines showing trends in the data. The left axis shows temperature change relative to the 1961 to 1990 average, and the right axis shows the estimated actual global mean temperature (°C). Linear trend fits to the last 25 (yellow), 50 (orange), 100 (purple), and 150 years (red) are shown; these correspond to 1981 to 2005, 1956 to 2005, 1906 to 2005, and 1856 to 2005, respectively. Notice that for shorter recent periods, the slope of the trend is steeper, indicating that warming has accelerated. The blue line is based on smoothed data to capture the decadal variations. To give an idea of whether the fluctuations are meaningful, decadal 5% to 95% (light gray) error ranges above and below that line are given (accordingly, annual values do exceed those limits). Results from climate models suggest that there was little change before about 1915 and that a substantial fraction of the early 20th-century change was contributed by naturally occurring influences including solar radiation changes, volcanism, and natural climate variability. From about 1940 to 1970 the industrialization following World War II increased pollution in the Northern Hemisphere, contributing to cooling by blocking sunlight. Increases in carbon dioxide and other greenhouse gases dominate the observed warming after the mid-1970s. (Top) Patterns of global temperature trends from 1979 to 2005 estimated at the surface (left) and for the troposphere (right; the lower atmosphere extending about 10 km [6.2 mi] above Earth's surface), from satellite data. Gray areas indicate incomplete data. Note the more spatially uniform warming in the satellite troposphere record, and the surface temperature changes more clearly relate to land and ocean.[73]

SOURCE: Climate Change 2007: The Physical Science Basis. Working Group I Contribution to the Fourth Assessment Report of the Intergovernmental Panel on Climate Change, Figure TS.6. Cambridge University Press.

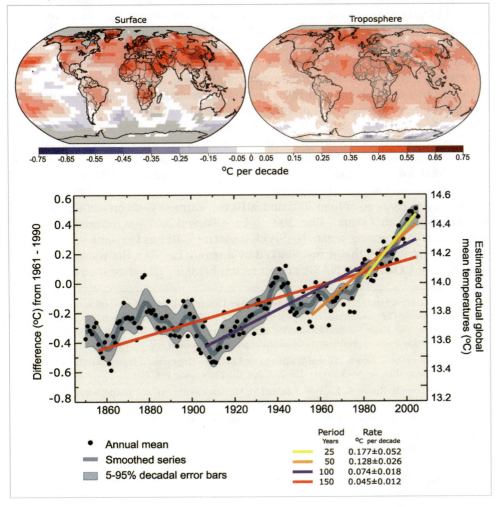

[73] Climate Change 2007: *The Physical Science Basis*. Working Group I Contribution to the Fourth Assessment Report of the Intergovernmental Panel on Climate Change, Figure TS.6. Cambridge University Press.

Sensors

Warming is documented by several types of independent sensors: Weather balloon measurements have found the global mean near-surface air temperature is warming by approximately 0.18°C (0.32°F) per decade,[74] satellite measurements of the lower atmosphere show warming of 0.16°C to 0.24°C (0.29°F–0.43°F) per decade since 1982,[75] continental weather stations document warming of approximately 0.2°C (0.36°F) per decade,[76] and ocean measurements using various types of sensors show persistent heating since 1970.[77] Notably, consistent with theory, satellite records of warming in the layers of the atmosphere near Earth's surface are matched by simultaneous cooling in the higher layers of the atmosphere (Figure 2.5). This makes perfect sense given that heat trapping near the surface would lead to cooling of the overlying air.

In a normal atmosphere where the amount of heat arriving from the Sun equals the amount that radiates back out to space, the temperature would not change and global warming would not be occurring. However, with increasing greenhouse gases, the amount of heat radiating back out to space is less than the amount arriving from the Sun, and the difference is trapped near Earth's surface by carbon dioxide and other anthropogenic gases.

By dissecting the temperature record of the past 160 years, researchers[78] have been able to define the components of temperature change that are the result of volcanic eruptions, the El Niño Southern Oscillation, variations in the Sun's energy, and warming due to increasing greenhouse gases. Finding that more than 90% of the excess heat trapped by greenhouse gases has been absorbed by the oceans, the study concludes that since 1850 and 1950, approximately 75% and 100%, respectively, of the observed global warming is due to human influences. In fact, it was determined that greenhouse gas emissions are responsible for 166% of the observed warming since 1950; that is, there would have been more greenhouse warming produced over the period, but it has been offset by aerosols (fine particles that reflect sunlight in the upper atmosphere, thus providing a cooling effect) produced by human manufacturing.

NASA collects global climate data including land and ocean measurements and provides an annual report and periodic updates (Figure 2.6). They report[79] that although 2008 was the coolest year of the decade because of strong cooling of the tropical Pacific Ocean, 2009 saw a return to near-record global temperatures, and 2010 tied with 2005 as the record high temperature over the period of monitoring. The year 2009 was only a fraction of a degree cooler than 2005 and 2010, the warmest years on record, and it tied with a cluster of other years—1998, 2002, 2003, 2006, and 2007—as the second-warmest year since recordkeeping began. NASA characterized 2011 as the ninth warmest year on record and noted[80] that it was marked by a strong La Niña, the warmest La Niña year on record. Officials at NASA predict record-breaking global average temperature

---

[74] J. K. Angell, "Global, Hemispheric, and Zonal Temperature Deviations Derived from Radiosonde Records." In *Trends Online: A Compendium of Data on Global Change.* (Oak Ridge, Tenn., Carbon Dioxide Information Analysis Center, Oak Ridge National Laboratory, U.S. Department of Energy, 2009), doi: 10.3334/CDIAC/cli.005; http://cdiac.esd.ornl.gov/trends/temp/angell/angell.html (accessed July 9, 2012).

[75] K. Y. Vinnikov and N. C. Grody, "Global Warming Trend of Mean Tropospheric Temperature Observed by Satellites," *Science* 302, no. 5643 (2003): 269-272, doi: 10.1126/science.1087910.

[76] J. Hansen, M. Sato, R. Ruedy, K. Lo, D.W. Lea, and M. Medina-Elizade, "Global Temperature Change," *Proceedings of the National Academy of Sciences* 103 (2006): 14288–14293, doi: 10.1073/pnas.0606291103, http://pubs.giss.nasa.gov/abstracts/2006/Hansen_etal_1.html (accessed July 9, 2012).

[77] S. Levitus, J. I. Antonov, T. P. Boyer, et al, "Global Ocean Heat Content 1955-2008 in Light of Recently Revealed Instrumentation Problems," *Geophysical Research Letters* 36 (2009): L07608, doi:10.1029/2008GL037155, http://www.agu.org/pubs/crossref/2009/2008GL037155.shtml (accessed July 9, 2012).

[78] M. Huber and R. Knutti, "Anthropogenic and Natural Warming Inferred from Changes in Earth's Energy Balance," *Nature Geoscience* 5 (2011): 31–36, doi:10.1038/ngeo1327.

[79] See the NASA GISS website for description of their methods of data collection and analysis: http://data.giss.nasa.gov/gistemp/ (accessed July 9, 2012) They publish annual global temperature summaries in December and January of each year.

[80] "NASA finds 2011 Ninth Warmest Year on Record," http://www.giss.nasa.gov/research/news/20120119/ (accessed July 9, 2012).

Figure 2.5. Map of temperature trends 1979 to 2010 (°C/decade) and 12 months running mean global temperature time series with respect to 1979 to 1998, in the lower troposphere (near surface) where temperature has risen in recent decades, and in the lower stratosphere (above the troposphere) where temperature has fallen because of heat trapping in the troposphere.[81]

SOURCE: From Wikipedia, http://en.wikipedia.org/wiki/File:RSS_troposphere_stratosphere_trend.png; see terms of use, http://wikimediafoundation.org/wiki/Terms_of_Use.

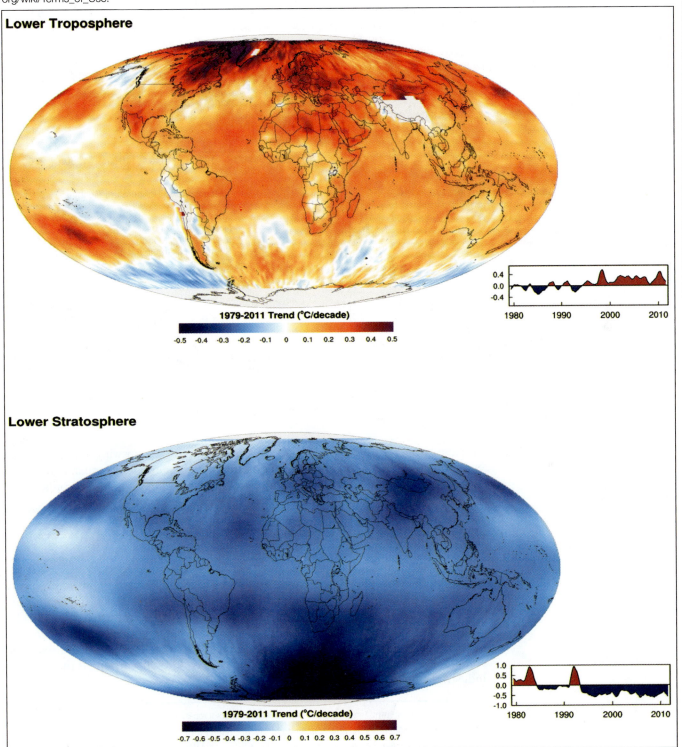

[81] See also C. A. Mears, F. J. Wentz, P. Thorne, and D. Bernie, "Assessing Uncertainty in Estimates of Atmospheric Temperature Changes from MSU and AMSU Using a Monte-Carlo Estimation Technique," *Journal of Geophysical Research* 116 (2011): D08112, doi:10.1029/2010JD014954.

Figure 2.6. Global temperature. **a,** Land and ocean trends: green indicates land; purple indicates ocean. **b,** Combined land and ocean are indicated by red. **c,** Temperature trends in the Northern (red) and Southern (blue) Hemispheres. Green bars represent uncertainty in the measurements. Temperature is graphed (on the vertical axis) as an "anomaly." This means it is the difference between the measured temperature of a year and a reference temperature. The reference is the global temperature averaged over the period 1951–1980. Individual annual means are plotted as single points joined by straight line segments; a bold line smoothes these annual measurements by plotting the five-year average. Note: 0.56°C = 1°F.

Source: NASA Goddard Institute for Space Studies, http://data.giss.nasa.gov/gistemp/graphs/.

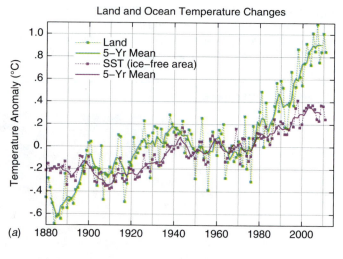

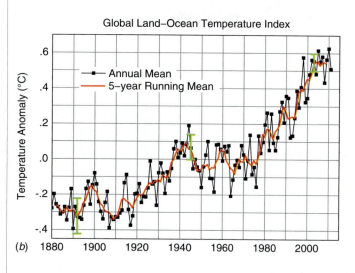

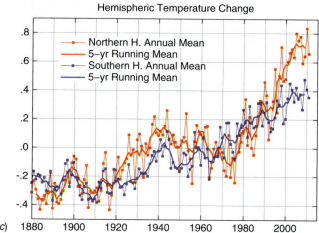

over the period 2012 to 2015 because solar activity is on the upswing and the next El Niño will increase tropical Pacific temperatures and influence global mean temerature. In summary, the decade 2002 to 2011 marked the warmest decade since recordkeeping began,[82] and, as documented,[83] the warmest in the past 1,000 years.

. . . AND HUMANS ARE THE CAUSE

The world is changing in significant ways that are most simply explained in the context of global warming. Taken together, the thousands of observations of changing ecosystems, environments, and natural processes that appear to be shifting in unusual fashion constitute a massive database pointing to the impacts of a warming world.[84] But what is the evidence that humans are the cause? Several principal observations define a "human fingerprint" on climate change that the majority of the scientific community find sufficient to support the hypothesis "Global warming is the result of industrial emissions of greenhouse gas."[85]

[82] See NASA analysis: http://www.nasa.gov/topics/earth/features/temp-analysis-2009.html (accessed July 9, 2012).

[83] R. D'Arrigo, R. Wilson, G. Jacoby, "On the Long-Term Context for Late Twentieth Century Warming," *Journal of Geophysical Research—Atmospheres* 111 (2011): D03103, doi:10.1029/2005JD006352, http://www.ncdc.noaa.gov/paleo/pubs/darrigo2006/darrigo2006.html (accessed July 9, 2012).

[84] J. R. Lanzante, T. C. Peterson, D. J. Seidel, and K. P. Shine, "Tropospheric Temperature Trends: History of an Ongoing Controversy." See also http://www.sciencedaily.com/releases/2010/11/101116080321.htm (accessed July 9, 2012).

[85] See "10 Indicators of a Human Fingerprint on Climate Change" at http://www.skepticalscience.com/10-Indicators-of-a-Human-Fingerprint-on-Climate-Change.html (accessed July 9, 2012).

Fingerprint #1: There Is More Industrial Carbon in the Atmosphere

One clear human fingerprint is found in the type of carbon (C) released into the air by industrial emissions. Carbon, and other elements, occurs naturally in forms known as isotopes.[86] In the case of carbon, there are three: ^{12}C, ^{13}C, and ^{14}C. While engaging in photosynthesis, plants prefer to absorb CO_2 wherein the carbon is composed of the lighter form, ^{12}C. Petroleum is composed largely of fossilized marine algae, and coal is composed of fossilized terrestrial wetland plants; thus both energy sources contain high abundances of light carbon (^{12}C). This can be measured using the ratio of the two types of carbon: $^{13}C/^{12}C$. In plants, $^{13}C/^{12}C$ is relatively low (that is, the amount of ^{12}C is high). Burning oil and coal releases this light carbon, which immediately combines with oxygen in the atmosphere to form $^{12}CO_2$; thus we should be able to detect a decrease in $^{13}C/^{12}C$ in the atmosphere as more fossil fuel emissions accumulate. Indeed, this is exactly what is found. Measurements of the ratio of $^{13}C/^{12}C$ in the air,[87] and in corals[88] and sponges[89] that take up atmospheric carbon (mixed in seawater), reveal a strong decrease in $^{13}C/^{12}C$ over the past 200 years, with a significant acceleration in the decrease since about 1960 to 1970. Thus, the growth of carbon dioxide in the atmosphere is wholly attributable to combustion of coal and petroleum; humans are raising the CO_2 level.

Fingerprint #2: Less Heat Is Escaping to Space

Direct evidence that more carbon dioxide causes warming is found in the fact that less heat is escaping into space. As discussed earlier, carbon dioxide traps infrared radiation (proved by laboratory experiments[90]) from the planet surface that would otherwise escape to space. A decrease in the infrared energy emitted by Earth from 1970 to 1997 has been detected by satellites[91] and has since been verified by additional measurements.[92] Because this heat is trapped in the lowest atmospheric layer, the troposphere, it is warming the air.

Fingerprint #3: Oceans Are Warming from the Top Down

The oceans are warming[93] in the only manner possible under an enhanced greenhouse: from the top down. Measurements of ocean warming show that the water temperature has a depth profile that varies widely by ocean; it cannot be explained by natural internal climate variability or solar and volcanic forcing. The pattern of

---

[86] Atoms of an element all have the same atomic number (the number of protons) but they may have different numbers of neutrons. The number of neutrons plus the number of protons is known as an atom's *mass number*. Atoms of the same element with differing mass numbers are known as *isotopes*. The mass number (identifying the isotope) is written as a superscript on the left side of an element's symbol.

[87] A. Manning, R. Keeling, "Global Oceanic and Land Biotic Carbon Sinks from the Scripps Atmospheric Oxygen Flask Sampling Network." *Tellus* 58 (2006): 95–116.

[88] G. Wei, M. McCulloch, G. Mortimer, W. Deng, L Xie, "Evidence for Ocean Acidification in the Great Barrier Reef of Australia," *Geochemica, Cosmochemica Acta* 73 (2009): 2332–2346.

[89] P. Swart, L. Greer, B. Rosenheim, et al., "The ^{13}C Suess Effect in Scleractinian Corals Mirror Changes in the Anthropogenic CO2 Inventory of the Surface Oceans," *Geophysical Research Letters* 37 (2010): L05604, doi:10.1029/2009GL041397.

[90] D. Burch, "Investigation of the Absorption of Infrared Radiation by Atmospheric Gases," Semi-Annual Technical Report, AFCRL, 1970, publication U-4784.

[91] J. Harries, H. Brindley, P. Sagoo, and R. Brantges, "Increases in Greenhouse Forcing Inferred from the Outgoing Longwave Radiation Spectra of the Earth in 1970 and 1997," 410 (2001): 355–357; doi:10.1038/35066553.

[92] J. Griggs and J. Harries, "Comparison of Spectrally Resolved Outgoing Longwave Data between 1970 and Present," Proceedings of SPIE, 5543 (2004): 164. See also C. Chen, J. Harries, H. Brindley, and M. Ringer, "Spectral Signatures of Climate Change in the Earth's Infrared Spectrum between 1970 and 2006." European Organization for the Exploitation of Meteorological Satellites (EUMETSAT), http://www.eumetsat.eu/Home/Main/Publications/Conference_and_Workshop_Proceedings/groups/cps/documents/document/pdf_conf_p50_s9_01_harries_v.pdf (accessed July 9, 2012). Talk given to the 15th American Meteorological Society (AMS) Satellite Meteorology and Oceanography Conference, Amsterdam, September 2007.

[93] T. Barnett, D. Pierce, K. Achutarao, et al., "Penetration of Human-Induced Warming into the World's Oceans," *Science* 309, no. 5732 (2005): 284–287.

warming is complex, but it has been captured by sensors that depict the upper layer of the oceans (varying from 500 to 75 m depth) warming in a way that is consistent with models simulating human production of greenhouse gases.

Fingerprint #4: Nights Are Warming Faster than Days

During the day, sunlight heats the air. At night the air cools by radiating heat out to space. Part of this heat is trapped by greenhouse gases. Thus, in a situation where the Sun has not increased its output, but the greenhouse effect is amplified, nights will become warmer faster than days. That is, if global warming were caused by the Sun we would expect to see that the days would warm faster than the nights. Observations[94] clearly show that nights are warming faster than days. Thus the detailed pattern of global warming is consistent with an amplified greenhouse effect (resulting from industrial exhaust of heat-trapping gases).

Fingerprint #5: More Heat Is Returning to Earth

Radiation works both ways; that is, infrared (IR) radiation can be measured moving upward from a warm Earth surface as well as moving downward from a warm atmosphere. With an enhanced greenhouse effect, where the molecules of CO_2, CH_4, CFCs, and other greenhouse gases are re-radiating heat (IR) in all directions, one would expect to observe an increase in downward IR radiation from the troposphere to the ground. As expected, this has been directly observed.[95] In fact, researchers state, "This experimental data should effectively end the argument by skeptics that no experimental evidence exists for the connection between greenhouse gas increases in the atmosphere and global warming." Thus, because of an amplified greenhouse effect due to industrial emissions, more heat is returning to Earth.

Fingerprint #6: Winter Is Warming Faster than Summer

Think about it: If the Sun were causing global warming, you would continue to see a seasonal effect in the warming pattern. But that is not what we see. Data show that winter is warming faster than summer. That is, temperature is becoming more uniform throughout the year. One way to think about this is to realize that in an atmosphere that is uniformly warming under an amplified greenhouse effect, a cool winter would be more out of equilibrium with the rising temperature than summer. Thus, one would expect winter to warm faster than summer. This is exactly what has been observed.[96]

Fingerprint #7: The Stratosphere Is Cooling

A corollary to fingerprint #2 (less heat is escaping to space) is that less heat is finding its way to the stratosphere. Because the amplified greenhouse effect is located in the troposphere near Earth's surface, heat that would otherwise find its way to the atmospheric layers above is being trapped by industrial greenhouse gases below. As a result, satellites and weather balloons are recording[97] cooling temperatures in the stratosphere simultaneous with warming in the troposphere (see Figure 2.5).

[94] L. Alexander, X. Zhang, T. Peterson, et al, "Global Observed Changes in Daily Climate Extremes of Temperature and Precipitation," *Journal of Geophysical Research* 111 (2006), doi:10.1029/2005JD006290.

[95] W. Evans and E. Puckrin, "Measurements of the Radiative Surface Forcing of Climate, 2006," P1.7, AMS 18th Conference on Climate Variability and Change. See also K. Wang and S. Liang, "Global Atmospheric Downward Longwave Radiation over Land Surface Under All-Sky Conditions from 1973 to 2008." *Journal of Geophysical Research* 114 (2009) (D19).

[96] K. Braganza, D. Karoly, T. Hirst, et al, "Indices of Global Climate Variability and Change: Part I. Variability and Correlation Structure," *Climate Dynamics* 20 (2003): 491–502. See also K. Braganza, D. Karoly, A. Hirst, et al, "Simple Indices of Global Climate Variability and Change: Part II: Attribution of Climate Change during the Twentieth Century," *Climate Dynamics* 22 (2004): 823–838, doi:10.007/s00382-004-0413-1.

[97] G. Jones, S. Tett, P. Stott, "Causes of Atmospheric Temperature Change 1960–2000: A Combined Attribution Analysis," *Geophysical Research Letters* 30 (2003): 1228. See also C. Mears, F. Wentz, "Construction of the Remote Sensing Systems, v. 3.2 Atmospheric Temperature Records from the MSU and AMSU Microwave Sounders," *Journal of Atmospheric and Oceanic Technology* 26 (2009): 1040–1056.

Fingerprint #8: Physical Models Require Human Greenhouse Gas Emissions

As we will see in Chapter 4, the fundamental laws of nature that explain the movement of heat, the behavior of molecules, and the physics of natural processes can be programmed to build computer models of climate. These are basically larger and more complex versions of computer models that are used to predict the weather on the TV news every day. When these models attempt[98] to simulate the past century of warming using only natural factors (e.g., variations in sunlight, volcanic eruptions, and other climate processes), they instead predict global cooling. But when human emissions of greenhouse gases are introduced to the models along with the natural factors, they faithfully reproduce the observed temperature record: global warming. In fact, researchers[99] have found that computer models are growing in sophistication and accuracy to the point that they are approaching direct observation of the planet as a reliable source of information.

Fingerprint #9: Multiple and Independent Lines of Evidence

Attributing global warming to industrial emissions is not only a consensus among the scientific community, it is the common explanation for multiple and independent lines of evidence. Direct observations show that

- There is more industrial carbon in the atmosphere and it is amplifying the greenhouse effect.
- Less heat is escaping to space.
- Oceans are warming from the top down.
- Nights are warming faster than days.
- More heat is returning to Earth.
- Winter is warming faster than summer.
- The stratosphere is cooling.
- Physical laws of nature predict global warming consistent with observations.

Most of the scientific community finds the authenticity and conformity of these observations sufficient to support the hypothesis "Global warming is the result of industrial emissions of greenhouse gas."

COULD IT BE ANY CLEARER?

In a warmer world, one would expect to observe certain changes including melting glaciers, a warming and acidifying ocean, sea-level rise, changes in ecosystems, and new patterns in the weather. These are all observed.

Briefly, for example, Greenland and Antarctic ice are melting at an accelerating rate.[100] Sensors on satellites have measured this over a sufficient period (a decade or so) that we not only know melting is a persistent annual trend, we also know the rate of melting is actually accelerating from one year to the next. The ocean is getting

---

[98] S. Solomon, D. Qin, M. Manning, et al. (eds.), *Contribution of Working Group I to the Fourth Assessment Report of the Intergovernmental Panel on Climate Change* (Cambridge, U.K., Cambridge University Press, 2007).

[99] T. Reichler and K. Junsu, "How Well Do Coupled Models Simulate Today's Climate?" *Bulletin of the American Meteorological Society* 89 (2008): 303–311, doi: http://dx.doi.org/10.1175/BAMS-89-3-303.

[100] See http://www.colorado.edu/news/r/f595fae00e6b451d4016ab9a43a049f8.html (accessed July 9, 2012). See also I. Velicogna, "Increasing Rates of Ice Mass Loss from the Greenland and Antarctic Ice Sheets Revealed by GRACE," *Geophysical Research Letters* 36 (2009): L19503, doi:10.1029/2009GL040222; and E. Rignot, I. Velicogna, M. van den Broeke, A. Monaghan, and J. Lenaerts, "Acceleration of the Contribution of the Greenland and Antarctic Ice Sheets to Sea Level Rise," *Geophysical Research Letters* 2011, doi:10.1029/2011GL046583.

warmer, and it is acidifying[101] as it mixes with an atmosphere that is enriched with excess carbon dioxide and the concentration of dissolved CO_2 in the ocean grows with each year. Not surprisingly, both warming and acidification are occurring at rates that are predicted by long-established chemical and physical theory.

Sea level is rising and the rate of rise has accelerated. This is an anticipated consequence of a warming world.[102] In a warming world it would be expected that the southern line of permanently frozen ground (permafrost) would begin to migrate to the north as warmer climate zones expand in the northern hemisphere. Indeed this is observed taking place in Canada.[103] Simultaneously, the boundaries of the tropics, defined by temperature, rainfall, wind, and ozone patterns, have shifted poleward by at least 2° latitude in the past 25 years.[104] Excess heating of the tropics has sped up the rate of evaporation and atmospheric circulation (the Hadley Cell) so that surface winds have accelerated nearly around the entire planet.[105] These and many other phenomena stand in testimony to the reality of warming and taken as a whole are consistently in keeping with expectations of how a warmer atmosphere would change the world.

Global warming is also changing the weather. In the decade 2000 to 2009 the United States experienced twice as many record daily high temperatures as record lows; that is, the hotter days are getting hotter and the colder days are getting hotter.[106] Throughout the spring, summer, and fall of 2010 thousands of daily high temperature records were set across the United States, outnumbering the daily record lows by as much as 6 to 1 (April), 5 to 1 (September), and 3.5 to 1 (August); this exceeded the 2 to 1 average of the previous decade.[107] In the summer of 2012 the U.S. experienced a prolonged heat wave that severely impacted agricultural production that year. In fact, extreme weather events are expected to grow in frequency and magnitude as the world continues to warm.[108] Ecosystem changes are occurring as well: Mild winters in British Columbia allow an infestation of the boring mountain pine beetle,[109] and warming oceans have led to coral bleaching, a problem in some hot months that has reached epidemic proportions.[110] There are many other examples of observed changes around the world that are consistent with expected impacts of warming, and these are highlighted throughout this book.

[101] C. Pelejero, E. Calvo, and O. Hoegh-Guldberg, "Palaeo-Perspectives on Ocean Acidification," *Trends in Ecology and Evolution* 25, no. 6 (2010): 332-344, doi: 10.1016/j.tree.2010.02.002. See also D. M. Murphy, S. Solomon, R. W. Portmann, et al. "An Observationally Based Energy Balance for the Earth since 1950," *Journal of Geophysical Research* 114 (2009): D17107, doi:10.1029/2009JD012105.

[102] M.A. Merrifield, S.T. Merrifield, and G.T. Mitchum, "An Anomalous Recent Acceleration of Global Sea Level Rise."

[103] S. Thibault and S. Payette, "Recent Permafrost Degradation in Bogs of the James Bay Area, Northern Quebec, Canada."

[104] D. J. Seidel, Q, Fu, W. J. Randel, and T. J. Reichler, "Widening of the Tropical Belt in a Changing Climate."

[105] G. Li and B. Ren, "Evidence for Strengthening of the Tropical Pacific Ocean Surface Wind Speed during 1979–2011," *Theoretical and Applied Climatology* 107, no. 1–2 (2012): 59–72, doi: 10.1007/s00704-011-0463-3, doi: 10.1007/s00704-011-0463-3. See also I. R. Young, S. Zieger, and A. V. Babanin, "Global Trends in Wind Speed and Wave Height," *Science* 332, no. 6028 (2011): 451–455, doi: 10.1126/science.1197219.

[106] G. A. Meehl, C. Tibaldi, G. Walton, and L. McDaniel, "The Relative Increase of Record High Maximum Temperatures Compared to Record Low Minimum Temperatures in the U.S.," *Geophysical Research Letters* 36 (2009): L23701, doi: 10.1029/2009GL040736.

[107] See analysis by CapitalClimate: http://capitalclimate.blogspot.com/2010/10/endless-summer-xii-septembers.html (accessed July 9, 2012).

[108] D. Medvigy and C. Beulieu, "Trends in Daily Solar Radiation and Precipitation Coefficients of Variation since 1984," *Journal of Climate* 25 (2011): 1330-1339, doi: 10.1175/2011JCL14115.1.

[109] W. A. Kurz, C. C. Dymond, G. Stinson, et al., "Mountain Pine Beetle and Forest Carbon Feedback to Climate Change," *Nature* 452 (2008): 987–990, doi: 10.1038/nature06777.

[110] J. Mao-Jones K. B. Ritchie, L. E. Jones, and S. P. Ellner, "How Microbial Community Composition Regulates Coral Disease Development," *PLoS Biology* 8, no. 3 (2010): e1000345, doi: 10.1371/journal.pbio.1000345. See also N. A. J. Graham S. K. Wilson, S. Jennings, et al., "Dynamic Fragility of Oceanic Coral Reef Ecosystems," *Proceedings of the National Academy of Science* 103 (2006): 8425–8429.

Yet, despite the strong evidence that the climate is changing and humans are the cause, countries across the globe continue to burn fossil fuels at rates that are greater than ever. Global carbon dioxide emissions from fossil fuel combustion and cement production grew 5.9% in 2010.[111] This is the highest total annual growth in carbon dioxide ever recorded, the highest annual growth rate since 2003, and prior to that the highest since 1979. Slowing, and eventually stopping, the production of greenhouse gases from industrial activities requires the directed energies of all the world's economies. Unfortunately, to date, efforts to succeed are having little obvious effect.

Community Resilience

Over the past five decades the average temperature of the atmosphere has increased at the fastest rate in recorded history. Under this trend the average temperature could be 1.8°C to 4.0°C (3.2°F–7.2°F) higher by the end of the century, and cities will be exposed to heat waves, extreme weather, crippling summer temperatures, water shortages related to drought, and high energy demand for air conditioning. Global warming is making life more dangerous. To adjust to this fact, cities, towns, and suburbs can take steps[112] to increase their resilience in the face of climate change.

Bring more vegetation into neighborhoods[113] in the form of green roofs, roadside plantings, vegetated swales, rain gardens, and other forms. These features improve storm water management, they lower the temperature, and they absorb carbon dioxide from the air.

Plant community gardens[114] such as urban orchards and vegetable patches as a food source. These help lower temperature, and growing food in our neighborhoods reduces the number of driving errands in a community.

Use drought-resistant landscaping as a way to save water because water shortages are likely to become more frequent with warmer temperatures.

Use light-colored pavement, roofing, and other surfaces. Dark colors absorb heat, but light surfaces reflect sunlight and lower the planet's temperature. For instance, on the hottest day of the New York City summer in 2011, a white roof was found to be 23°C (42°F) cooler than a traditional black roof. This lowers electricity demand for air conditioning, which in turn reduces carbon emissions from power plants.

Stop building on coastlines.[115] Sea-level rise is real. It will accelerate, and storms, tsunamis, high waves, and high winds cause more damage when the ocean is higher. What used to be the storm of the century will eventually become the storm of the decade. Communities can adapt to sea-level rise, but planning needs to begin in advance of the problem.

Save older buildings because new construction generates heat, requires large volumes of water, disrupts vegetation, and adds to the carbon dioxide in the atmosphere.

Follow new "Original Green Building Practices"[116] when building. Especially in a warmer climate, it is important that buildings be constructed and sited

[111] G. Peters, G. Marland, and C., Le Quere, et al., "Rapid Growth in CO_2 Emissions after the 2008–2009 Global Financial Crisis," *Nature Climate Change* 2 (2012): 2–4.

[112] See Kaid Benfield, "Think Progress," http://thinkprogress.org/romm/2012/04/03/450059/nine-low-tech-steps-for-community-resilience-in-a-warming-climate/ (accessed July 9, 2012).

[113] See "Green Infrastructure," http://switchboard.nrdc.org/blogs/kbenfield/how_green_infrastructure_for_w.html (accessed July 9, 2012).

[114] See "City Gardens that Respect the Urban Fabric," http://switchboard.nrdc.org/blogs/kbenfield/city_gardens_that_respect_the.html (accessed July 9, 2012).

[115] See "Are You in the Zone? New Tool Helps Communities Prepare for Surging Seas," http://switchboard.nrdc.org/blogs/dlashof/are_you_in_the_zone_new_tool_h.html (accessed July 9, 2012).

[116] See the site of the Natural Resources Defense Council on Original Green: http://switchboard.nrdc.org/blogs/kbenfield/they_dont_makeem_like_they_use.html (accessed July 9, 2012).

to take advantage of natural processes. These practices include building front porches and planting deciduous trees on the south side where they provide shade in summer and allow sun in the winter. Plant evergreens on the side that will benefit by protection from winter winds. Use close-to-the-source materials and naturally insulating design. Place new buildings in walkable settings with everyday conveniences nearby.

Keep the community footprint small and well connected. One characteristic of urban sprawl is that it is vehicle dependent, and transportation is the main source of carbon dioxide. Walkable and bikeable destinations, effective mass transit, and small-scale commuting and errands promote a low carbon footprint among communities.

Update zoning and building codes to promote resilience and a low carbon footprint.

As you read through the following chapters, keep in mind steps that you can take to lower your contribution to global warming and to increase your own safety in a warming world.

ANIMATIONS AND VIDEOS

General Circulation, http://kingfish.coastal.edu/marine/Animations/Hadley/hadley.html

Plate Tectonics, http://demo.wiley.ru:30011/fletcher_Geology_1e_anim/ch03/plate_tectonics/index.html

Heat circulation within Earth, http://www.pbs.org/wnet/savageearth/animations/hellscrust/index.html

NASA, Climate Change Visualization 1880–2010, http://www.youtube.com/watch?v=X8XqHwSrakA

Conversation with climatologist Dr. James Hansen, Director of NASA's Goddard Institute for Space Studies, and Climate Crisis Coalition Coordinator Tom Stokes on May 10, 2008, http://www.youtube.com/watch?v=eLBDVZO-8xM&feature=channel_video_title

COMPREHENSION QUESTIONS

1. What is the relationship between global warming and climate change?

2. Describe some of the scientific evidence that increased surface temperature is having measurable impacts on human communities and natural ecosystems.

3. Describe the principal features of the graph showing global temperature anomalies (Figure 2.1).

4. What evidence supports the conclusion that humans are the primary cause of global warming?

5. Describe the primary human activities causing the problem of global warming and climate change.

6. If global warming is real, why is the stratosphere cooling?

7. Temperature records show that climate varies strongly from one year to the next. What does this mean in terms of interpreting the data for the presence or absence of global warming?

8. What is the IPCC?

9. What can we expect Earth to be like in the future if the climate crisis is not addressed?

10. Describe the climate changes and impacts observed in the United States.

THINKING CRITICALLY

1. Which aspect of climate change worries you the most? Why?

2. Suppose a scientist reported that climate had been cooling for several decades in one county in the central United States. What questions would you ask before accepting this information? And once you accept these data as true, what impact would they have on your understanding of global warming?

3. As mayor of a small town in Florida, what steps are you considering with regard to the problem of climate change?

4. Solar output over the period 2008–2010 was low, and scientists are predicting that this trend will continue for another decade or so before the Sun's heat recovers to normal levels. Describe the impact that low solar output could have on global warming both over the next decade and after.

5. As a homeowner planning on staying in your new home for at least 30 years, what proactive steps will you consider to adapt your house to climate change?

6. What steps would you like to see the U.S. President and Congress take with regard to climate change?

7. What effects could heat waves have on a large city?

8. What is the average rate of global warming?

9. Describe a study designed to test the theory that humans are causing global warming.

10. Why is weather very likely to change as climate changes?

CLASS ACTIVITIES (FACE TO FACE OR ONLINE)

ACTIVITIES

1. Explore the websites below. What evidence is provided by each of these that climate is changing and humans are the most important cause?

   a. NOAA Climate Service http://www.noaa.gov/climate.html.

   b. NASA Goddard Institute of Space Science http://www.giss.nasa.gov.

   c. U.S. Global Change Research Program http://www.globalchange.gov.

2. NASA uses satellites to gather information about factors and processes contributing to the increase in Earth's temperature. Which factors and processes that satellites use for this purpose are mentioned in the NASA video "The Temperature Puzzle," http://www.nasa.gov/multimedia/videogallery/index.html?media_id=11886846.

3. Watch Al Gore, "15 Ways to Avert a Climate Crisis," http://www.youtube.com/watch?v=rDiGYuQicpA. What are the 15 steps that former Vice President Al Gore describes that each of us can take to help avoid the climate problem?

HOW DO WE KNOW THAT HUMANS ARE THE PRIMARY CAUSE OF GLOBAL WARMING?

FIGURE 3.0. The average global land temperature for May 2012 was the warmest on record and marks the 327th consecutive month with a global temperature above the 20th-century average. The average temperatures across the United States for the first six months of 2012, and the 12-month period ending in June 2012, were the warmest on record. The map above shows May temperatures relative to average across the globe. Red shading indicates above-average temperatures and blue shading indicates below-average temperatures.

IMAGE CREDIT: NOAA map by Dan Pisut, based on Global Historical Climatology Network data from the National Climatic Data Center (NCDC).

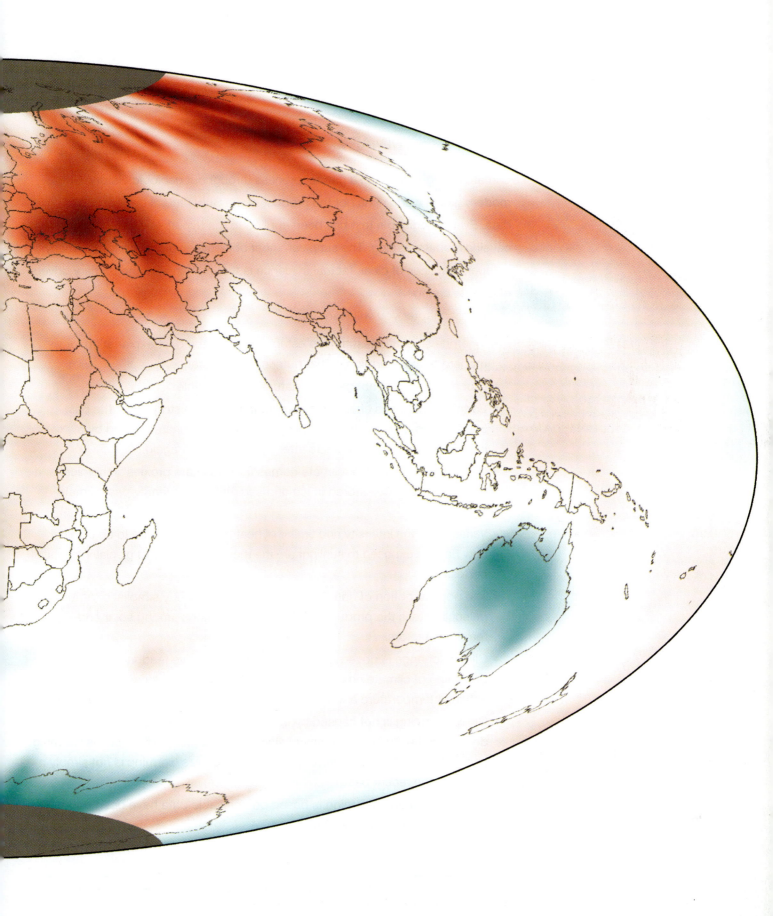

CHAPTER SUMMARY

Climate change has been a natural process throughout geologic history. But modern global warming is not the product of the Sun, natural cycles, or bad data. Every imaginable test has been applied to the hypothesis that humans are causing global warming. The simplest, most objective explanation for the many independent lines of clear, factual evidence is that humans are the primary drivers of climate change.

In this chapter you will learn that:

- Vigorous scientific testing has established that the best explanation for modern global warming is that humans are the cause.
- Global warming is not part of a natural climate cycle.
- Earth's climate has changed throughout geologic history. Over the past 500,000 years or so, the climate system has been characterized by swings in temperature, from glacial to interglacial and back again.
- Our knowledge of paleoclimate comes from "climate proxies": chemical and other types of clues stored in ice and sediment that identify past climate change.
- Variations in the intensity and timing of heat from the Sun due to changes in how Earth is exposed to sunlight are the most likely cause of glacial/interglacial cycles.
- The popular notion of natural "climate cycles" is overly simplistic. Actual paleoclimate is the product of complex interactions among solar and terrestrial factors.
- Climate change is governed by positive and negative feedbacks that make the timing of climate changes irregular. These feedbacks can suppress or enhance temperature and other climate processes in unpredictable ways.
- Global warming is not caused by the Sun, it did not stop during the first decade of the 21st century, scientists do not disagree that climate is warming and that humans are the primary cause, and today's warming is not simply a repeat of the recent past.
- Claims that global warming is based on unreliable data have been rigorously tested[1] and are simply not true.

---

[1] M. Menne, J. Williams, and M. Palecki, "On the Reliability of the U.S. Surface Temperature Record," *Journal of Geophysical Research* 115 (2010): D11108, doi: 10.1029/2009JD013094.

Learning Objectives

Despite rigorous testing, scientists are unable to identify a natural process that is responsible for modern global warming. The accumulation of heat-trapping greenhouse gases in the atmosphere resulting from various human activities has been shown through modeling, theoretical calculations, empirical evidence, and natural chemical and physical processes to be the cause of global warming.

PUBLIC DISCUSSION

We saw in Chapter 2 that global warming poses serious threats to ecosystems and human communities;[2] thus, there has been public discussion of managing the impacts of warming by limiting the production of greenhouse gases and adapting to the inevitable consequences caused by gases that have already been released. Limiting greenhouse gas production and adaptation will be costly, however, and will require significant changes to human behavior.[3]

The public discourse on this issue includes personal worldviews and political ideologies; consequently, the theory that humans are responsible for global warming is vigorously tested not only in science circles using the peer-review process, but also in the nonscientific court of public opinion. This chapter discusses some of the more widely cited tests.

Is Global Warming Part of a Natural Climate Cycle?

This is an important question that deserves careful explanation.[4] If global warming is part of a natural climate cycle, then that suggests that changing human behavior to stop burning fossil fuels is essentially pointless; we might as well avoid the expensive disruption to our economy that switching to noncarbon fuels will entail. If global warming is a natural phenomenon, then humanity must prepare for a future characterized by a warmer world over which we have little control. If, however, global warming is not part of a natural process, if it is the result of releasing greenhouse gases into the atmosphere by human activities, then humans have an obligation to future generations to counteract the problem with significant, potentially expensive, and probably disruptive changes in our behavior. For our own good, and for the safety of our children and the global environment, we need to change our carbon-burning (e.g., oil, coal, natural gas) habits to prevent the most dangerous effects of climate change.

In either case, the answers involve making decisions that risk hundreds of billions of dollars of economic activity, global ecosystems, and the livelihoods of the entire human race. Thus, the field of past climate history, known as paleoclimatology, is a key point around which the entire discussion of climate change revolves. Knowing past climate history can improve our understanding of the question "Is global warming part of a natural climate cycle?" To start, let's investigate paleoclimate in geologic history.

---

[2] For more on this topic see the University Corporation for Atmospheric Research (UCAR) website "Understanding Climate Change," http://www2.ucar.edu/news/backgrounders/understanding-climate-change-global-warming (accessed July 10, 2012).

[3] See report on limiting greenhouse gas production, and congressional reaction to it: http://www.nytimes.com/2011/05/13/science/earth/13climate.html (accessed July 10, 2012).

[4] See the animation "Sir David Attenborough: The Truth about Climate Change" at the end of this chapter.

PALEOCLIMATE

Earth's climate has changed throughout geologic history. In fact, it might surprise you to learn that the past half million years or so marks one of the coolest phases of Earth history (Figure 3.1). One of the great challenges in studying past climate is that geologic materials that record Earth's climate history are very difficult to find as you go farther back in time. Thus, our knowledge of how climate has changed over the course of many millions of years is based on rare evidence that has to be interpreted and tested by many researchers before it becomes widely accepted.

The past 542 million years of Earth history is known as the Phanerozoic Eon. Our understanding of Phanerozoic climate history comes from the interpretation of fossils and chemical clues in rocks and sediments from this time. Although it is entirely likely that short-term annual and interannual climate processes have operated in various forms throughout geologic history, the evidence from hundreds of millions of years ago rarely preserves the details needed to define these processes. Instead, the ancient record of climate shows the effects of longer-term processes:

- Plate tectonics causing the clustering of continents in the high latitudes (promoting cooling) or on the equator (promoting warming)
- Prolonged periods of volcanic outgassing (again the result of plate tectonics) that change climate
- Chemical weathering of Earth's crust that draws down carbon dioxide concentrations in the atmosphere
- Changes in ocean circulation owing to shifting continental positions
- Release of large quantities of frozen methane (a powerful greenhouse gas) by warming ocean water
- Impacts by large meteorites that change climate
- Positive and negative feedbacks that amplify the effects of the above processes

Figure 3.1 presents global surface temperature history relative to the 20th century's average temperature, over Phanerozoic time. Notice the compressed time scale to the left (early Phanerozoic) and the expanded time scale to the right (late Phanerozoic,

Figure 3.1. Paleoclimate history of the Phanerozoic Eon, the past 542 million years. The vertical axis plots temperature change [ΔT (°C)] difference from the average global temperature of the mid 20th century (shown as "0" here). Abbreviations for, and names of, geologic periods of time are shown along margins of the graph (e.g., EO stands for the Eocene Epoch, lasting from approximately 56 to 34 million years ago). Geologic research indicates that throughout much of Earth's history, global temperature was significantly warmer than present until approximately 40 to 50 million years ago, when global temperature began a long and slow decline.

Source: Figure from "Paleoclimatology," *Wikipedia,* http://en.wikipedia.org/wiki/Paleoclimatology.

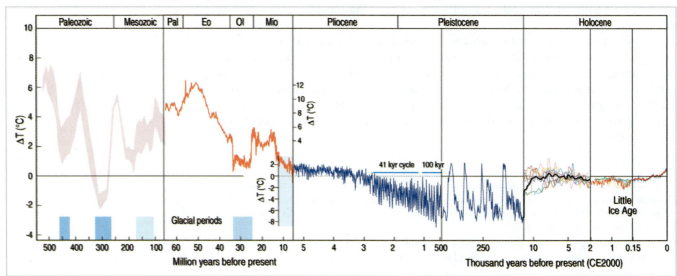

modern). Throughout this history, global temperature was largely warmer than at present. However, by approximately 40 to 50 million years ago global average temperature started a long gradual cooling that led to a series of ice ages and warm periods that have characterized the past 500,000 years or so. One hypothesis for this gradual global cooling is that formation of the Himalayan Mountain system during the Paleocene (Pal) and Eocene (Eo) epochs caused an increase in chemical reactions between newly exposed rock of the mountain system and the atmosphere that reduced atmospheric carbon dioxide levels (known as the Uplift Weathering Hypothesis[5]).

As Figure 3.1 shows, Earth's climate is not steady.[6] This is especially true of the past 500,000 years or so, a time when the climate system has experienced great swings in temperature, from extreme states of cold (glacials) to dramatically warmer periods (interglacials). Glacials (simply known as ice ages[7]) are typified by the growth of massive continental ice sheets reaching across North America and Northern Europe. At their maximum, these glaciers were more than 4 km (2.5 miles) thick in places; today the ice sheets on Greenland and Antarctica are remnants of the most recent ice age. Accompanying the spread of ice sheets was dramatic growth in mountain glaciers, many of which expanded into ice caps that covered large areas of mountainous territory.

We currently live in the latest interglacial, known as the Holocene Epoch, which began about 10,000 years ago. The last ice age, occurring at the end of the Pleistocene Epoch, began approximately 75,000 years ago and peaked between 20,000 and 30,000 years ago. Formed over a period of 50,000 years, the landscape of formerly glaciated areas is widespread and characterized by myriad glacial landforms that document this episode.

Climate Proxies

Fossilized plankton, coral, and other geologic indicators document past climate changes. Known as paleoclimate, past climate change is one way to separate natural climate change from human-caused, or anthropogenic, climate change.

We also know this history of climate because scientists can measure chemical telltales (known as climate proxies) of past climate in samples obtained by drilling in continental ice sheets and mountain glaciers,[8] as well as in sediment composed of fossilized plankton on the seafloor[9] (Figure 3.2).

In ice cores, the past carbon dioxide content of the atmosphere is measured directly from air that was trapped during the formation of glacial ice. The longest ice cores (more than 3 km [1.8 miles] in length) come from Greenland and Antarctica. In this case, CO_2 is used as a climate proxy because it is directly related to the heat-trapping ability of the atmosphere. But cores provide other measures of past climate as well. For instance, fossil snow also contains information about the temperature of the atmosphere and the amount of sunlight-blocking dust, and deep-sea cores can record changes in ocean chemistry that reveal the history of global ice volume.

Deep-sea sediment is composed of the microscopic shells of fossil plankton. The chemistry of these shells—for instance, tiny plankton from the phylum Foraminifera (shown in Figure 3.2B)—provides chemical clues to the climate prevailing when they were formed. Cores of these sediments offer a record of climate history extending hundreds of thousands to millions of years back through time.

Foraminifera use dissolved compounds and ions in seawater to precipitate microscopic shells of $CaCO_3$. Both the calcite ($CaCO_3$) of a foraminifer's skeleton and a

[5] M. Raymo, W. Ruddiman, and P. Froelich, "Influence Of Late Cenozoic Mountain Building on Ocean Geochemical Cycles," *Geology* 16 (1988): 649–653. See also M. Raymo and W. Ruddiman, "Tectonic Forcing of Late Cenozoic Climate," *Nature* 359 (1992): 117–122.

[6] See the animation "Ice Stories: Working to Reconstruct Past Climates" at the end of this chapter.

[7] See "Ice Age," *Wikipedia*, http://en.wikipedia.org/wiki/Ice_age (accessed July 10, 2012).

[8] See National Science Foundation, Office of Polar Programs: http://www.nsf.gov/news/news_summ.jsp?cntn_id=115495&org=OPP&from=news.

[9] See the Integrated Ocean Drilling Program: http://www.iodp.org/ (accessed July 10, 2012).

Figure 3.2. Cores of ice and deep-sea sediments provide geologic samples that contain evidence of past climate. **a,** Scientists from several nations have established collaborative drilling programs on the Green-land and Antarctic ice sheets, as well as on high-elevation ice caps in mountains. **b,** The Integrated Ocean Drilling Program is funded by a consortium of nations interested in using sea-floor sediments to improve our understanding of Earth's history and natural processes.

IMAGE CREDIT: Figure 3.2a (top left) Marc Steinmetz/Aurora Photos Inc., (top right) Carlos Muñoz-Yagüe/ Photo Researchers, (bottom left) Marc Steinmetz/Aurora Photos Inc., (bottom center) Pasquale Sorrentino/ Photo Researchers, (bottom right) Marc Steinmetz/Aurora Photos Inc.; Figure 3.2b (top left) Courtesy of J. Farrell(c) ECORD/IODP, (top right) ©AP/Wide World Photos, (bottom left) HO/AFP/NewsCom, (bottom right) PASIEKA/Science Photo Library/Getty Images, Inc.

(a)

(b)

Figure 3.3. The $H_2{}^{18}O$ water molecule does not evaporate as readily as the $H_2{}^{16}O$ molecule, and once in the atmosphere, water vapor composed of $H_2{}^{18}O$ tends to condense and precipitate more readily in cooling air than a molecule composed of $H_2{}^{16}O$. Because most water vapor originates from the tropical ocean, by the time it travels to high latitudes and high elevations where glaciers form, it is enriched in $H_2{}^{16}O$ relative to seawater. Hence, snow and ice are also relatively enriched in the $H_2{}^{16}O$ molecule.

SOURCE: Fletcher, *Physical Geology: The Science of Earth,* 2012.

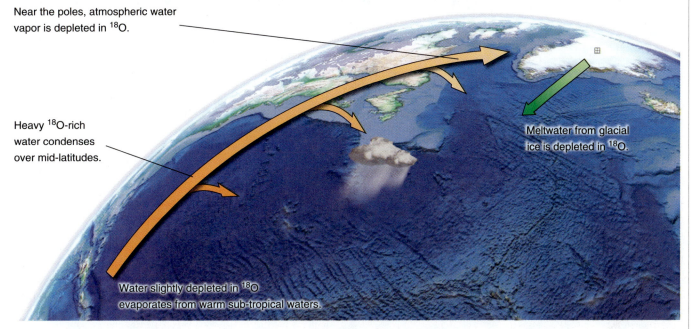

Near the poles, atmospheric water vapor is depleted in ^{18}O.

Heavy ^{18}O-rich water condenses over mid-latitudes.

Meltwater from glacial ice is depleted in ^{18}O.

Water slightly depleted in ^{18}O evaporates from warm sub-tropical waters.

molecule of water (H_2O) in seawater contain oxygen (O). In nature, oxygen occurs most commonly as the isotope ^{16}O, but it is also found as ^{17}O and ^{18}O. (Isotopes[10] are atoms with a different mass number than other atoms of the same element.) Water (H_2O) molecules composed of the heavier isotope ^{18}O do not evaporate as readily as those composed of the lighter isotope ^{16}O. Likewise, in atmospheric water vapor, heavier water molecules with ^{18}O tend to precipitate (as rain and snow) more readily than those composed of lighter ^{16}O (Figure 3.3).

Both evaporation and precipitation of oxygen isotopes occur in relation to temperature. $H_2{}^{18}O$ tends to be left behind (but not entirely) when water vapor is formed during the evaporation of seawater, and it tends to be the first molecule to condense when rain and snow are forming. Hence, because most water vapor in the atmosphere is formed by evaporation in the tropical ocean, by the time it travels the long distance to the high latitudes and elevations where ice sheets and glaciers are located, it is relatively depleted of $H_2{}^{18}O$ and enriched in $H_2{}^{16}O$. This means that during an ice age vast amounts of $H_2{}^{16}O$ are locked up in global ice sheets for thousands of years. At the same time, the oceans are relatively enriched in $H_2{}^{18}O$. Because the ratio of ^{18}O to ^{16}O in the shells of foraminifera mimics the ratio of these isotopes in seawater, the oxygen isotope content of these shells provides a record of global ice volume through time.

Oxygen isotopes in fossil foraminifera provide a record of global ice volume, and in ice cores oxygen isotopes provide a record of changes in air temperature above the glacier. Because the atmosphere is so well mixed, the isotopic content of air above a glacier indicates the temperature of the atmosphere; hence, the isotopic content of snow is useful as a proxy for global atmospheric temperature. At the poles, as an air

---

[10] Every atom in the known universe is a tiny structural unit consisting of electrons, protons, and (usually) neutrons. An atom's center, or nucleus, is composed of protons (large, heavy, and having a positive electrical charge, +) and neutrons (large, heavy, and having no electrical charge). The number of protons plus the number of neutrons makes the mass number. Some atoms of a given element can have a different mass number because they have a different number of neutrons. These are called *isotopes*. For example, carbon atoms normally contain 6 protons (the number of protons is called the *atomic number,* and it's what defines an element) and 6 neutrons; some, however, contain 7 or 8 neutrons. Hence, carbon always has an atomic number of 6, but its mass number may be 12, 13, or 14. These variations in mass number create isotopes of carbon that are written like this: ^{12}C, ^{13}C, ^{14}C.

TABLE 3.1. Proxies for Air Temperature and Ice Volume in Geologic History

Measure	Sample Analyzed	Target of Analysis	Phenomenon for Which Evidence is a Proxy	Finding
Air temperature	Trapped air and the chemistry of samples from ice cores	Abundance of oxygen isotopes in ice	Air temperature	The history of changing air temperature reflected in ice cores strongly correlates to the history of changing global ice volume in marine sediments.
Global ice volume	Foraminifera in deep-sea cores	Abundance of oxygen isotopes in foraminifera	Global ice volume	History of global ice volume correlates well with history of air temperature in ice cores.

mass cools and water vapor condenses to snow, molecules of $H_2{}^{18}O$ condense more readily than do molecules of $H_2{}^{16}O$, depending on the temperature of the air. Typically, above a glacier, the condensation falls out of a cloud as snow. Thus, the oxygen isotopic content of snow (measured as the ratio of ^{18}O to ^{16}O) is a proxy for air temperature; hence, cores of glacial ice record variations in air temperature through time.

Paleoclimate Patterns

Because global ice volume and air temperature are related, the records of oxygen isotopes in foraminifera and in glacial ice show similar patterns.[11] These records provide researchers with two independent proxies for the history of global climate. Many researchers[12] have tested and verified the history of global ice volume preserved in deep-sea cores and the history of temperature preserved in glacial cores from every corner of the planet. Past episodes of cooler temperature reflected in ice cores strongly correlate to periods of increased global ice volume in marine sediments. Likewise, past episodes of warmer climate correlate well to periods of decreased ice volume (Table 3.1).

An example of this record is shown in Figure 3.4. The variation in abundance of oxygen isotopes in ice (a proxy for atmospheric temperature) and in marine foraminifera (a proxy for global ice volume) do indeed display strong agreement and provide researchers with a global guide for interpreting past climate patterns and events.

These natural archives show that global climate change is characterized by alternating warm periods and ice ages occurring approximately every 100,000 years. This history of cooling and warming has several important features:

- Major glacial and interglacial periods are repeated approximately every 100,000 years.
- Numerous minor episodes of cooling (called stadials) and warming (called interstadials) are spaced throughout the entire record.
- Global ice volume during the peak of the last interglacial (known as the Eemian[13]), approximately 125,000 years ago, was lower than at present, and global climate was warmer.

[11]Discussion based on C. H. Fletcher, *Physical Geology: The Science of Earth* (Hoboken, N.J., Wiley, 2011).

[12]See various types of paleoclimate indicators used to reconstruct past climate at NOAA's Paleoclimatology Program website: http://www.ncdc.noaa.gov/paleo/recons.html (accessed July 10, 2012).

[13]See "Eemian," *Wikipedia*, http://en.wikipedia.org/wiki/Eemian_Stage (accessed July 10, 2012).

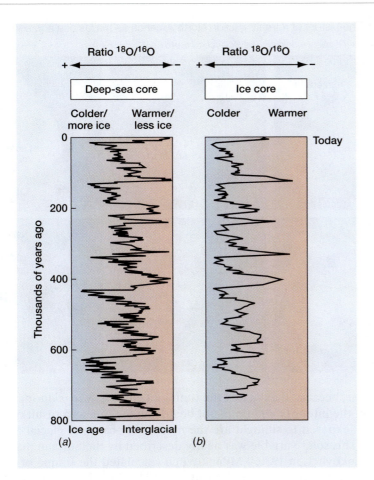

Figure 3.4. a, The ratio of ^{18}O to ^{16}O in deep-sea cores of fossil foraminifera provides a proxy for global ice volume. **b,** The ratio of ^{18}O to ^{16}O in cores of glacial ice documents changes in atmospheric temperature, confirming that decreased ice volume in deep-sea cores correlates to times of warmer atmosphere, whereas increased ice volume recorded in deep-sea cores correlates to times of cooler atmosphere.

Source: Fletcher, *Physical Geology: The Science of Earth*, 2012.

- Following the last interglacial, global climate deteriorated in a long, drawn-out cooling phase, culminating approximately 20,000 to 30,000 years ago with a major glaciation.

- The present interglacial has lasted approximately 10,000 years.

The marine oxygen isotope record suggests that over the past 500,000 years, each glacial–interglacial period has lasted about 100,000 years; hence, there have been approximately five glaciations in this period. During the length of a typical 100,000-year period, climate gradually cools and ice slowly expands until it reaches a peak. At the peak, glaciers cover Iceland, Scandinavia, the British Isles, and most of Canada southward to the Great Lakes. In the Southern Hemisphere, part of Chile is covered, and the ice of Antarctica covers part of what is now the Southern Ocean. In mountainous regions the snowline lowers by about 1,000 m (3,280 feet) in altitude from the warmest to the coldest portions of a period.

Once ice cover reaches a maximum during a glacial episode, within a couple of thousand years global temperature rises again and the glaciers retreat to their minimum extent and volume (Figure 3.5). The last ice age culminated about 20,000 to 30,000 years ago, and by approximately 5,000 years ago most of the ice had melted (except for remnants on Greenland and Antarctica). Since then, glaciers have generally retreated to their smallest extent, with the exception of short-term climate fluctuations, such as the Little Ice Age, a cool period that lasted from about the 16th to the 19th century (discussed later).

Orbital Parameters

Scientists are still uncertain about all the factors that drive variations in paleoclimate. There is, however, agreement that regular and predictable differences in Earth's exposure to solar radiation over the past half-million years must play an

Figure 3.5. The history of retreating ice in North America as the last ice age ended.
SOURCE: Fletcher, *Physical Geology: The Science of Earth*, 2012.

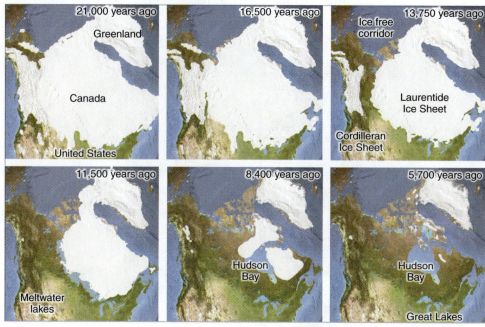

important role, because they match the timing of climate swings during that period. Variations in the intensity and timing of heat from the Sun as a result of changes in how Earth is exposed to sunlight are the most likely cause of glacial and interglacial periods. This solar variable was neatly described by the Serbian mathematician Milutin Milankovitch in 1930.[14] Milankovitch calculated the timing of three major components (parameters) of Earth's orbit around the Sun that contribute to changes in global climate: eccentricity, obliquity, and precession.

Figure 3.6. Earth's seasons are the result of a 23.5° tilt in the planet's axis. Because of this tilt, different parts of the globe are oriented toward the Sun at different times of the year. Summer is warmer than winter (in each hemisphere) because the Sun's rays hit Earth at a more-direct angle than during winter and because days are much longer than nights. During winter, the Sun's rays hit Earth at a less-direct angle, and days are very short.
SOURCE: Fletcher, *Physical Geology: The Science of Earth*, 2012.

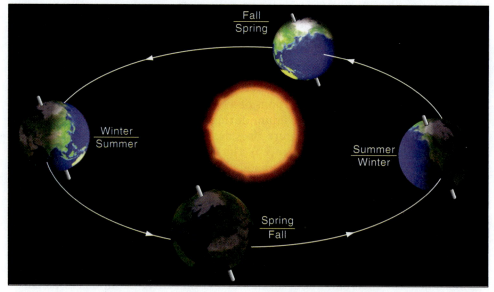

[14]See "Milankovitch Cycles," *Wikipedia*, http://en.wikipedia.org/wiki/Milankovitch_cycles (accessed July 10, 2012).

Figure 3.7. The primary orbital parameters driving climate changes over the past half-million years are eccentricity, obliquity, and precession. These parameters regulate the intensity of insolation reaching Earth's surface, triggering changes in atmospheric temperature.

SOURCE: Fletcher, *Physical Geology: The Science of Earth*, 2012.

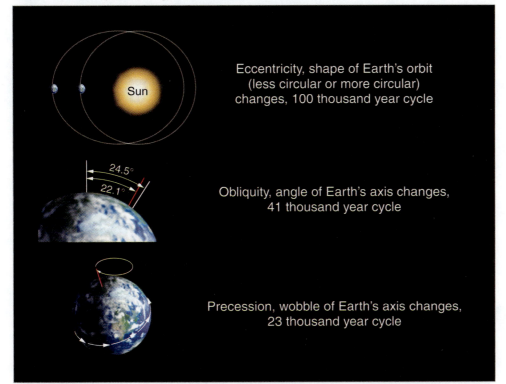

Eccentricity, shape of Earth's orbit (less circular or more circular) changes, 100 thousand year cycle

Obliquity, angle of Earth's axis changes, 41 thousand year cycle

Precession, wobble of Earth's axis changes, 23 thousand year cycle

INSOLATION   To understand these orbital parameters, we first must appreciate the effect of Earth's tilted axis on insolation, the amount of solar radiation received at Earth's surface through the year. Earth's axis is tilted an average of 23.5° from the vertical (Figure 3.6). As Earth orbits the Sun, this tilt means that during one portion of the year (summer) the Northern Hemisphere is tilted toward the Sun and receives greater insolation, whereas 6 months later (winter) it is tilted away from the Sun and receives less insolation. The reverse applies to the Southern Hemisphere: when it is summer in the Northern Hemisphere, it is winter in the Southern Hemisphere, and when it is winter in the Northern Hemisphere, it is summer in the Southern Hemisphere. These annual extremes in insolation create the seasons.

FIRST ORBITAL PARAMETER: ECCENTRICITY   Earth orbits the Sun on a flat plane called the ecliptic; however, three aspects of the geometry of this orbit change in a regular pattern under the influence of the combined gravity of Earth, the Moon, the Sun, and the other planets. These orbital parameters dictate the insolation reaching Earth's surface over time, which in turn regulates climate (Figure 3.7).

The shape of Earth's orbit changes from a nearly perfect circle to more elliptical and back again in a 100,000-year cycle and a 400,000-year cycle. The change in the shape of Earth's orbit is known as eccentricity. Eccentricity affects the amount of insolation received at the point in its orbit at which Earth is farthest from the Sun (aphelion) and at the point in its orbit at which Earth is closest to the Sun (perihelion). Eccentricity shifts the seasonal contrast in the Northern and Southern Hemispheres. For example, when Earth's orbit is more elliptical (less circular), one hemisphere has hot summers and cold winters while the other has moderate summers and moderate winters. When Earth's orbit is more circular, both hemispheres have similar contrasts in seasonal temperature.

SECOND ORBITAL PARAMETER: OBLIQUITY The angle of Earth's axis of spin changes its tilt between 22.1° and 24.5° on a 41,000-year cycle. Obliquity describes the changing tilt of Earth's axis. Changes in obliquity cause large changes in the seasonal distribution of sunlight at high latitudes and in the length of the winter dark period at the poles. They have little effect on low latitudes.

THIRD ORBITAL PARAMETER: PRECESSION Finally, Earth's axis of spin slowly wobbles. Like a spinning top running out of energy, the axis wobbles toward and away from the Sun over the span of approximately 23,000 years. This wobbling or changing of Earth's tilt as it spins is known as precession. Precession affects the timing of aphelion and perihelion, and this has important implications for climate because it affects the seasonal balance of radiation. For example, when perihelion falls in January, winters in the Northern Hemisphere and summers in the Southern Hemisphere are slightly warmer than the corresponding seasons in the opposite hemispheres. The effects of precession on the amount of radiation reaching Earth are closely linked to the effects of obliquity (changes in tilt). The combined variation in these two factors causes radiation changes of up to 15% at high latitudes, greatly influencing the growth and melting of ice sheets.

Short, Cool Summers

The small variations in Earth–Sun geometry related to eccentricity, obliquity, and precession change how much sunlight each hemisphere receives during Earth's year-long journey around the Sun. They also determine the time of year at which the seasons occur and the intensity of seasonal changes. Milankovitch theorized that ice ages occur when orbital variations cause lands in the region of 65° North latitude (the approximate latitude of central Canada and northern Europe) to receive less sunshine in the summer. Why the Northern Hemisphere? Because most of the continents are located in the Northern Hemisphere and glaciers form on land, not water.

When orbital parameters combine to create short, cool summers in the Northern Hemisphere, some snow is likely to last into the following winter and thereby accumulate from year to year, leading to the formation of glaciers. In addition, as snow builds up, its bright, white surface reflects more radiation back into space, thus enhancing the cooling trend toward an ice age. Based on this reasoning and his calculations, Milankovitch predicted that the ice ages would peak every 100,000 and 41,000 years, with additional significant variations every 19,000 to 23,000 years.

Figure 3.8 plots variations in all three orbital parameters across the past one million years. Indeed, according to the ocean sediment record of global ice volume, Milankovitch's predictions were accurate. Ice ages and interglacials occur roughly every 100,000 years, and the timing of stadials and interstadials varies on the more-rapid schedule approximated by his predictions of 41,000, 19,000, and 23,00 years, although that exact timing is not preserved in the sediment or ice core records. The fact that the exact timing of obliquity and precession are not preserved in geologic records has to do with the influence of Earth's surface environments on how the climate system changes over time.

If you compare the total solar forcing in Figure 3.8 (yellow line) to the paleoclimate record in marine sediments (black line), you will notice that the timing and magnitude of the two do not exactly match. For instance, the solar forcing that led to the Eemian is considerably greater than the forcing that is producing the modern interglacial. Also, the drop in insolation following the Eemian is greater than the low at the culmination of the last ice age 20,000 to 30,000 years ago, yet the last ice age was much colder. What creates these disparities? The answer is that Earth's climate is not driven solely by insolation. It is also influenced by climate feedbacks.

Understanding the history of Earth's past climate, and thus addressing the question of whether modern global warming is a natural process, requires a familiarity with orbital parameters and climate feedbacks.

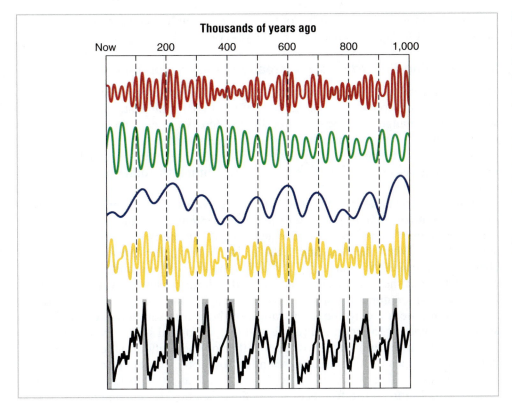

Figure 3.8. This plot shows the relative insolation (also known as solar forcing) due to precession (red), obliquity (green), and eccentricity (blue). The cumulative insolation at 65° North latitude caused by the combined influence of the three orbital parameters is plotted in yellow. Black shows a history of global ice volume, with interglacials (vertical gray bars) showing regular variations in climate approximately every 100,000 years and more often, just as Milankovitch predicted.

SOURCE: Fletcher, *Physical Geology: The Science of Earth,* 2012.

THE IMPORTANT ROLE OF CLIMATE FEEDBACKS

When scientists first tried to build computer models to simulate paleoclimate, they could not get them to reproduce past climate change unless they added changes in carbon dioxide levels to accompany the changes in insolation caused by orbital parameters. This was an indication that it takes more than insolation alone to predict changes in the climate system. Although scientists are still trying to understand what causes natural changes in carbon dioxide levels, most believe that past episodes of climate warming were initiated by orbital forcing, and then enhanced and extended by the rise of greenhouse gases. In other words, warming led to CO_2 rise, not the other way around. Because deep sea cores reveal that carbon dioxide levels are much higher today than at any time in the past 15 million years,[15] pinning down the cause-and-effect relationship between carbon dioxide and climate change continues to be a focal point of modern research. In the case of paleoclimate, the cause and effect was related to climate feedbacks.

A climate feedback is a process taking place on Earth that amplifies (a positive feedback) or minimizes (a negative feedback) the effects of insolation. Earth's environmental system generates positive and negative climate feedbacks that can enhance or suppress the timing and intensity of the Earth–Sun geometry. Climate feedbacks are responsible for the difference between the influence of orbital parameters and Earth's actual climate. That is, the climate is not solely controlled by sunlight; feedbacks are an equally powerful process, and it is the combined influence of feedbacks and orbital parameters that determines the climate.

---

[15]A. Tripati, D. Roberts, and R. Eagle, "Coupling of CO_2 and Ice Sheet Stability over Major Climate Transitions of the Last 20 Million Years," *Science* 326, no. 5958 (2009): 1394–1397, http://www.sciencemag.org/cgi/content/abstract/1178296.

Let's examine two case studies of how feedbacks influenced Earth's climate: the end of the last ice age, and the Younger Dryas cold spell.

Why Warming Preceded CO$_2$ Increase at the End of the Ice Age

Ice cores record past greenhouse gas levels as well as temperature; hence, they allow researchers to compare the history of the two. In the past, when the climate warmed, the change was accompanied by an increase in greenhouse gases, particularly carbon dioxide. However, increases in temperature preceded increases in carbon dioxide. This pattern is opposite to the present pattern, in which industrial greenhouse gases are causing increases in temperature. The difference is related to climate feedbacks.

One idea for understanding the role of feedbacks in paleoclimate was developed by scientists analyzing a core of marine sediments from the ocean floor near the Philippines.[16] That area of the Pacific contains foraminifera that live in tropical surface water. When they die and settle to the bottom, they preserve a record of changing tropical air temperatures. But different types of foraminifera living on the deep seafloor at the same location are bathed in bottom waters (water that travels along the seafloor, not at the surface) fed from the Southern Ocean near Antarctica. These foraminifera record the temperature of those cold southern waters. The fossils of both types of foraminifera (those that live in tropical surface water and those living in bottom waters) are deposited together on the seafloor. Upon radiocarbon dating[17] both types of foraminifera, scientists found that water from the Antarctic region warmed before waters in the topics—as much as 1,000 to 1,300 years earlier. The explanation for this difference, they believe, is a positive climate feedback that enhanced warming caused by orbital parameters.

First, predictable variations in Earth's eccentricity and obliquity increased the amount of sunlight hitting high southern latitudes during spring in the Southern Hemisphere. That increase warmed the Southern Ocean. As a result, sea ice shrank back toward Antarctica, uncovering and warming ocean waters that had been isolated from the atmosphere for millennia. As the Southern Ocean warmed, it was less able to hold dissolved carbon dioxide, and great quantities of CO$_2$ escaped into the atmosphere (warm water can hold less dissolved gas than cold water). The released gas proceeded to warm the global climate system. This process was responsible for driving climate out of its glacial state and into an interglacial state at the end of the last ice age. It explains how small temperature changes caused by orbital parameters led to a positive feedback in global carbon dioxide that warmed the world.

Rapid Climate Change: The Younger Dryas[18]

When scientists first analyzed paleoclimate evidence in marine and glacial oxygen isotope records, they discovered that the Milankovitch theory predicted the occurrence of ice ages and interglacials with remarkable accuracy. But they also found something that required additional explanation: climate changes that appeared to have occurred very rapidly and that were not predicted by orbital parameters. Because Milankovitch's theory tied climate change to slow and regular variations in Earth's orbit, it was assumed that climate variations would also be slow and regular. The discovery of rapid changes was a surprise. Here again, the answer lay in the climate feedback system.

[16]L. Stott, A. Timmermann and R. Thunell, "Southern Hemisphere and Deep Sea Warming Led Deglacial Atmospheric CO$_2$ Rise and Tropical Warming," *Science* 318, no. 5849 (2007): 435–438, 2007 doi: 10.1126/science.1143791.

[17]Radiocarbon dating is a laboratory technique for determining how old a geologic sample is by using the constant rate of radioactive decay of the isotope carbon-14 in organic materials. The technique is typically only applicable to samples younger than 50,000 years old. See "Radiocarbon Dating," *Wikipedia*, http://en.wikipedia.org/wiki/Radiocarbon_dating (accessed July 10, 2012).

[18]See the video "Richard Alley's Global Warming" at the end of this chapter.

Figure 3.9. The Younger Dryas climate event was a dramatic cooling that lasted approximately 1,300 years during the transition between glacial and interglacial states. The return to warm conditions was equally rapid, occurring within the span of a single human lifetime.

SOURCE: Fletcher, *Physical Geology: The Science of Earth,* 2012.

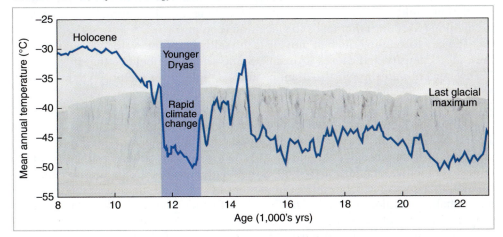

Cores[19] show that although it took thousands of years for Earth to totally emerge from the last ice age and warm to today's balmy climate, fully one third to one half of the warming—about 10°C (18°F) at Greenland—occurred within mere decades, at least according to ice records in Greenland (Figure 3.9). At approximately 12,800 years ago, following the last ice age, temperatures in most of the Northern Hemisphere rapidly returned to near-glacial conditions and stayed there during a climate event called the Younger Dryas (named after the alpine flower Dryas octopetala). The cool episode lasted about 1,300 years, and by 11,500 years ago temperatures had warmed again. Ice core records show that the recovery to warm conditions occurred with startling rapidity, less than a human generation. Changes of this magnitude would have a huge impact on modern human societies, and there is an urgent need to understand and predict such abrupt climate events.

A look at marine sediments confirms that this pattern is present and may be a global characteristic of climate change. Hence, scientists conclude that although the general timing and pace of climate change is set by the orbital parameters, some feedback process must play an important role in its precise timing and magnitude. What might that process be? Global thermohaline circulation is thought to play a key role in the case of the Younger Dryas.

The Conveyor Belt Hypothesis

As we discussed in relation to Figure 1.5, today warm water near the equator in the Atlantic Ocean is carried to the north on the Gulf Stream, which flows from southwest to northeast in the western Atlantic. The Gulf Stream releases heat into the atmosphere through evaporation, and the heat in turn moves across Northern Europe and moderates the climate. As a result, the Gulf Stream becomes cool and salty as it approaches Iceland and Greenland. This makes it very dense; as a result, it sinks deep into the North Atlantic (making a deep current called the North Atlantic Deep Water) before it can freeze. From there it is pulled southward toward the equator. The Gulf Stream continuously replaces the sinking water, warming Europe and setting up a global oceanic conveyor belt.

Thermohaline circulation transports heat around the planet and hence plays an important role in global climatology. Acting as a conveyor belt carrying heat from the equator into the North Atlantic, it raises Arctic temperatures, discouraging the expansion of ice sheets. However, influxes of freshwater from melting ice on the

[19] R. G. Fairbanks, "The Age and Origin of the 'Younger Dryas Climate Event' in Greenland Ice Cores," *Paleoceanography* 5, no. 6 (1990): 937–948, doi: 10.1029/PA005i006p00937.

lands that surround the North Atlantic (such as Greenland) can slow or shut down the circulation by preventing the formation of deep water. This process leads to cooling in the Northern Hemisphere, thereby regulating snowfall in the crucial region where ice sheets shrink and grow (65°N). Hence, a shutdown of the thermohaline circulation could play a role in a negative climate feedback pattern, beginning with ice melting (warming) that leads to glaciation (cooling).

The key to keeping the belt moving is the saltiness of the water, which increases the water's density and causes it to sink. Some scientists[20] believe that if too much freshwater entered the North Atlantic—for example, from melting Arctic glaciers and sea ice—the surface water would freeze before it could become dense enough to sink toward the bottom. There is, in fact, observational evidence[21] that thermohaline circulation has slowed 20% over the decade 2000 to 2009. If the water in the north did not sink, the Gulf Stream would eventually stop moving warm water northward, leaving Northern Europe cold and dry. Modeling[22] of this problem suggests that a low rate of meltwater addition to the North Atlantic would not significantly alter the circulation. However a moderate or high rate of Greenland melting could make the thermohaline circulation weaken further. This further weakening would not necessarily make the global climate in the next two centuries cooler than in the late 20th century, but it would instead lessen the warming, especially in the northern high latitudes.

This hypothesis of rapid climate change is called the conveyor belt hypothesis, and the paleoclimate record found in ocean sediment cores is beginning to support it. Paleoclimate studies have shown that in the past, when heat circulation in the North Atlantic Ocean slowed, Northern Europe's climate changed. Although the last ice age peaked about 20,000 to 30,000 years ago, the warming trend that followed it was interrupted by cold spells at 17,500 years ago and again at 12,800 years ago (the Younger Dryas). These cold spells happened just after melting ice had diluted the salty North Atlantic water, slowing the ocean conveyor belt. It is this idea that led to the movie The Day After Tomorrow. In the movie, global warming results in freshwater from melting ice stopping the thermohaline circulation, which in turn produces deadly cooling in the North Atlantic—an unlikely scenario.

Hence, we have seen two types of climate feedback:

- A **positive climate feedback** in the Antarctic that ended the last ice age. Predictable variations in Earth's tilt and orbit caused warming, which triggered the withdrawal of sea ice in the Southern Ocean. This led to additional warming of ocean water, reducing its ability to hold dissolved carbon dioxide. The carbon dioxide escaped into the atmosphere and warmed the planet beyond the temperatures that would have been achieved by orbital parameters alone.

- A **negative climate feedback** late in the transitional phase between the last ice age and the modern interglacial. Warming at approximately 12,800 years ago produced abundant freshwater in the North Atlantic that diluted the salty Gulf Stream. This put a temporary end to the global thermohaline circulation and triggered rapid cooling in the Northern Hemisphere. Later, after a period of cooling lasting approximately 1,300 years, the Younger Dryas, thermohaline circulation once again became a source of heat transport throughout the world's oceans. This renewed circulation triggered warming, apparently very rapidly, that would not have been predicted by orbital parameters alone.

[20]W. S. Broecker, "Was the Younger Dryas Triggered by a Flood?" *Science* 312, no. 5777 (2006): 1146–1148, doi: 10.1126/science.1123253.

[21]U. Send, M. Lankhorst, and T. Kanzow, "Observation of Decadal Change in the Atlantic Meridional Overturning Circulation Using 10 Years of Continuous Transport Data," *Geophysical Research Letters* 38 (2011): L24606, doi: 10.1029/2011GL049801.

[22]A. Hu, G. Meehl, W. Han, J. Yin, "Effect of the Potential Melting of the Greenland Ice Sheet on the Meridional Overturning Circulation and Global Climate in the Future," *Deep Sea Research* II 58 (2011): 1914–1926.

These climate feedbacks, and others that are still being discovered, work in parallel with orbital parameters to determine the nature of Earth's climate. Interestingly, Milankovitch cycles predict that today global climate should be cooling. For instance, one famous study[23] about Milankovitch cycles concluded that "this model predicts that the long-term cooling trend which began some 6,000 years ago will continue for the next 23,000 years."

Another study[24] showed that the influence of Milankovitch cycles predicts that a new continental ice sheet should be forming in northeast Canada. The fact that Earth has not cooled over this interglacial as predicted has led to the anthropogenic hypothesis, which proposes that human agriculture (involving deforestation, rice wetland production, and animal husbandry) has controlled global climate for several thousand years.[25] It is clear that Milankovitch cycles, the major natural paleoclimate process, did not anticipate the global warming problem we face today.

Dansgaard–Oeschger Events

In some cases, global climate follows semicyclic patterns of warming and cooling that are more frequent than Milankovitch cycles. As we have seen already, the popular notion of "natural cycles" is overly simplistic, and actual climate is the product of complex interactions among solar and terrestrial processes. Climate change in fact is governed by feedback processes that make cycle timing irregular and can suppress or enhance temperature and other climate processes in unpredictable ways.

An example of how feedback processes make cycle timing irregular and affect temperature and other climate processes unpredictably is the sequence of Dansgaard–Oeschger (DO) events that occurred during the last ice age in the North Atlantic (Figure 3.10). DO events are recorded in Greenland ice cores and in North Atlantic seafloor sediments. These rapid climate events have led some to pose an important scientific question: "Is present global warming part of a natural DO event?"[26]

DO events are rapid climate fluctuations, occurring 25 times during the last glacial age, that are revealed in ice core and marine sediment records in the Northern Hemisphere. They take the form of rapid warming episodes, typically in a matter of decades, each followed by gradual cooling over a longer period. The pattern in the Southern Hemisphere is different, with slow warming and much smaller temperature fluctuations. However, orbital geometry does not predict these events.

Several explanations have been promoted to explain DO events, but their exact origin is still unclear. It has been hypothesized that they are the result of periodic collapses of thick glacier ice in Canada (ice buildup eventually collapses under its own weight) or changes in Atlantic thermohaline circulation triggered by an influx of freshwater.[27] The question of whether DO events extend into the present interglacial is controversial, the last clear candidate for a DO event was 11,500 years ago (the Younger Dryas),[28] and it has been questioned if this event resulted in climate

---

[23]J. Imbrie and J. Z. Imbrie, "Modeling the Climatic Response to Orbital Variations," *Science* 207 (1980): 943–953, doi: 10.1126/science.207.4434.943.

[24]A. E. Carlson, A. N. Legrande, D. W. Oppo, et al., "Rapid Early Holocene Deglaciation of the Laurentide Ice Sheet," *Nature Geoscience* 1 (2008): 620–624.

[25]W. Ruddiman, "The Anthropogenic Greenhouse Era Began Thousands of Years Ago," *Climatic Change* 61 (2003): 261–293; W. Ruddiman, "Cold Climate During the Closest Stage 11 Analog to Recent Millennia," *Quaternary Science Reviews* 24 (2005): 1111–1121. W. Ruddiman, *Plows, Plagues, and Petroleum: How Humans took Control of Climate* (Princeton: Princeton University Press, 2007).

[26]See "Bond Events," *Wikipedia*, http://en.wikipedia.org/wiki/Bond_event (accessed July 10, 2012).

[27]M. Maslin, D. Seidov, and J. Lowe, "Synthesis of the Nature and Causes of Rapid Climate Transitions during the Quaternary," *Geophysical Monograph* 126 (2001): 9–52, http://www.essc.psu.edu/~dseidov/pdf_copies/maslin_seidov_levi_agu_book_2001.pdf (accessed July 10, 2012).

[28]See discussion at http://en.wikipedia.org/wiki/1500-year_climate_cycle (accessed July 10, 2012).

Figure 3.10. Climate change over 140,000 years: Yellow and red are from Antarctic ice cores; blue and purple are from Greenland ice cores. Greenland ice cores use ^{18}O as a proxy for temperature, and Antarctic ice cores use an isotope of hydrogen, ^{2}H. Note the rapid climate changes in the Greenland ice core during the glacial age, about 80,000 to 15,000 years ago, which barely register in the corresponding Antarctic record. These are Dansgaard–Oeschger events.

Source: Figure from "Dansgaard–Oeschger Event," *Wikipedia*, http://en.wikipedia.org/wiki/Dansgaard-Oeschger_event (accessed July 10, 2012).

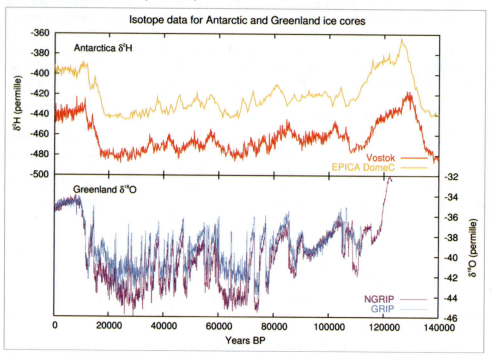

change that was truly global in extent.[29] In fact, some high-resolution records of climate extending 50,000 years into the past find no such cycle.[30]

Bipolar Seesaw

The fundamental problem with assigning modern global warming to cyclic-type DO events is the bipolar seesaw. Weak DO events are found in Antarctic ice cores, but their effect in the Southern Hemisphere is opposite in timing to the Northern Hemisphere. That is, cooling in the north is accompanied by warming in the south, and vice versa.[31] This seesaw[32] is thought to be related to thermohaline circulation.

The bipolar seesaw hypothesis goes like this: DO slow cooling may be associated with an influx of fresh water to the North Atlantic, which reduces the strength of the thermohaline circulation. As a result, there is excess heat in the tropics, available for oceanic currents to transfer toward the Southern Ocean. Thus warming is recorded in Antarctic ice core records. Eventually, warming in Antarctica releases fresh water, which weakens Southern Hemisphere circulation. This weakening allows the North

---

[29]T. Barrows, S. Lehman, L. Fifield, P. De Deckker, "Absence of Cooling in New Zealand and the Adjacent Ocean During the Younger Dryas Chronozone," *Science* 318, no. 5847, (2007): 86–89.

[30]D. Fleitmann, H. Cheng, S. Badertscher, et al., "Timing and Climatic Impact of Greenland Interstadials Recorded in Stalagmites from Northern Turkey," *Geophysical Research Letters* 36 (2009): L19707, doi: 10.1029/2009GL040050.

[31]S. Barker, P. Diz, M. Vautravers, et al., "Interhemispheric Atlantic Seesaw Response During the Last Deglaciation," *Nature* 457, 7233 (2009):1097–1102, doi: 10.1038/nature07770. See discussion by J. P. Severinghaus, "Climate Change: Southern See-Saw Seen," *Nature* 457, no. 7233 (2009): 1093–1094.

[32]B. Stenni, D. Buiron, M. Frezzotti, et al., "Expression of the Bipolar Seesaw in Antarctic Climate Records During the Last Deglaciation," *Nature Geoscience* 4 (2011): 46–49, doi: 10.1038/ngeo1026.

Atlantic thermohaline circulation to suddenly switch on, producing a rapid DO warming event in the North Atlantic. Warming in the south is contemporaneous with cooling in the north; and vice versa.

Today, global warming is occurring simultaneously around the planet; hence it cannot be tied to a modern-day DO event.[33] This was proved[34] by detailed study of ice cores that capture a record of atmospheric temperature in both the Northern and Southern Hemispheres as well as by cores of ocean sediments;[35] no period of simultaneous change in the north and south is seen over the past 20,000 years. This means that a fundamental characteristic of modern global warming has never occurred since the last ice age: Natural warming alternates between the hemispheres (the seesaw effect), but anthropogenic warming occurs across the entire planet simultaneously.

Here is another way to think of it: DO events lead to no net change in Earth's heat budget, because of offsetting trends in the Northern and Southern Hemispheres. Global warming is global and represents a significant net increase in Earth's heat budget. In any case, DO events have not been shown to exist in the Holocene, nor have they been documented as global in extent.[36]

This rather lengthy treatment of paleoclimate was necessary to fully examine the hypothesis that global warming today is the result of a natural process. Despite testing the full list of options of climate processes known to exist in the recent past, no climate cycle candidates emerge with the characteristics to account for modern warming. In a few pages we test the last persistent issue, that medieval time was warmer than present. But first let's ask, "Is global warming caused by the Sun?"

IS GLOBAL WARMING CAUSED BY THE SUN?

In recent centuries there has been a steady increase in solar radiation (Figure 3.11), and certainly the Sun is the dominant factor governing Earth's climate. The most prominent feature of the Sun's activity is the 11-or-so-year sunspot cycle. This is associated with natural increases and decreases in solar radiation. The record of regular and frequent sunspot observations extends to about 1750; prior to those, less-frequent observations were made.[37]

These observations document that the Sun shows considerable variability. Between 1650 and 1700 the Maunder Minimum was a period of greatly reduced sunspot activity, and climate on Earth was cooler than average. This period corresponded to the coldest portion of the Little Ice Age, during which Europe and North America experienced bitterly cold winters.[38] The less severe Dalton Minimum lasted from 1790 to 1830 and also corresponded to a period of lower than average temperatures. The past 50 years is known as the Modern Maximum. This is a period of relatively high solar activity that began around 1900. The Modern Maximum reached a

---

[33]D. Seidov and M. Maslin, "Atlantic Ocean Heat Piracy and the Bipolar Climate See-Saw during Heinrich and Dansgaard-Oeschger events," *Journal of Quaternary Science*, v. 16.4, (2001): 321–328.

[34]S. Björck, "Current Global Warming Appears Anomalous in Relation to the Climate of the Last 20,000 Years," *Climate Research* 48, no. 1 (2011): 5 doi: 10.3354/cr00873.

[35]I. Hessler, S. Steinke, J. Groeneveld, L. Dupont, and G. Wefer, "Impact of Abrupt Climate Change in the Tropical Southeast Atlantic during Marine Isotope Stage (MIS) 3," *Paleoceanography*, 26 (2011): PA4209, doi: 10.1029/2011PA002118.

[36]See discussion at http://www.realclimate.org/index.php/archives/2006/11/revealed-secrets-of-abrupt-climate-shifts/ (accessed July 10, 2012). Also see T. Blunier and E. J. Brook, "Timing of Millennial-Scale Climate Change in Antarctica and Greenland during the Last Glacial Period," *Science* 291 (2001): 109–112; T. Blunier, J. Chappellaz, J. Schwander, A. et al., "Asynchrony of Antarctic and Greenland Climate Change during the Last Glacial Period," *Nature* 394 (1998): 739–743.

[37]See "Solar Variation," *Wikipedia*, http://en.wikipedia.org/wiki/Solar_variation (accessed July 10, 2012).

[38]IPCC, "Observed Climate Variability and Change," In *Climate Change 2001: Working Group I: The Scientific Basis* (Cambridge, U.K., Cambridge University Press, 2001). 2.3.3: "Was There a 'Little Ice Age' and a 'Medieval Warm Period'?" http://www.grida.no/publications/other/ipcc_tar/?src=/climate/ipcc_tar/wg1/070.htm (accessed July 10, 2012).

Figure 3.11. Since about 1750 there has been continuous measurement of sunspot activity. Prior to that there were occasional observations that have been interpreted to extend the record to about 1600.

SOURCE: Figure from "Solar Variation," *Wikipedia*, http://en.wikipedia.org/wiki/Solar_variation (accessed July 10, 2012).

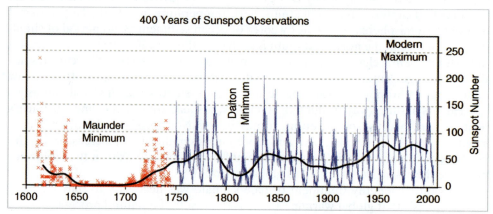

double peak, once in the 1950s and again in the 1990s. The causes for solar variations are not well understood, but because sunspots affect the brightness of the Sun, solar radiation is lower during periods of low sunspot activity.

The Modern Maximum is partly responsible for global warming, especially the temperature increases between 1900 and 1950. One study[39] argues that warming due to the relatively high level of solar activity since 1950 is responsible for 16% to 36% of recent warming; however, the authors state, "Most warming over the last 50 years is likely to have been caused by increases in greenhouse gases." Other researchers place the amount of recent warming due to the Sun at lower estimates. For instance, Benestad and Schmidt[40] state, "the most likely contribution from solar forcing to global warming is $7 \pm 1\%$ for the 20th century and is negligible for warming since 1980."

Solar radiation has not increased over the period when global warming has been strongest—since the 1970s (Figure 3.12). One study concludes that sunspot number[41] and global temperature data strongly correlate until the last 30 years, but "during these last 30 years the solar total irradiance, solar UV irradiance and cosmic ray flux have not shown any significant secular trend, so that at least this most recent warming episode must have another source."[42] In fact, several studies come to essentially the same conclusion. For example: "Solar forcing has declined over the past 20 years, while surface air temperatures have continued to rise."[43]

Another aspect of questioning the Sun's role in climate change involves realizing that if global warming were caused by a more-active Sun, researchers would expect to see warmer temperatures in all layers of the atmosphere. Instead, satellites have measured cooling in the stratosphere (the upper atmosphere)[44] and warming in the

[39]P. A. Stott, G. S. Jones, and J. F. B. Mitchell, "Do Models Underestimate the Solar Contribution to Recent Climate Change?" *Journal of Climate* 16, no. 24 (2003): 4079–4093.

[40]R. E. Benestad and G. A. Schmidt, "Solar Trends and Global Warming," *Journal of Geophysical Research* 114 (2009): D14101, doi: 10.1029/2008JD011639.

[41]"Sunspot number" refers to the number of dark areas on the Sun's surface that are counted daily by observatories. A high sunspot number means the Sun is active and producing more heat. A low sunspot number means the Sun is producing less heat. Sunspots tend to come and go on an 11-year cycle (shown in Figures 3.11 and 3.12).

[42]I. Usoskin, M. Schussler, S. Slanki, and K. Mursula, "Solar Activity over the Last 1150 Years: Does It Correlate with Climate?" Proceedings of the 13th Cool Stars Workshop, Hamburg, 5–9 July, 2004. See: http://www.mps.mpg.de/dokumente/publikationen/solanki/c153.pdf, last accessed 11/11/09.

[43]M. Lockwood and C. Fröhlich, "Recent Oppositely Directed Trends in Solar Climate Forcings and the Global Mean Surface Air Temperature. II. Different Reconstructions of the Total Solar Irradiance Variation and Dependence on Response Time Scale," *Proceedings of the Royal Society A* 464 (2008): 1367–1385, doi: 10.1098/rspa.2007.0347.

[44]D. T. Shindell, "Climate and Ozone Response to Increased Stratospheric Water Vapor," *Geophysical Research Letters* 28 (2001): 1551–1554.

Figure 3.12. In the past 30 years global temperature and solar radiation show little correlation. The Sun's energy has been measured by satellites since 1978, and it has followed its natural 11-year cycle of small ups and downs, but with no net increase. Over the same period, global temperature has strongly increased.

SOURCE: Figure from U.S. Global Change Research Program: http://www.globalchange.gov/resources/gallery (accessed July 10, 2012).

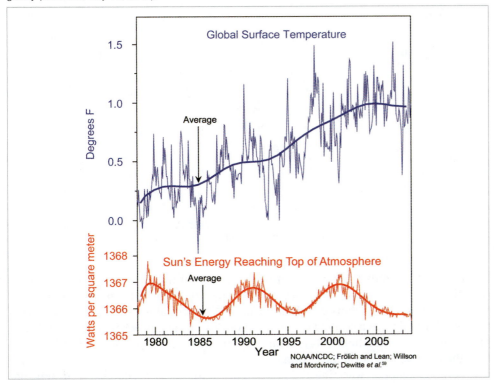

troposphere (at the surface). That's because greenhouse gases are trapping heat in the lower atmosphere and the Sun is not playing a significant role in climate change today.[45]

DID GLOBAL WARMING END AFTER 1998?

Did global warming end after 1998? This test of the global warming hypothesis surfaced because 2008 was dramatically cooler than previous years. Figure 3.13 shows a plot of the NASA global climate data from 1970 to 2011 and from 1997 to 2011. It is clear in both cases that warming has continued. Apparently the idea that global warming ended originated simply because eyeballing the data suggests the absence of a trend since 1998. But a linear regression,[46] the blue line in both plots, is a more legitimate method of determining a trend. A regression of these data reveals that it was a mistake to think that warming had stopped; an upward-sloping trend showing that warming has continued is clearly evident.

As discussed in Chapter 2, the Associated Press conducted a blind test. They gave unidentified temperature data to four independent statisticians and asked them to look for trends. The experts found no true temperature declines over time. One climate scientist said the following: "To talk about global cooling at the end of the hottest decade the planet has experienced in many thousands of years is ridiculous."[47]

---

[45]See two reports that allude to this logic: http://climate.nasa.gov/causes/; and T. R. Karl, J. M. Melillo, and T. C. Peterson, (eds.), *Global Climate Change Impacts in the United States* (Cambridge, U.K., Cambridge University Press, 2009).

[46]Linear regression is a mathematical technique that calculates a straight line from data as a best estimate of the trend.

[47]The Associated Press article describing the statistical testing contains this quote and can be found here: http://news.yahoo.com/s/ap/20091026/ap_on_bi_ge/us_sci_global_cooling (accessed July 10, 2012).

Figure 3.13. Average annual surface temperature (red; ocean and land, NASA[48]) and trend (blue). The vertical axis plots temperature change from a standard (the average temperature of 1951–1980). Large plot shows the trend for 1970–2011, and the inset graphs show the trend for 1997–2011.

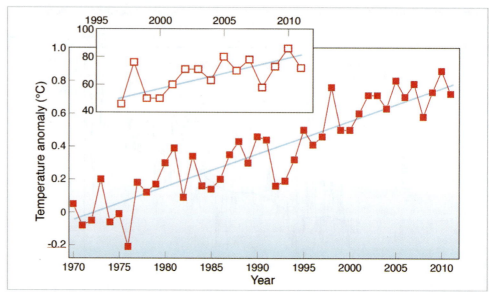

In another study,[49] scientists showed that naturally occurring periods of no warming or even slight cooling can easily be a part of a longer-term pattern of global warming. These researchers conclude, "Claims that global warming is not occurring that are derived from a cooling observed over short time periods ignore natural variability and are misleading."[50] It is clear that global warming did not end in 1998 and that the climate has not been cooling.

Although global warming deniers might continue to argue that warming has not been strong over the past decade, a look at the temperature of the ocean reveals that warming has not relented over the past decade, not even a little; it's just that the excess heat in Earth's climate system is being stored in the ocean. We learned earlier in the chapter that 93% of the heat trapped by increasing greenhouse gases goes into warming the ocean, not the atmosphere. So taking the ocean's temperature is the most comprehensive way to monitor global warming. A group of National Oceanic and Atmospheric Administration (NOAA) scientists has revised and updated their decade-old compilation of temperature measurements from the upper 2,000 m (6560 ft) of the world's oceans.[51] Ocean warmth (Figure 3.14) has steadily increased since 1990, and the upper ocean has warmed so much in the past 50 years that its additional heat, if released, would be enough to warm the lower atmosphere by about 36°C (65°F).

Although it is no surprise that the rate of atmospheric warming declined somewhat after 1998, scientists have struggled to learn why. The answer lies with climate processes that are complex and not well understood: reflection of sunlight by aerosols in the stratosphere, and processes related to the storage of heat by ocean circulation. In 2010 researchers[52] determined that the concentration of aerosols in the stratosphere over the previous decade has been somewhat higher than assumed, and they calculated that the

[48]NASA-GISS dataset is available at http://data.giss.nasa.gov/gistemp/tabledata/GLB.Ts.txt (accessed July 10, 2012).

[49]D. Easterling and M. Wehner, "Is the Climate Warming or Cooling?" *Geophysical Research Letters* 36 (2009): L08706.

[50]See the NASA website on this study: http://climate.nasa.gov/news/index.cfm?FuseAction=ShowNews&News ID=175 (accessed July 10, 2012).

[51]S. Levitus, J. Antonov, T. Boyer, et al., "World Ocean Heat Content and Thermosteric Sea Level Change (0-2000), 1955–2010," *Geophysical Research Letters* 39 (2012): L10603, doi: 10.1029/2012GL051106.

[52]S. Solomon, J.S. Daniel, R. R. Neely, J. P. Vernier, and E. G. Dutton, "The Persistently Variable 'Background' Stratospheric Aerosol Layer and Global Climate Change," *Science* 333, no 6044 (2011): 866–870, doi: 10.1126/science.1206027.

Figure 3.14. The world ocean has warmed steadily since 1990.

SOURCE: Figure from "*Science*Shot: No Letup in World's Warming," *Science*, http://news.sciencemag.org/sciencenow/2012/04/scienceshot-no-letup-in-worlds.html?rss=1.

CREDIT: Adapted from S. Levitus et al., Geophys. Res. Letts.; © AGU 2012.

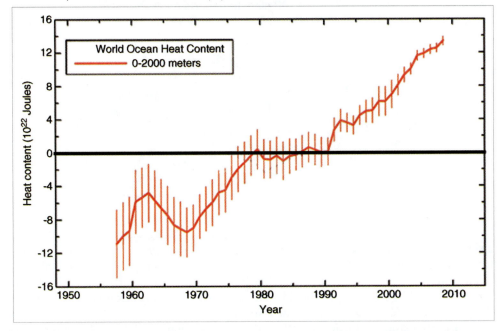

reflection of sunlight by these particles might have had a cooling effect about 20% more than would be expected without them. Aerosol particles are produced by burning coal, wood, and animal dung. There is an especially high production of aerosols by power plants in Asia burning sulfur-rich coal whose influence had not previously been recognized. Aerosols are also produced by volcanic eruptions, and relatively small volcanic eruptions such as the 2006 eruptions of Soufriere Hills in Montserrat and Tavurvur in Papua New Guinea may have contributed more aerosols than previously realized.

Another team of researchers[53] looked at heat storage in the uppermost ocean (0–700 m, 0–2300 ft) during the period 2003–2010 and found it had not gained any heat. By using an ensemble of computer climate models to trace heat budget variations, they learned that an eight-year period without upper ocean warming is not unusual. They probed the history of El Niño, which releases heat to the atmosphere, and the thermohaline circulation, which buries heat deep in the oceans. Models suggested that both processes combined starting in 2003, preventing excess energy from accumulating in the shallow ocean as usual. Approximately 45% of the missing heat was instead released to space, and 35% was stored below 700 m depth in the North Atlantic Ocean. How long will this pattern continue? The researchers point to recently observed changes in these two modes of climate variability and predict an upward trend in upper ocean heat content.

DO SCIENTISTS DISAGREE ON GLOBAL WARMING?

According to the vast majority of climate scientists,[54] the planet is heating up. In a survey[55] of 3,146 earth scientists, among the most highly qualified climatologists (those who wrote more than 50% of their peer-reviewed publications in the past five

[53]C. A. Katsman and G. J. van Oldenborgh, "Tracing the Upper Ocean's 'Missing Heat'", *Geophysical Research Letters*, 38 (2011): L14610, doi: 10.1029/2011GL048417.

[54]W. R. L. Anderegg, J. W. Prall, J. Harold, and S.H. Schneider, "Expert Credibility in Climate Change," *Proceedings of the National Academy of Sciences* 107, no. 27 (2010): 12107–12109, doi: 10.1073/pnas.1003187107.

[55]P. T. Doran and M. K. Zimmerman, "Examining the Scientific Consensus on Climate Change," *Eos* 90, no. 3 (2009) http://tigger.uic.edu/~pdoran/012009_Doran_final.pdf (accessed July 10, 2012).

Figure 3.15. Response of scientists who publish in earth science to the question "Do you think human activity is a significant contributing factor in changing mean global temperatures?"
SOURCE: Figure from SkepticalScience.com.

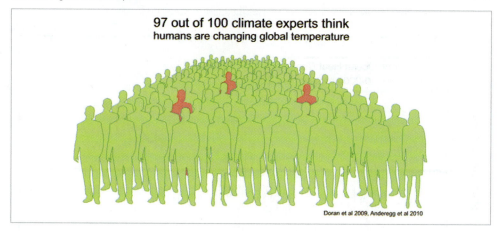

years on the subject of climate change) more than 95% agreed "human activity is a significant contributing factor in changing mean global temperatures." The survey found that as the level of active research and peer-reviewed publication in climate science increases, so does agreement that humans are significantly changing global temperatures (Figure 3.15). The divide between expert climate scientists (97.4%) and the general public (58%) is particularly striking.

An earlier study, published in 2004,[56] came to a similar conclusion: There is strong scientific consensus on global warming, and there is agreement that humans are the primary cause. The study analyzed all peer-reviewed scientific papers between 1993 and 2003 using the keyword phrase "climate change." Here is what its authors concluded: "The 928 papers were divided into six categories: explicit endorsement of the consensus position, evaluation of impacts, mitigation proposals, methods, paleoclimate analysis, and rejection of the consensus position. Of all the papers, 75% fell into the first three categories, either explicitly or implicitly accepting the consensus view, and 25% dealt with methods or paleoclimate, taking no position on current anthropogenic climate change. Remarkably, none of the papers disagreed with the consensus position."

It is strikingly obvious, from any rational point of view, that there is very strong scientific consensus that global warming is real and that humans are the primary cause.

ARE CLIMATE DATA FAULTY?

Climate measurements come from many sources. Satellites measure Earth's temperature from a distance, weather balloons measure vertical profiles of the atmosphere as they ascend, and ground-based weather stations located at thousands of sites across the globe measure the temperature of the lower troposphere and the ground. The agencies that collect temperature data take pains to remove factors that might skew the data artificially, relative to the true temperature. One such factor is known as the urban heat island effect; basically, urbanized regions tend to be hotter than the adjacent countryside.

In assembling climate data, NASA compares long-term urban temperature trends to nearby rural trends. They then adjust the urban trend so it matches the rural trend. The methodology that NASA Goddard Institute for Space Studies uses in assembling its dataset is explained in detail on their website.[57] Contrary to

---

[56]N. Oreskes, "Beyond the Ivory Tower: The Scientific Consensus on Climate Change," *Science* 306, no. 5702 (2004): 1686, http://www.sciencemag.org/cgi/content/full/306/5702/1686# (accessed July 10, 2012).

[57]See the methodology explained here: http://pubs.giss.nasa.gov/docs/2001/2001_Hansen_etal.pdf. This methodology is periodically updated and discussed at this site: http://data.giss.nasa.gov/gistemp/updates/ (accessed July 10, 2012).

popular belief, because most urban climate stations are located in parks and other nonindustrial areas, NASA found that 42 percent of city trends are cooler relative to their country surroundings.

This is consistent with a study[58] finding that no statistically significant impact of urbanization could be identified in annual temperatures. Researchers found that industrial sections of towns may well be significantly warmer than rural sites, but urban meteorological observations are more likely to be made within parks that are cool islands compared to industrial regions. Another study[59] analyzed 50-year records of temperatures on calm nights and on windy nights. It concluded "temperatures over land have risen as much on windy nights as on calm nights, indicating that the observed overall warming is not a consequence of urban development." The reasoning for this conclusion is that windy nights will circulate air from cool surroundings into the warm city, and thus warming should be suppressed on those nights if it is due to the urban effect.

Claims that global warming is based on unreliable data have been rigorously tested[60] and are simply not true. In any case, satellite data and ocean measurements confirm that global warming is not an artifact of ground-based measurements.

IS TODAY'S WARMING SIMPLY A REPEAT OF THE RECENT PAST?

It has been claimed that two periods in recent geologic history were warmer than today, and therefore today's warming is no big deal and probably a natural event. One of these periods, known as the Eemian Interglacial, was indeed warmer as a result of the orbital parameters at the time, and the other, called the Medieval Climate Anomaly (MCA), might have been warmer at various times in various places within the North Atlantic region, but this does not mean it was a global phenomenon. Let's look into the MCA first.

Medieval Climate Anomaly (MCA)

It has been claimed that the MCA was a time of warmer temperatures prior to industrial greenhouse gas production.[61] However, the statistical validity of proxy temperature reconstructions of the climate during medieval times has been questioned by a number of authors,[62] and the IPCC AR4 concludes that "it is likely that the 20th century was the warmest in at least the past thirteen hundred years."[63] (Figure 3.16). The MCA has not been well documented outside of the North Atlantic region, and one cannot assume it was a global phenomenon. Additionally, even in the North Atlantic, modeling and geologic proxy data used by the IPCC indicate that temperatures were very unlikely to have been higher than present temperatures and that they rose and fell at different times in different places. This scientific evidence makes the event fundamentally different from today's global warming pattern, which is nearly everywhere synchronous and statistically significant.

[58]T. C. Peterson, "Assessment of Urban versus Rural in situ Surface Temperatures in the Contiguous United States: No Difference Found," *Journal of Climate* 16, no 18 (2003): 2941–2959.

[59]D. Parker, "A Demonstration that Large-Scale Warming Is Not Urban," *Journal of Climate* 19, no. 12, (2006): 2882–2895.

[60]M. Menne, C. Williams, and M. Palecki, "On the Reliability of the U.S. Surface Temperature Record," *Journal of Geophysical Research* 115 (2010): D11108, doi: 10.1029/2009JD013094.

[61]See the video "Climate Denial Crock of the Week: The Medieval Warming Crock" at the end of this chapter.

[62]See the review in the IPCC AR4, Chapter 6, Box 6.4 "Hemispheric Temperatures in the Medieval Warm Period" p. 468, available at http://www.ipcc.ch/pdf/assessment-report/ar4/wg1/ar4-wg1-chapter6.pdf (accessed July 10, 2012).

[63]E. Jansen, J. Overpeck, K.R. Briffa, et al., "Palaeoclimate." In S. Solomon, D. Qin, M. Manning, et al (eds), *Contribution of Working Group I to the Fourth Assessment Report (AR4) of the Intergovernmental Panel on Climate Change* (Cambridge, U.K., Cambridge University Press, 2007).

Figure 3.16. Reconstructions of global temperature (using climate proxies) reveal that global temperatures today are the highest in the past 1,300 years. Scientists have shown that local temperatures from one place to another may have been as warm as or even warmer than today, but these do not represent a global trend, only a local pattern. The black line is the record provided by modern reliable instruments. The level of brown shading represents the percentage of statistical probability in the temperature reconstruction.

Source: Climate Change 2007: The Physical Science Basis. Working Group I Contribution to the Fourth Assessment Report of the Intergovernmental Panel on Climate Change, Figure 6.10 (c). Cambridge University Press.

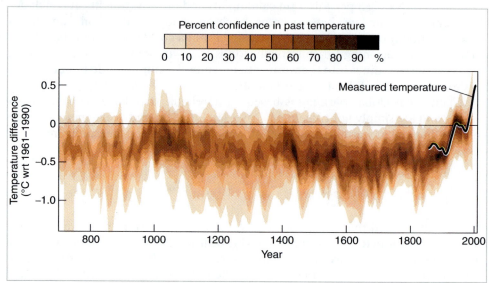

An important study[64] was published in 2006 on the MCA and it concludes:

- Dramatic warming has occurred since the 19th century.
- The record warm temperatures in the last 15 years are indeed the warmest temperatures in at least the last 1,000 years and possibly in the last 2,000 years.
- Comparisons between modern global warming and the MCA can only be made at the local and regional scale.

Researchers[65] examined tree ring data recording land temperatures and found clear MCA (warm), Little Ice Age (cool), and recent (warm) episodes preserved in North American and Eurasian tree ring records. They conclude that MCA temperatures were nearly 0.7°C (1.26°F) cooler than in the late 20th century, with an amplitude difference of 1.14°C (2.05°F) from the coldest (1600–1609) to warmest (1937–1946) decades. The paper also stresses "that presently available paleoclimate reconstructions are inadequate for making specific inferences, at hemispheric scales, about MCA warmth relative to the present anthropogenic period and that such comparisons can only still be made at the local/regional scale." This means that it is wrong to use the MCA in a discussion of global warming, because the MCA was not likely a global event.

The Little Ice Age was a period of cooling that occurred after the MCA. The event has been depicted[66] as having three maxima beginning about 1650, 1770, and 1850, each separated by slight warming intervals. Scientific consensus[67] is that this cooling occurred (like the MCA) with varying degrees of intensity, at different times, in different places. Paleoclimatologists no longer expect to agree on either the start

---

[64]R. D'Arrigo, R. Wilson, and G. Jacoby, "On the Long-Term Context for Late Twentieth Century Warming," *Journal of Geophysical Research–Atmospheres*, 111, no. D3, (2006): D03103, doi: 10.1029/2005JD006352, see http://www.ncdc.noaa.gov/paleo/pubs/darrigo2006/darrigo2006.html (accessed July 10, 2012).

[65]R. D'Arrigo, R. Wilson, and G. Jacoby, "On the Long-Term Context for Late Twentieth Century Warming."

[66]See NASA Glossary: http://earthobservatory.nasa.gov/Glossary/?mode=alpha&seg=l&segend=n (accessed July 10, 2012).

[67]S. Solomon, D. Qin, M. Manning, et al., eds, *AR4*.

Figure 3.17. The last interglacial consisted of five stadials and interstadials, named MIS5a–e. The last glacial consisted of two stadials, MIS4 and MIS2, as well as one interstadial, MIS3. The present interglacial is MIS1, also known as the Holocene Epoch. The acronym MIS stands for Marine Isotopic Stage, because these detailed records were first identified using oxygen isotopes in seafloor sediments.

SOURCE: Fletcher, *Physical Geology: The Science of Earth,* 2012.

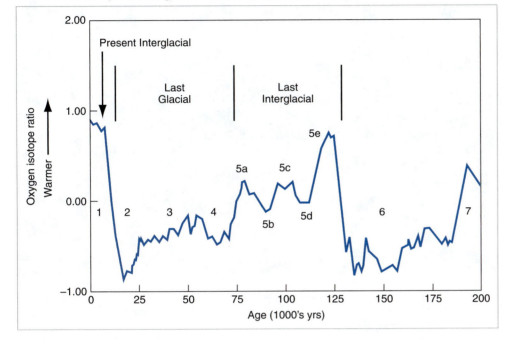

or end dates of this event, which varied according to local conditions.[68] Another researcher[69] studying the MCA established, "The Medieval period is found to display warmth that matches or exceeds that of the past decade in some regions, but which falls well below recent levels globally."

Eemian Interglacial

There is one period of recent geologic history that is generally agreed to have been warmer than today: 125,000 years ago, the Eemian Interglacial. Paleoclimatologists have calculated that it was warmer during Eemian time because orbital parameters favored greater warmth. The Eemian was a natural episode of global warming and one that is intensely studied[70] to improve understanding of what we might expect with continued global warming later this century. One study[71] asserts that during Eemian time, global sea level peaked 5.5 to 9 m (18 to 30 ft) above present despite temperatures estimated to be only 2°C (3.6°F) above pre-industrial levels.

The last interglacial, broadly defined, occurred between approximately 130,000 and 75,000 years ago. Climate during this 55,000-year period was not continuously warm. Rather, researchers have identified five major phases consisting of three interstadials (warmings) and two stadials (coolings). These show up clearly in the ice-core records as well as the deep-sea record. Figure 3.17 shows these phases, using the scientific naming system for climate stages of this time. The last interglacial is named after the oxygen isotope proxy that was used to first identify it in cores of

[68]See "Little Ice Age," *Wikipedia*, http://en.wikipedia.org/wiki/Little_Ice_Age (accessed July 10, 2012).

[69]M. E. Mann, Z. Zhang, S. Rutherford, et al., "Global Signatures and Dynamical Origins of the Little Ice Age and Medieval Climate Anomaly," *Science* 326, no. 5957 (2009): 1256–1260, doi: 10.1126/science.1177303.

[70]D. R. Muhs, K. R. Simmons and B. Steinke, "Timing and Warmth of the Last Interglacial Period: New U-series Evidence from Hawaii and Bermuda and a New Fossil Compilation for North America," *Quaternary Science Reviews* 21 (2002): 1355–1383.

[71]A. Dutton and K. Lambeck, "Ice Volume and Sea Level during the Last Interglacial." *Science*, 337 (2012): 216–219.

Figure 3.18. This rocky shoreline in Hawaii is composed of limestone formed by a fossil reef that grew under higher-than-present sea levels during MIS5e.[72]

Image Credit: Courtesy of Chip Fletcher.

Large coral head from MIS5e

marine sediments: Marine Isotopic Stage 5 (MIS5), and the stadials and interstadials are labeled MIS5a–e. Of these, MIS5e (the Eemian) was the warmest, and most like the current Holocene Epoch.

MIS5e offers a geologically recent example of a warm period with characteristics similar to those of the Holocene; however, it differs from the Holocene in that the warmth at the time was driven by orbital parameters that were very different from Holocene time. Because it is also a relatively recent event, many rocks and sediments that record climate conditions from that time have not been lost to erosion.

MIS5e lasted approximately 12,000 years, from 130,000 to 118,000 years ago, and the average age of fossil corals from around the world that grew at that time is 125,000 years. For example, Figure 3.18 shows a fossil reef on the Hawaiian island of Oahu that illustrates another important feature of MIS5e: Sea level was higher than present, estimated by different authors to have been between 4 to 6 m[73] (13 to 20 ft) and 5.5 to 9 m[74] (18 to 30 ft). Researchers have therefore concluded that because climate was warmer, melted ice contributed to the higher sea level. This conclusion is supported by deep cores of ice in Greenland and Antarctica that preserve temperature records from MIS5e, showing that it was warmer, with lower ice volume, than the present-day climate.

The Eemian has been cited as a possible analogue for a future climate[75] under increased global warming. Studies have shown that CO_2 concentrations in the atmosphere were relatively high[76] (though not as high as they are today owing to the contribution of industrial greenhouse gases), temperatures were higher than at present, and sea level was higher.[77] Scientists study MIS5e to improve understanding of the

[72]C. H. Fletcher, C. Bochicchio, C. L. Conger, et al., "Geology of Hawaii Reefs." In *Coral Reefs of the U.S.A.* (New York: Springer, 2008), pp. 435–488.

[73]R. E. Kopp, F. J. Simons, J. X. Mitrovica, A. C. Maloof, and M. Oppenheimer, "Probabilistic Assessment of Sea Level during the Last Interglacial Stage," *Nature* 462 (2009): 863–868, doi: 10.1038/nature08686.

[74]A. Dutton and K. Lambeck, "Ice Volume and Sea Level during the Last Interglacial." *Science*, 337 (2012): 216–219.

[75]P. U. Clark and P. Huybers, "Interglacial and Future Sea Level," *Nature* 462 (2009): 856–857.

[76]U. C. Müller, "Cyclic Climate Fluctuations during the Last Interglacial in Central Europe," *Geology* 33, no. 6 (2005): 449–452.

[77]M. T. McCulloch and T. Esat, "The Coral Record of Last Interglacial Sea Levels and Sea Surface Temperatures." *Chemical Geology* 169 (2000): 107–129.

Figure 3.19. Computer models that simulate climate during MIS5e indicate that melting of the Greenland ice sheet was responsible for a global sea-level rise of approximately 4 m (13 ft).

Source: "Arctic, Antarctic Melting May Raise Sea Levels Faster than Expected," http://www.ucar.edu/news/releases/2006/melting.shtml (accessed July 10, 2012).

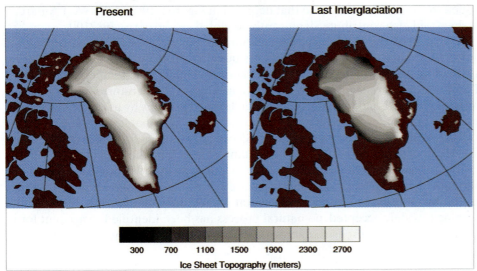

duration of the last interglacial period and global characteristics at the time. Both of these goals are intended to provide a basis for testing and advancing computer models that can be used to predict future climate.

Computer models of climate change during MIS5e indicate that sea-level rise started with melting of the Greenland ice sheet and not the Antarctic ice sheet.[78] Research also suggests that ice sheets across both the Arctic and Antarctic could melt more quickly than expected this century because temperatures are likely to rise higher than they did during MIS5e, especially in the Polar Regions. If these predictions are correct, by 2100 the Arctic could warm by 3°C to 5°C (5.4°F to 9°F) in summer. During MIS5e, meltwater from Greenland and other Arctic sources raised sea level by as much as 4 m (13 ft). However, because global sea level actually rose significantly higher, researchers have concluded that Antarctic melting and thermal expansion of warm seawater must have produced the remainder of the rise in sea level.

The rise in sea levels produced by Arctic warming and melting could have floated, and thus destabilized ice shelves at the edge of the Antarctic ice sheet and led to their collapse, a positive feedback to sea level rise. If such a process occurred today, it would be accelerated by global warming year round. In the last few years, sea level has begun rising more rapidly, and now it is rising at a rate of more than 3 cm per decade (12 in per century). Recent studies have also found accelerated rates of annual melt of both the Greenland and Antarctica ice sheets.[79]

During MIS5e, the amount of global warming needed to initiate this melting was less than 3.5°C (6.3°F) above modern summer temperatures, similar to the amount that is predicted to occur by mid-century if CO_2 levels continue to rise unchecked. The amount of Greenland ice sheet melting that produced higher sea levels is shown in Figure 3.19. This reconstruction predicts that sea level rose at a rate exceeding 1.6 m (5.3 ft) per century, a rate that would be potentially catastrophic for coastal

[78]J. T. Overpeck, B. L. Otto-Bliesner, G. H. Miller, et al., "Paleoclimatic Evidence for Future Ice-Sheet Instability and Rapid Sea-Level Rise," *Science* 311 (2006): 1747–1750.

[79]I. Velicogna, "Increasing Rates of Ice Mass Loss from the Greenland and Antarctic Ice Sheets Revealed by GRACE," *Geophysical Research Letters* 36 (2009): L19503, doi: 10.1029/2009GL040222. See also E. Rignot, I. Velicogna, M. R. van den Broeke, A. Monaghan, and J. Lenaerts, "Acceleration of the Contribution of the Greenland and Antarctic Ice Sheets to Sea Level Rise," *Geophysical Research Letters* 38 (2011): LO5503, doi: 10.1029/2011GL046583.

R 3   How Do We Know that Humans Are the Primary Cause of Global Warming?

101

communities worldwide if it were to happen today. Computer modeling[80] predicts several other features of the Eemian that could portend global conditions by the end of the 21st century: Global carbon dioxide rose by 1% per year (half the current rate of rise); 2100 will be significantly warmer than MIS5e, so Greenland is already headed toward a state similar to that depicted in Figure 3.19; and the West Antarctic ice sheet will also contribute significantly to global sea-level rise by 2100.

The end of the Eemian was characterized by precipitous changes in global climate. Although the period has been studied intensively, global climate during MIS5a–d is poorly understood. Researchers speculate that temperatures during MIS5d and 5b were significantly cooler than present temperatures, that global ice volume expanded, and that global sea level dropped perhaps by as much 25 m (82 ft) below the present level. These were, in effect, mini ice ages that lasted a few thousand to 10,000 years each. The interstadials MIS5a and 5b were likely periods during which global temperature was cooler and ice volume greater than they are today.

The origin of global warming has been vigorously tested for decades. There are still many details to work out, and there are occasional surprises in ongoing work, but the hypothesis that human production of heat-trapping gas has led to global warming is widely accepted, no natural process has been identified to account for it, and it represents a consensus opinion of the scientific community.

ANIMATIONS AND VIDEOS

"Ice Stories: Working to Reconstruct Past Climates," http://www.youtube.com/watch?v=r81MtDimgSg

David Attenborough, "The Truth about Climate Change," http://www.youtube.com/watch?v=S9ob9WdbXx0&feature =related

"Richard Alley's Global Warming," http://www.youtube.com/ watch?v=T4GThA35s1s&feature=relmfu

Climate Denial Crock of the Week, "The Medieval Warming Crock," http://www.youtube.com/watch?v=vrKfz8NjEzU

COMPREHENSION QUESTIONS

1. What is a climate proxy? Identify two climate proxies and describe how they work.

2. Explain how the orbital parameters influence climate.

3. Do ice cores and ocean cores tell the same story about paleoclimate?

4. What was the role of carbon dioxide in Earth's climate system at the end of the last ice age?

5. How do climate feedbacks work? Describe one positive and one negative feedback.

6. Is global warming caused by the Sun?

7. Describe the Medieval Climate Anomaly.

8. Describe natural climate cycles. Are these responsible for modern global warming?

9. Did global warming end after 1998? Why or why not?

10. What is the Eemian? Why do researchers study the Eemian?

THINKING CRITICALLY

1. What climate processes are recorded in ice cores and deep sea cores and why are they related?

2. You are asked to appear before a congressional hearing into climate change. Explain how paleoclimate improves our understanding of certain aspects of the global warming issue.

3. What is a Dansgaard–Oeschger event and what role does it play in the discussion of modern climate change?

4. The bipolar seesaw has been used to explain why Dansgaard–Oeschger events are not responsible for global warming. Elaborate.

5. How do oxygen isotopes reveal paleoclimate patterns?

6. How do we know that orbital parameters are not responsible for modern global warming?

7. Explain how climate feedbacks play a critical role in understanding the origin of modern climate change.

8. The Medieval Climate Anomaly has been used to explain global warming as a natural event. What is the logic behind this and why is it wrong?

9. Describe why attributing global warming to natural climate cycles is not supported by the evidence.

10. What aspects of the Eemian make it useful for understanding the impacts of global warming?

[80]J. T. Overpeck, B. L. Otto-Bliesner, G. H. Miller, et al., "Paleoclimatic Evidence for Future Ice-Sheet Instability and Rapid Sea-Level Rise."

CLASS ACTIVITIES (FACE TO FACE OR ONLINE)

ACTIVITIES

1. Visit the "Powers of 10" website http://www.ncdc.noaa.gov/paleo/ctl/index.html and answer the following questions.

   a. How is climate related to the water cycle?

   b. Describe the time scales of climate change.

   c. What is a climate proxy and what do different proxies tell us about climate variability?

   d. Compare and contrast climate variability on the time scale of 100 years versus 10,000 years.

2. Watch the video "Climate Denial Crock of the Week: That 1500 Year Thing" http://www.youtube.com/watch?v=G0HGFSUx2a8&feature=view_all&list=PL029130BFDC78FA33&index=55 and answer the following questions.

   a. How are paleoclimate data used to address the question of natural cycles as a cause of global warming?

   b. How does heat play a role in this issue?

   c. Describe the methods used by climate deniers as outlined in this video and the potential impacts.

   d. Describe the difference between 1500-year climate cycles and modern global warming.

3. Watch the video "Climate Denial Crock of the Week: The Urban Heat Island" http://www.youtube.com/watch?v=B7OdCOsMgCw&feature=view_all&list=PL029130BFDC78FA33&index=54 and answer the following questions.

   a. Describe the urban heat island effect.

   b. Explain why the urban heat island effect is not a real source of error in global warming data. Use information from the video as well as from this chapter.

   c. What are some of the impacts of global warming on the natural world?

   d. Describe the methods used by "Climate Denial Crock of the Week" to improve understanding of climate change issues.

HOW DO SCIENTISTS PROJECT FUTURE CLIMATE?

FIGURE 4.0. The NASA Center for Climate Simulation (NCCS) has provided supercomputing resources to NASA scientists and engineers for over 25 years. This model visualization depicts specific atmospheric humidity on June 17, 1993, during the Great Flood that hit the Midwestern United States.

IMAGE CREDIT: Research: Michele Rienecker, Max Suarez, Ron Gelaro, Julio Bacmeister, Ricardo Todling, Larry Takacs, Emily Liu, Steve Pawson, Mike Bosilovich, Siegfried Schubert, Gi-Kong Kim, NASA/Goddard; Visualization: Trent Schindler, NASA/Goddard/UMBC

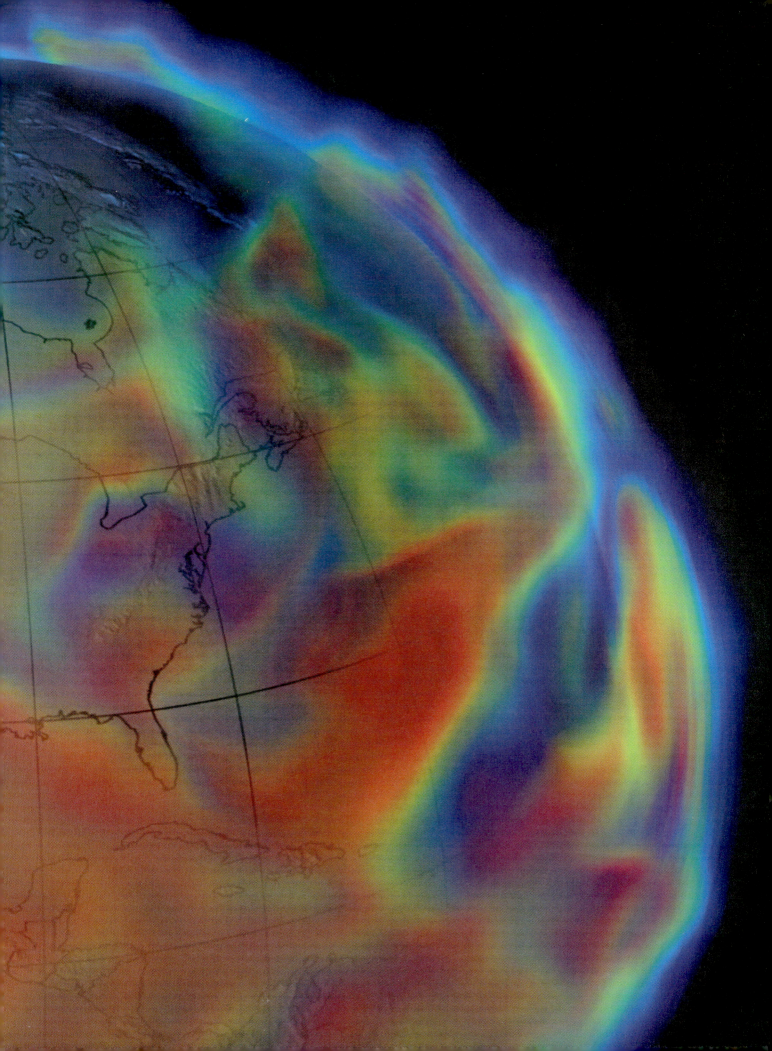

CHAPTER SUMMARY

Climate models successfully reproduce the past 100 years of climate change, but only when greenhouse gases, produced by human activities, are included. Models published by the International Panel on Climate Change use a range of potential future scenarios of greenhouse gas emissions to predict that surface air warming in the 21st century will likely (better than 66% probability) range from a low of 1.1°C to a high of 6.4°C (2.0°F to 11.5°F). Climate models provide important results for understanding future global climate, but their ability to project regional and localized climate is still limited.

In this chapter you will learn that:

- Climate processes interact with one another over different lengths of time, sometimes enhancing and sometimes suppressing each other's effects. Global circulation models (GCMs) are computer models attempting to make order out of this complexity.

- Global climate is influenced by explosive volcanic eruptions, the ice-albedo effect, clouds, and variations in solar radiation; models must take these into account.

- Confidence in climate model projections is strengthened because of the agreement between model simulations of the past and actual observed temperature increases.

- If greenhouse gas and aerosol concentrations were kept at year 2000 levels, climate models project that a temperature rise of about 0.1°C (0.18°F) per decade would be expected for the next two decades.

- Climate models project a temperature rise of about 0.2°C (0.36°F) per decade for the next two decades for all potential future scenarios of greenhouse gas emissions.

- The best estimate for a low scenario of surface air warming in the 21st century is 1.8°C (3.24°F), with a likely range of 1.1 to 2.9°C (2.0 to 5.2°F).

- The best estimate for a high scenario of surface air warming in the 21st century is 4.0°C (7.2°F), with a likely range of 2.4°C to 6.4°C (4.3°F to 11.5°F).

- Climate models project that these increases in global surface air temperature will likely cause increased drought, sea-level rise, frequency of warm spells, heat waves, heavy rainfall events, intensity of tropical cyclones (including hurricanes), extreme high tides, reductions of sea ice, and other physical impacts.

Learning Objectives

Researchers use models to simulate the complex behavior and interaction of climate processes that operate among the oceans, land, biosphere, and atmosphere. These models provide useful large-scale predictions of future climate and its impacts.

Earth's climate system is very complex. Some climate processes operate on cycles (such as seasons and glacial–interglacial cycles), some occur with irregular timing (such as ENSO and Dansgaard–Oeschger events), and some are essentially unpredictable far in advance but it is possible to say that they are "likely" or "unlikely" (such as hurricanes, snowfall, rain, and extremely hot days). In reality, climate processes all interact with one another over different lengths of time, sometimes enhancing and sometimes suppressing each other's effects. Climate complexity is enormous. Several fundamental processes are not well understood, and getting any one of them wrong could mean the success or failure of a climate model.

For example, in Chapter 3 we discussed the relative slowdown in warming that occurred during the first decade of this century. Subsequent research on this problem has resulted in several hypotheses to explain the pattern. One group of researchers[1] noted that the upper ocean stopped accumulating heat following 2003. They concluded that some of the missing energy had been lost to space, and the remainder had been stored deep in the thermohaline circulation system of the North Atlantic. Another group[2] concluded that the lack of strong warming resulted from stratospheric aerosol particles that "persistently varied" rather than staying stable as researchers had assumed. They identified Asian power plants that burn sulfur-rich coal and moderately small volcanic eruptions as the source of these particles. The message is that modeling climate complexity[3] continues to be a difficult aspect of predicting the impacts of global warming.

Attempting to make order out of this complexity is the role of computer models called global circulation models (GCMs).[4] GCMs must be able to reproduce the influence that a number of climate-regulating processes exert on climate variability. Processes that influence climate include large-scale volcanic eruptions that emit light-scattering aerosols into the stratosphere, the Pacific climate process known as the El Niño Southern Oscillation (ENSO), variations in solar radiation such as the sunspot cycle, the role of greenhouse gas buildup resulting from human activities, the growth and development of clouds, and others.

---

[1] C. A. Katsman and G. van Oldenborgh, "Tracing the Upper Ocean's 'Missing Heat'". *Geophysical Research Letters* 38 (2011): L14610, doi:10.1029/2011GL048417.

[2] S. Solomon, J. Daniel, R. Neely, J. Vernier, and E. Dutton, "The Persistently Variable 'Background' Stratospheric Aerosol Layer and Global Climate Change," *Science* 333, no. 6044 (2011): 866–870, doi: 10.1126/science.1206027.

[3] C. Schultz, Interview with De-Zheng Sun, co-editor of "Climate Dynamics: Why Does Climate Vary?" *Eos* 92, no. 34 (2011): 285–286.

[4] See this link for a description of a global climate model: http://www.sciencedaily.com/releases/2004/06/040623082622.htm (accessed July 10, 2011).

CLIMATE MODELS

Researchers use global circulation models to study climate change. These are computer-based mathematical programs that simulate the behavior and interaction of Earth's oceans, land, and atmosphere. GCMs consist of thousands of mathematical calculations that solve the equations of fluid dynamics on supercomputers. The equations of fluid dynamics are used to calculate the properties of fluids (gas, and thus Earth's atmosphere, is considered a fluid) such as velocity, pressure, density, temperature, and how they change over space and time. These equations are used as laws to predict the behavior of the atmosphere and the ocean in various useful and applied settings. Because both gases and liquids behave as fluids, fluid dynamics can predict water flow in a pipe, circulation of the atmosphere, airplane flight, acoustics, ocean currents, and other very relevant phenomena. The models describe how air temperature and pressure, winds, clouds, various types of gases, and other characteristics of the climate (Figure 4.1) all respond to heating by the Sun and other phenomena that drive climate.[5] GCMs include equations that predict how greenhouse gases influence the climate.

Figure 4.1. General circulation models are used for weather forecasting, simulating climate, and predicting climate change. Models must take many factors into account such as how the atmosphere, the oceans, the land, ecosystems, ice, topography, and energy from the Sun all affect one another and Earth's climate.

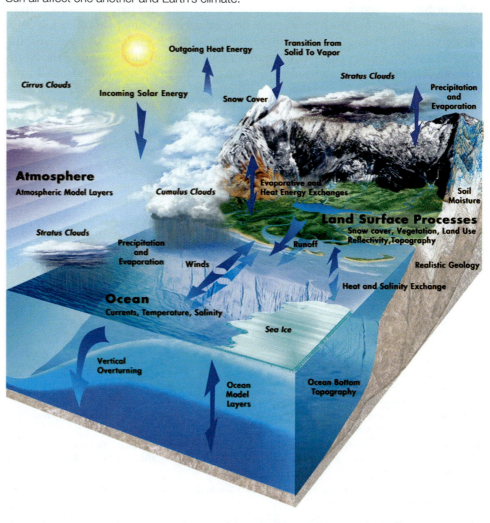

[5]See the video "Recipe for a better climate model" at the end of the chapter.

GCMs receive input in the form of data on ocean currents and seawater temperature, the concentration of CO_2 and other greenhouse gases in the atmosphere; the amount of sunlight; the cover of vegetation, ice, and snow; the development of clouds; and other factors that affect the heating of Earth's surface. These inputs are used to guide how the equations treat the various factors that influence climate. There are various types of models; separate atmospheric GCMs and oceanic GCMs treat different parts of the climate system. Atmospheric and ocean GCMs can be united to form a coupled general circulation model, and if other components such as a model of sea ice or a model of evapotranspiration over land are added, the GCM can become a full climate model. The most actively researched use of climate models in recent years has been to project temperature changes resulting from increases in atmospheric concentrations of greenhouse gases.

Climate models are designed to simulate climate on a range of scales, from global to regional (hundreds of kilometers). But few models regularly tackle climate changes at the local level (tens of kilometers). Most break up the atmosphere into ten to thirty vertical levels between Earth's surface and outer space, where climate phenomena are represented by complex calculations. A model's output might include predictions of long-term precipitation patterns, or an estimate of changes in global mean sea level 100 years from now, or an approximation of future global temperatures at a certain concentration of greenhouse gases.

Model Operations

The resolution of a model determines whether the climate predictions it makes are specific to an area the size of the continental United States, the size of New England, or the size of Manhattan. Spatial resolution governs the size of grid cells in a model (in degrees of latitude and longitude or in kilometers or miles), and temporal resolution refers to how often (in "model time") the model recalculates climate factors (every half hour, 6 hours, every week, etc.). Models use grids of cells to establish the locations at which to execute calculations and thus make estimates of climate traits such as temperature, wind speed, and others that are of concern to a researcher.

A typical climate model might have horizontally spaced grid cells of 100 km^2 (62 mi^2). This is equivalent to saying "Calculate the temperature at a point, then move 100 km west and calculate temperature again, then another 100 km west and repeat. Once you've gone all the way around the globe, move 100 km north and repeat the process; and so on."[6] In effect, the model is creating virtual weather stations at 100-km intervals around the planet surface and reporting calculated climate characteristics at each of them. Climate models must also calculate atmospheric characteristics using three-dimensional grids that extend upward through the atmosphere. Modern models typically have about 30 atmospheric layers and a horizontal cell spacing of only a few kilometers or so.

One of the problems with dividing the atmosphere up into lots of little grid boxes is that there are many climate processes that are smaller than a box: cloud formation, rainfall, the effect of topography on winds, storms, and others. This problem could potentially mean that individual clouds, which play an important role in the climate system, might not be represented in modeling, or they might be mischaracterized. Somehow the processes that form clouds and other small-scale climate factors (and their consequences) must be represented.

Researchers address this problem with expert estimations based on the fundamentals of climate science. For example, cloudiness and rainfall can be estimated based on knowledge of the temperature and humidity in a box. Raindrops require a very small solid particle in the air to precipitate, and thus modelers must also estimate how much dust (aerosol) is in the box. This process of estimation is called parameterization, and

---

[6] After R. M. Russell, *Resolution of Climate Models*, 2011, http://eo.ucar.edu/staff/rrussell/climate/modeling/climate_model_resolution.html (accessed July 10, 2011).

most models run many parameterization schemes to approximate many climate processes. Some of these schemes are well defined and thought to be quite reliable, but others are far less well understood, and confidence in them is not high.

Models must include a calculation of time in their operations, for instance allowing a realistic exchange of heat between the ocean surface and the atmosphere or from one part of the globe to another. When a model starts, it begins with a set of initial conditions for the atmosphere and ocean and then calculates how they will have changed after one time step, say, half an hour. The time step must be chosen with care. For example, if you want to run a model through 50 years as quickly as possible, you want to use as large a time step as possible. However, past some critical threshold the time step is so large that air (or, more accurately, energy) can travel farther than one grid box in one time step, and it becomes impossible to accurately determine how various elements of the climate develop. Some aspects of the climate can change more rapidly than others, and so they need to be calculated more frequently. For example, the movement of the air needs to be calculated every half hour, but the balance of incoming and outgoing radiation can be calculated less frequently. In the ocean, the ratio of the horizontal grid size to the length of a time step must not exceed the largest flow speed of water in the ocean.

Weather is chaotic, meaning that it does obey the laws of physics (every effect has a cause), but there are so many possible causes affecting weather that it is impossible to know about all of them. To address this problem in climate modeling, researchers need to get an idea of all the possible ways that climate could change, and the likelihood, or probability, of each possible way. The probability of a certain climate outcome is developed by running ensembles (groups) of global circulation models,[7] each of which uses different parameterizations (expert estimates of key processes) and makes different types of assumptions. An ensemble is a collection of model runs designed to identify the most probable future climate.

There are several ways to conduct ensemble experiments. One common method is to run several different models to discover all of their answers to the same question. For instance, let's say we wanted to know "How will central Pacific tropical sea-surface temperature (SST) change if global mean air temperature increases by 2°C?" To get at the answer we might choose 20 different climate models, each with slightly different parameterizations of various climate processes (e.g., ENSO, cloudiness, aerosols). If 15 out of the 20 (75%) agree that central Pacific tropical SST will rise between 1.7°C and 2.3°C, and the rest offer answers that fall outside this range, we have confidence that the answer to our question is that SST will rise between 1.7°C and 2.3°C. It is still important to assess the 25% of answers that fall outside this range and improve our understanding of why they do not agree, but we can conclude that the most agreed upon range of temperature in this case has the highest statistical probability of being correct. Often, in cases like this, an ensemble mean is reported, which is the mean prediction of all 20 model runs. The range of model outcomes around the mean allows researchers to calculate the probability of an answer's falling within some range; scientists typically use the 95th percentile or "There is a 95% chance that the answer falls within a certain range of values."

Grid cells (Figure 4.2) can be made smaller (for higher resolution), but this requires more computing time, which, on supercomputers, can be very expensive. To pay for this, researchers typically seek special government grants. As a general rule, increasing the resolution of a model by a factor of two (say going from a cell size of 100 km² to one of 50 km²) means that about ten times more computing power will be needed (the model will take ten times as long to run on the same computer).

---

[7] H.-M. Kim, P. Webster, and J. Curry, "Evaluation of Short-Term Climate Change Prediction in Multi-Model CMIP5 Decadal Hindcasts," *Geophysical Research Letters* 39 (2012): L10701, doi:10.1029/2012GL051644.

Figure 4.2. The resolution of climate models has increased over time. **a,** In the 1990s, models used the T42 grid, where temperature, moisture, and other processes were simulated in grid boxes about 200 by 300 km (120 × 180 mi). For the IPCC 2007 report, models like the Community Climate System Model[8] (one of the world's leading GCMs) at the National Center for Atmospheric Research used the T85 resolution, with grid boxes about 100 by 150 km (60 × 90 mi). As models improve, better resolution allows more realistic climate processes, which makes regional (T170 and T340) climate projections more accurate. **b,** Computer models reach high into a virtual atmosphere and deep into the ocean. They simulate climate by dividing the world into three-dimensional grid boxes, measuring physical processes such as temperature at each grid point. Such models can be used to simulate changes in climate over years, decades, or even centuries.

SOURCE: Figure from University Corporation for Atmospheric Research (UCAR) http://www2.ucar.edu/climate/faq/aren-t-computer-models-used-predict-climate-really-simplistic#mediaterms (accessed July 10, 2012). Copyright University Corporation for Atmospheric Research, NCAR/CGD, Figure by Gay Strand.

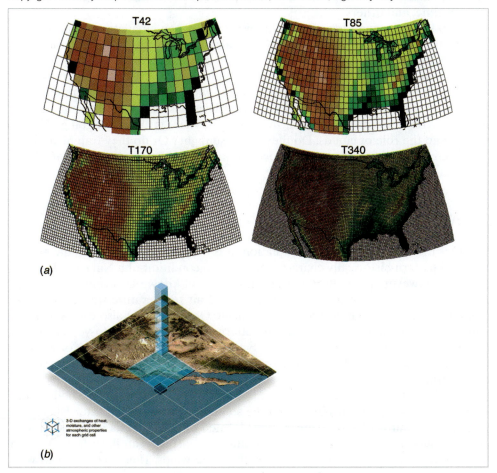

Even with improved resolution, the ability of models to depict climate in the place where you live is limited. Global models are called that for a reason: Their output is averaged over time and over space because achieving even a global projection is a major chore for the fastest supercomputer on the planet. A commonly used model, the Community Climate System Model (at the National Center for Atmospheric Research), is so complex it requires about 3 trillion computer calculations to simulate a single day of global climate.

One way that researchers use models to provide more localized information is through a regional climate model (RCM).[9] RCMs need a partner global

---

[8] See the model home page at http://www.cesm.ucar.edu/ (accessed July 10, 2012).

[9] See description in *Third Assessment Report* (TAR) of the IPCC, 2001 http://grida.no/publications/other/ipcc_tar/?src=/climate/ipcc_tar/wg1/380.htm (accessed July 10, 2012).

model to specify boundary conditions; boundary conditions are an existing set of climate parameters such as water vapor, winds, temperature, greenhouse gas content, and others that set the stage for the RCM calculations. For instance, a researcher can use a global model to project various climate parameters under a scenario of doubled CO_2 content. Then, using an RCM with a finer grid size over the New England region, the global climate parameters are used as boundary conditions to calculate climate in the finer cells of the RCM. The benefit is in the finer resolution of the RCM. For instance, in the global model perhaps only a dozen data points (12 cells) represent the topography of the White Mountains, the Berkshires, the Green Mountains, the Catskills, and other local topographic features that influence circulation, clouds, precipitation, and air temperature. In the RCM, perhaps this complex topography is represented with 60 to 120 cells, thus improving the simulation of winds, snow cover, rainfall, biological systems, the coastline, and others.

[10]Models do a good job of simulating air and ocean circulation, solar heating, and the role of greenhouse gases. There are also factors whose influence on global climate can be modeled, but when they will occur is essentially unpredictable: major volcanic eruptions that throw sulfur compounds into the high atmosphere that absorb and scatter sunlight, ENSO events with magnitudes that are unknowable beforehand, and of course the political decisions that will determine future fossil fuel consumption and land surface changes. There is also an ongoing effort to understand the role of clouds: Do they block sunlight? Or do they trap heat? Let's examine each of these factors and how scientists are working to understand their effect on climate.

WHAT IS ENSO?

ENSO is the El Niño Southern Oscillation. It is a large-scale, quasiperiodic meteorological pattern historically characterized by two conditions: La Niña and El Niño. Recently, however, a new third pattern has emerged known as a central-Pacific El Niño.[11] These conditions govern sea-surface and air temperature trends as well as rainfall patterns throughout the tropical Pacific Ocean. ENSO also exerts a global influence on weather and temperature patterns. In fact, the year-to-year variation in global average temperature shown in Figure 4.3 is largely a reflection of whether the year was dominated by La Niña (years tend to be cool) or El Niño (years tend to be warm).

ENSO[12] is related to the atmospheric pressure difference between a body of dry air (a high pressure system) located in the southeast Pacific over Easter Island and a body of wet air (a low pressure system) located over Indonesia in the southwest Pacific. Under normal conditions in the southern hemisphere, air flows from the high pressure to the low pressure and creates the trade winds. These blow east to west across the surface of the Pacific and drive a warm surface current of water into the western Pacific. The resulting accumulation of warm tropical water in the western Pacific is known as the warm pool; it extends well below the surface and it has the highest sea-surface temperatures on the planet. Seawater in the warm pool evaporates readily and produces lush rainfall throughout Southeast Asia, India, Africa, and other areas associated with the monsoon, the rainy season storms that nourish food production and ecosystems from Indonesia to Africa. In the eastern Pacific, this displaced seawater is replaced by nutrient-rich cold ocean water that rises from the deep sea, a process called upwelling. The upwelling current is loaded with nutrients fueling an important fishing industry off the coast of South America.

---

[10] See the animation "Supercomputing the Climate" at the end of the chapter.

[11] T. Lee and M. McPhaden, "Increasing Intensity of El Niño in the Central Equatorial Pacific," *Geophysical Research Letters* 37 (2010): L14603, doi: 10.1029/2010GL044007.

[12] See "El Niño Explained" at the end of the chapter.

Figure 4.3. Monthly (thin lines) and 12-month running mean (thick lines or filled colors in the case of Niño 3.4 index) global land–ocean temperature anomaly, global land and sea surface temperature, and El Niño index. All have a base period 1951–1980. Data are through January 2012. Year-to-year global mean temperature variability largely correlates to the prevalence of El Niño (warm) and La Niña (cool). Some temperature variability is dominated by large-scale volcanic emissions that tend to produce cooling of a year or two. Shown are three volcanic eruptions with global impact; from left to right, Mount Agung (Indonesia, 1963), El Chichon (Mexico, 1982), and Mount Pinatubo (Philippines, 1991).

SOURCE: Figure from http://www.columbia.edu/~mhs119/Temperature/T_moreFigs/

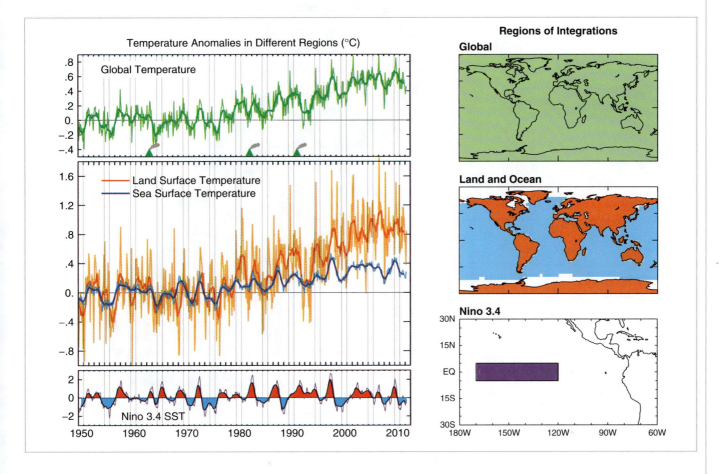

On occasion, the pressure difference between the two centers decreases and the trade winds respond by weakening (Figure 4.4). This condition is known as El Niño. As a result, the warm pool of the west Pacific surges to the east; it shallows and spreads across the surface, releasing its heat to the atmosphere and causing a broad area of the Pacific to experience warmer, wetter conditions than normal. The ocean surface in the central and eastern Pacific warms sufficiently to heat the lower troposphere and thereby temporarily raise global mean temperature[13] for any year characterized by El Niño.

El Niño can have devastating social and economic consequences. The eastern movement of the tropical warm pool takes with it a critical source of rainfall; as a result, seasonal rains in Indonesia collapse, leading to drought, famine, and forest fires in Southeast Asia. The monsoon, the life-sustaining, crop-nourishing rains that overcome the summer drought in India, is known to fail in the onset year of an El Niño event, thus leading to famine and water shortages. Precipitation in the east Pacific increases with the arrival of the warm seawater, causing higher (often catastrophic)

---

[13] D. Thompson, M. Wallace, P. Jones, and J. Kennedy, "Identifying Signatures of Natural Climate Variability in Time Series of Global-Mean Surface Temperature: Methodology and Insights," *Journal of Climate* 22 (2009): 6120–6141, doi:10.1175/2009JCLI3089.1.

Figure 4.4. The El Niño-Southern Oscillation (ENSO) is a large-scale meteorological pattern characterized by two conditions: the La Niña and the El Niño. These govern temperature and rainfall trends in the Pacific and Indian oceans and they exert a global influence on weather patterns and annual temperature. Winds blowing to the west are weak during El Niño, and sea-surface temperatures rise in the central and eastern Pacific.[14]

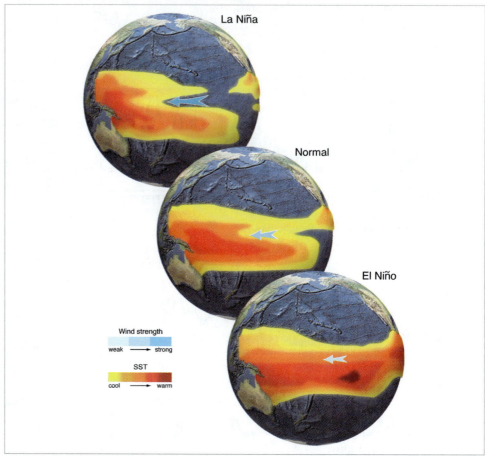

rainfall on the coasts of both North and South America. In the northern United States, winters are warmer and drier than average and summers are wetter than average. Torrential rains and damaging floods can occur across the southern United States. There is also a clear statistical relationship between El Niño and hurricanes. Whereas the number of hurricanes in the Atlantic basin tends to drop (~50%) during El Niño years, the number of hurricanes tends to increase in the Pacific.

A new type of El Niño has emerged in the past two decades, one that has its warmest waters in the tropical central Pacific Ocean rather than in the eastern Pacific. This new type of El Niño is known by several names, including central-Pacific El Niño, warm-pool El Niño, dateline El Niño, and El Niño Modoki (Japanese for "similar but different"). The intensity of these central Pacific El Niño events has nearly doubled in 20 years. They have been observed in 1991–92, 1994–95, 2002–03, and 2004–05, and the most intense occurred in 2009–10.[15] Researchers have found that many climate models predict that such events will become more frequent as global warming continues, suggesting that climate change may already be affecting El Niño by shifting the center of action from the eastern to the central Pacific.

[14] Temperature patterns modified from: http://www.cpc.noaa.gov/products/analysis_monitoring/ensocycle/ensocycle.shtml (accessed July 10, 2012).

[15] T. Lee and M. McPhaden, "Increasing Intensity of El Niño in the Central-Equatorial Pacific," *Geophysical Research Letter* 37, no. 14 (2010): L14603, doi: 10.1029/2010GL044007.

How will global warming affect ENSO? The answer to this question is still not settled.[16] It seems that one common trait among some climate models is the indication that global warming could result in a more general average state of the climate that is El Niño–like. However, that does not seem to be what is currently happening in the Pacific. Trade winds have been seen to strengthen,[17] and, counter to the predictions of some climate models, a persistent La Niña–like state has emerged over the past two decades. In fact, this is consistent with studies of ENSO variability over the past 1000 years.[18] Researchers found that during the medieval climate anomaly, megadroughts that occurred in western North America might have been the result of stronger or more frequent La Niña than El Niño. In Chapter 6 we will see that drought is once again visiting the western United States.

ADDITIONAL COMPLEXITIES IN GLOBAL CLIMATE

A number of natural factors influence global climate. These include large-scale volcanic eruptions of dust particles and sulfur compounds that scatter and absorb sunlight and cause temporary global cooling; solar radiation, changes in the amount of sunlight entering the atmosphere that influence Earth's radiation budget; changes in ice-albedo, which provides a positive feedback that enhances warming in the arctic; and clouds, poorly understood phenomena that can either reflect sunlight or trap heat, depending on their characteristics. These factors must be accurately incorporated into climate models to produce believable projections of future climate.

Explosive Volcanic Eruptions

Volcanic eruptions come in all sizes and shapes. Eruptions that are relatively passive, in which the lava quietly effuses from a volcanic vent and flows across the ground, are not known to influence the climate. But some eruptions are explosive. These can send thick columns of sulfuric compounds and ash high into the atmosphere (Figure 4.5). If they reach the stratosphere, these materials can scatter and absorb incoming sunlight and change Earth's radiation balance, temporarily offsetting warming with global cooling.

The style of any volcanic eruption depends largely on the chemistry of the molten rock, or magma, that fuels a volcano. In many volcanoes, the rock is rich in the element silicon (Si), which has the property of readily bonding with oxygen (O). These two elements make up more than 75% of the average chemistry of Earth's crust, and so much of the magma is involved in bonding. When silicon and oxygen bond they form a solid compound (SiO_2, silica). As a result, the magma becomes partially crystallized and, like a hot version of a Slurpee, it is thick and viscous and therefore hard for a volcano to expel. The magma acts as a plug and it builds up pressure from gases trying to escape the volcano.

Eventually the build-up of gas pressure in a silica-rich magma overcomes the viscous forces and the result is a massive, explosive eruption of such force that it can punch through the troposphere and penetrate deep into the stratosphere. These explosive eruptions are driven skyward by a column of rising gas that, no longer confined, expands exponentially. Erupting magma is ripped apart by enlarging pockets of gas, a mixture that geologists compare to the froth at the top of a beer that is poured too rapidly. Only in this case, the froth is made of rapidly solidified bits of glass. Glass is what forms when the atoms in magma do not have time to organize

[16] M. Collins, S. L. An, W. Cai, et al., "The Impact of Global Warming on the Tropical Pacific Ocean and El Niño. Review Abstract," *Nature Geoscience* 3 (2010): 391–397, doi:10.1038/ngeo868, http://www.nature.com/ngeo/journal/v3/n6/abs/ngeo868.html (accessed July 10, 2012).

[17] I. R. Young, S. Zieger, and A. V. Babanin, "Global Trends in Wind Speed and Wave Height," *Science* 332, no. 6028 (2011): 451–455, doi: 10.1126/science.1197219, http://www.sciencemag.org/content/332/6028/451.abstract (accessed July 10, 2012).

[18] D. Khider, L. Stott, J. Emile-Geay, R. Thunell, and D. Hammond, "Assessing El Niño Southern Oscillation Variability during the Past Millennium," *Paleoceanography* 26 (2011): PA3222, doi:10.1029/2011PA002139.

Figure 4.5. Explosive eruption of Sarychev Peak, Kuril Islands (northeast of Japan), June 12, 2009; photo taken from the International Space Station. According to NASA, "Ash from the multi-day eruption has been detected 2,407 kilometers east-southeast and 926 kilometers west-northwest of the volcano, and commercial airline flights are being diverted away from the region to minimize the danger of engine failures from ash intake." The eruptive column was measured at a height of more than 8 km (5 mi). Volcanic ash is not like the soft, fluffy ash produced by burning vegetation. It is tiny, abrasive particles of glass and rock that pose serious hazards to aircraft engines.

IMAGE CREDIT: M. Justin Wilkinson, NASA-JSC.

into true minerals characterized by a crystalline lattice work of atoms; glass is made of atoms that are randomly arranged. In particularly large eruptions, these bits of glass (and rock) the size of ash particles may be ejected deep into the stratosphere and can stay there, circulating around the entire planet on high-altitude winds, for over a year before they fall back to Earth.

Volcanic particles blasted into the stratosphere scatter and absorb incoming sunlight and cause temporary cooling. The amount of cooling depends on the total volume of particles and how long they stay in the air. Larger particles the size of sand grains fall out of the atmosphere in a few minutes and have little effect on the climate. Tiny ash particles thrown into the troposphere stay aloft for hours or days, causing darkness and cooling directly beneath the ash cloud, but these are soon washed out of the air by rain. Particles erupted into the stratosphere, which lacks rainfall and is dry, can remain for weeks to months, affecting sunlight and causing some cooling over large areas. In some cases, massive eruptions can produce particles that circle the globe and cause global cooling for a year or more.

Volcanoes that release large amounts of sulfur compounds affect the climate more strongly than those that eject ash alone. Once sulfur compounds reach the stratosphere they combine with water to form a haze of tiny droplets of sulfuric acid. These droplets absorb and scatter a great deal of sunlight for their size, and although they eventually grow large enough to fall to Earth, the stratosphere is so dry that it takes months or even years to happen. Hence, reflective hazes of sulfur droplets can cause significant cooling for as long as two years, and it is major sulfur-rich eruptions that cause the greatest global effects. For example, Mount Pinatubo (Philippines) erupted in 1991 and ejected almost 15 million tons of sulfur dioxide into the

Figure 4.6. In 1991, Mount Pinatubo in the Philippines erupted millions of tons of sulfur dioxide into the stratosphere, where it formed a layer of sulfuric acid droplets that scattered and absorbed incoming sunlight. Over the next 15 months, scientists measured a drop in the average global temperature of about 0.6°C (1°F). The Pinatubo eruption increased aerosol optical depth (a measure of how much light airborne particles prevent from passing through a column of atmosphere) in the stratosphere by a factor of 10 to 100 times normal levels (blue indicates clear air, red indicates hazy air).

Source: NASA Earth Observatory, http://earthobservatory.nasa.gov/IOTD/view.php?id=1510 (accessed July 10, 2012).

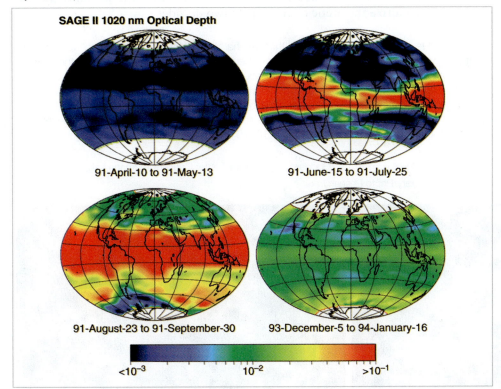

stratosphere. For many months a satellite tracked the sulfur cloud produced by the eruption as it lowered average global temperature by about 0.6°C (1°F) (Figure 4.6).

Research[19] has revealed that the Little Ice Age, a period of regional cooling in the North Atlantic, North America, and Europe (and perhaps beyond) that lasted for hundreds of years until the late 19th century, may have been triggered by an unusual 50-year-long episode of four massive tropical volcanic eruptions between A.D. 1275 and 1300. The stratospheric aerosols associated with this sequence of eruptions apparently produced persistent cold summers that generated a positive feedback in the form of expanding sea ice and weakened Atlantic Ocean currents that transport heat. Computer simulations paint a picture of a climate system being hit time and time again by cold conditions over a short period, all leading to a cumulative cooling effect that culminated in the start of the Little Ice Age. The study relied on a convergence of data from ice and sediment cores, patterns of dead vegetation (from the start of the Little Ice Age) that were recently revealed by receding Arctic ice, and computer simulations of climate feedbacks. The research indicates that the start of the cold period was prevalent throughout the North Atlantic region and involved major components of the climate system through a series of feedbacks that amplified the original impacts of explosive volcanic aerosols.

---

[19] G. Miller, J. Southon, C. Anderson, et al., "Abrupt Onset of the Little Ice Age Triggered by Volcanism and Sustained by Sea-Ice-Ocean Feedbacks." *Geophysical Research Letters* 39 (2012): L02708, doi: 10.1029/2011GL050168.

Ice-Albedo Feedback

Worldwide attention is paid to the annual summer retreat, and persistent long-term decline, of arctic sea ice.[20] Scientists devote great effort to understanding the behavior of arctic sea ice because as the summer extent of the ice pack decreases, white, sunlight-reflecting ice and snow is replaced by dark, heat-absorbing seawater (Figure 4.7). This switch constitutes a positive climate feedback that amplifies global warming called the ice-albedo feedback. In fact, it is this arctic amplification of global warming that is viewed as the cause for the dramatic warming that has come to characterize the arctic over the past two decades.[21]

Figure 4.7. Arctic sea ice has been declining for at least 30 years according to the National Snow and Ice Data Center.[22] **a,** Arctic sea ice extent typically reaches its low point each year in September. The ice extent in September 2012 was the lowest in the satellite record. Shrinking a dramatic 18% in 2012, scientists described the event as "unprecedented" and "uncharted territory" for the Arctic. **b,** Arctic ice reflects sunlight, helping to cool the planet. As this ice begins to melt, less sunlight gets reflected into space. **c,** Sunlight is instead absorbed into the oceans and land, raising the overall temperature and fueling further melting. This results in a positive feedback loop called ice-albedo feedback, which causes the loss of the sea ice to be self-compounding. The more it disappears, the more likely it is to continue to disappear. **d,** Graph of Arctic sea ice extent (October 15, 2012). 2012 in blue, 2011 in orange, and earlier years in other colors.

SOURCE: Figure 4.7a: National Snow and Ice Data Center. Figure 4.7b: NASA/Goddard Space Flight Center Conceptual Image Lab.

IMAGE CREDIT: Figure 4.7a and Fig. 4.7c: Science Photo Library/Photo Researchers

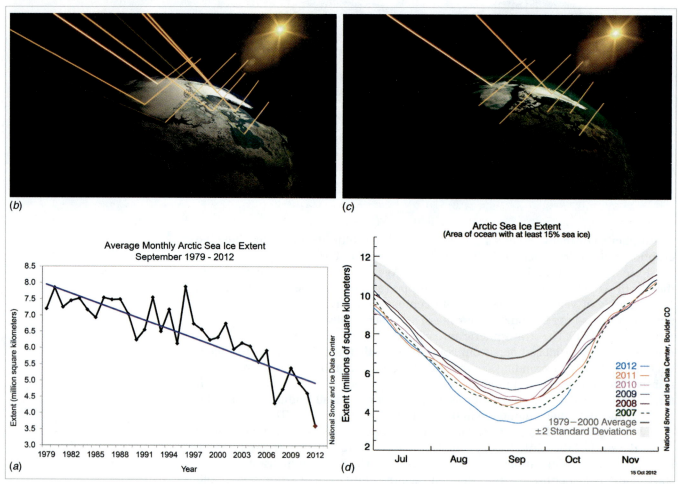

[20] See the news page of the National Snow and Ice Data Center, where sea ice and other cryosphere science are reported and frequently updated http://nsidc.org/arcticseaicenews/ (accessed July 10, 2012).

[21] D. Ghatak, A. Frei, G. Gong, J. Stroeve, and D. Robinson, "On the Emergence of an Arctic Amplification Signal in Terrestrial Arctic Snow Extent," *Journal of Geophysical Research* 115 (2010): D24105, doi:10.1029/2010JD014007. D. Perovich, K. Jones, B. Light, et al., "Solar Partitioning in a Changing Arctic Sea-Ice Cover," *Annals of Glaciology* 52, no. 57 (2011): 192–196.

[22] See the NSIDC home page here: http://nsidc.org/ (accessed July 10, 2012).

In addition to the excess warming produced by the albedo switch,[23] researchers also worry that continued decline of arctic sea ice will reach a tipping-point[24] where so much heat-absorbing water has been exposed that the ice decline[25] becomes self-amplified and unstoppable. The tipping-point[26] idea goes like this: With less sea ice, more sunlight is absorbed by the dark open water of the Arctic Ocean. Warm ocean water leads to additional sea ice melting, thus producing more open water, which absorbs even more heat and melts even more ice, and so forth until this becomes a self-fulfilling process that cannot be stopped. Researchers have long thought that this feedback loop can in principle become self-operating and independent of prevailing climate conditions.

The sea-ice story is focused on summer pack ice. Winter in the arctic is dark and cold and there is little worry that ice would not continue to form each winter. But as global warming and arctic amplification advance, the winter ice is increasingly characterized by thin annually forming ice that readily melts the following summer, rather than the thicker multiyear ice that seemingly is more stable. Eventually researchers fear the arctic will become ice free[27] in the summer months, and that the summer ice-free months will expand into early spring and late fall.[28] By opening the Arctic Ocean to resource exploitation such as oil drilling, fishing, sea floor dredging, and others, this ice-free condition can put at risk fragile arctic ecosystems.

Surprisingly, studies[29] indicate that the oldest and thickest Arctic sea ice is disappearing at a faster rate than the younger and thinner ice at the edges of the Arctic Ocean's floating ice cap. Researchers have documented that the average thickness of the Arctic sea ice cover is declining because it is rapidly losing its thick component, the multi-year ice. At the same time, the surface temperature in the Arctic is going up, which results in a shorter ice-forming season. It would take a persistent cold spell for most multi-year sea ice and other ice types to grow thick enough in the winter to survive the summer melt season and reverse the trend.

There are global implications for the rapid and geologically unusual warming that is happening today.[30] Because of the ice-albedo feedback, a future small increase in temperature could lead to larger warming over time, making the polar regions the most sensitive areas to climate change on Earth. The ice-albedo feedback has the potential to turn a small climate change into a big climate change. Why does this matter? There are three reasons:

- Human populations in the arctic depend on their ecosystems for their food and other basic needs. Rapid ecosystem change threatens the very survival of these human communities. Without options they become dependent on government support, displaced, and bereft of their traditional culture and identity.

---

[23] See the animation "Arctic Sea Ice Decline" at end of chapter.

[24] C. M. Duarte, T. Lenton, P. Wadhams, and P. Wassmann, "Abrupt Climate Change in the Arctic," *Nature Climate Change* 2 (2012): 60–63.

[25] See "Still Hope for Arctic Sea Ice" http://www.sciencedaily.com/releases/2011/02/110204092149.htm (accessed July 10, 2012).

[26] P. Wassmann and T. Lenton, "Arctic Tipping Points in an Earth System Perspective," *AMBIO* 41, no 1 (2012): 1–9.

[27] M. Wang and J. Overland, "A Aea Ice Free Summer Arctic within 30 Years?" *Geophysical Research Letters* 36 (2009): L07502, doi: 10.1029/2009GL037820.

[28] T. Markus, J. Stroeve, and J. Miller, "Recent Changes in Arctic Sea Ice Melt Onset, Freezeup, and Melt Season Length," *Journal of Geophysical Research* 114 (2009): C12024, doi:10.1029/2009JC005436. D. Ghatak, A. Frei, G. Gong, J. Stroeve, and D. Robinson, "On the Emergence of an Arctic Amplification Signal in Terrestrial Arctic Snow Extent," *Journal of Geophysical Research* 115 (2010): D24105, doi:10.1029/2010JD014007. D. Perovich, K. Jones, B. Light, et al., "Solar Partitioning in a Changing Arctic Sea-Ice Cover," *Annals of Glaciology* 52, no. 57 (2012): 192–196.

[29] D. Hall, J. Comiso, N. DiGirolamo, et al., "A Satellite-Derived Climate-Quality Data Record of the Clear-Sky Surface Temperature of the Greenland Ice Sheet," *Journal of Climate* 2012, doi: 10.1175/JCLI-D-11-00365.1, http://ntrs.nasa.gov/search.jsp?R=20120009049 (accessed July 10, 2012).

[30] Y. Axford, J. P. Briner, C. A. Cooke, et al., "Recent Changes in a Remote Arctic Lake are Unique within the Past 200,000 Years." *Proceedings of the National Academy of Sciences* 106, no. 44 (2009): 18443–18446; doi: 10.1073/pnas.0907094106.

- Ecosystem destruction anywhere on the planet reduces the diversity, interconnectedness, and complexity of living communities and thereby weakens the whole biological kingdom.

- Melting Greenland ice leads to global sea-level rise, and excessive meltwater may be capable of slowing the thermohaline circulation.

Greenland (Figure 4.8) is the largest island in the world, and it is covered in ice left over from the last ice age, which peaked about 20,000 to 30,000 years ago, covering approximately 27% of the world's land area with ice (compared with 10.4% today). Although Greenland ice has largely resisted the warm temperatures of the Holocene Epoch (from 10,000 years ago to present), it is succumbing now to anthropogenic global warming.[31] Since 1979, scientists have tracked the extent of summer

Figure 4.8. Map of changes in the percentage of light reflected by the Greenland ice sheet in summer (June, July, August) 2011 compared to the average from 2000 to 2006. Virtually the entire surface has grown darker owing to surface melting, dust and soot on the surface, and temperature-driven changes in the size and shape of snow grains. Previously, the bright surface of the ice reflected more than half of the sunlight that fell on it. This helped keep the ice sheet stable, as less absorbed sunlight meant less heating and melting. However, in the past decade satellites have observed a decrease in Greenland's reflectiveness. This darker surface now absorbs more sunlight, which accelerates melting.

SOURCE: R. Lindsey "Greenland Ice Sheet Getting Darker," NOAA Climate Watch Magazine, 2011, http://www.climatewatch.noaa.gov/article/2011/greenland-ice-sheet-getting-darker-2 (accessed July 10, 2012).

300 miles

Difference from average
reflectiveness (percent)

−18   0   18

[31] M. van den Broeke, J. Bamber, J. Ettema, et al., "Partitioning Recent Greenland Mass Loss," *Science* 326, no. 5955 (2009): 984–986, doi: 10.1126/science.1178176.

melting of the Greenland Ice Sheet. In 2007, the extent of melting broke the record set in 2005 by 10%, making it at the time the largest season of melting ever recorded. That record was broken in the 2010 melt season, in which melting started earlier, ended later, and peaked with more melting than any previous melt season.[32] Melting in 2011 did not reach the extent of the previous year, but nonetheless it was one of just three years since 1979 where melt area exceeded 30%.

Melting on portions of Greenland rose 150% above the long-term average, with melting occurring on 25 to 30 more days in 2010 than the average for the previous 19 years. In the past decade, the total mass deficit (the annual difference between snowfall and melting) tripled, and the amount of ice lost in 2008 was nearly three times the amount lost in 2007. In 2009, scientists announced[33] that Greenland's ice was melting at a rate three times faster than it was only five years earlier. This melting turns to water that flows into the North Atlantic and raises sea level. Additionally, this freshwater has the potential to slow the delivery of heat from the tropics via the North Atlantic Gyre, an important arm of the thermohaline circulation.

As we learned in Chapter 3, thermohaline circulation transports heat around the planet and hence plays an important role in global climatology. Acting as a conveyor belt carrying heat from the equator into the North Atlantic, the Gulf Stream raises Arctic temperatures, but Greenland ice has been in equilibrium with Gulf Stream heating for millennia, allowing Greenland ice to remain largely stable. Today, however, global warming is causing Greenland to melt, and the influx of freshwater from melting ice has the potential to slow or shut down thermohaline circulation by preventing the formation of deep water.

A shutdown of the thermohaline circulation[34] could play a role in a negative climate feedback pattern beginning with ice melting (warming) that ironically leads to glaciation (cooling). The key to keeping the circulation moving is the saltiness of the water. Saltier water increases in density and sinks. Many scientists believe that if too much freshwater enters the ocean—for example, from melting Arctic glaciers and sea ice—the surface water would freeze before it could become dense enough to sink toward the bottom. If the water in the north did not sink, the Gulf Stream eventually would stop moving warm water northward, leaving Northern Europe cold and dry within a single decade. This hypothesis of rapid climate change is called the conveyor belt hypothesis, and the paleoclimate record found in ocean sediment cores appears to support it.[35] Paleoclimate studies have shown that in the past, when heat circulation in the North Atlantic Ocean slowed, the climate of northern Europe changed.

Although the last ice age peaked about 20,000 to 30,000 years ago, the warming trend that followed it was interrupted by cold spells at 17,500 years ago and again at 12,800 years ago. These cold spells happened just after melting ice had diluted the salty North Atlantic water, slowing the ocean conveyor belt. It is this idea that led to the movie The Day After Tomorrow,[36] in which global warming results in freshwater from melting ice stopping the thermohaline circulation, which in turn produces deadly cooling (an unlikely scenario) in the North Atlantic. Ultimately, scientists fear that amplified warming in the arctic can have ripple effects that pose severe impacts to the world's coastal cities and the weather of Europe and North America.

[32] M. Tedesco, X. Fettweis, M. R. van den Broeke, et al., "The Role of Albedo and Accumulation in the 2010 Melting Record in Greenland," Environmental Research Letters 6 (2011): 014005, doi: 10.1088/1748-9326/6/1/014005.

[33] S. Mernild, G. Liston, C. Hiemstra, et al., "Greenland Ice Sheet Surface Mass-Balance Modelling and Freshwater Flux for 2007, and in a 1995–2007 Perspective," Hydrological Processes 2009, doi: 10.1002/hyp.7354.

[34] S. Rahmstorf, "The Concept of the Thermohaline Circulation," Nature 421, no. 6924 (2003): 699, doi:10.1038/421699a.

[35] R. B. Alley, "Wally Was Right: Predictive Ability of the North Atlantic 'Conveyor Belt' Hypothesis for Abrupt Climate Change," Annual Review of Earth and Planetary Sciences 35 (2007): 241–272, doi: 10.1146/annurev.earth.35.081006.131524

[36] See "The Day After Tomorrow," Wikipedia, http://en.wikipedia.org/wiki/The_Day_After_Tomorrow (accessed July 10, 2012).

Clouds

It is apparent to anyone who has been outside that clouds can exert influence over the climate.[37] A sunny moment can change to a cool one as a cloud passes overhead. Clouds interact with solar radiation and reflect incoming sunlight in significant amounts, causing the albedo (reflectivity) of the entire Earth to be about twice what it would be in the absence of clouds.[38] Clouds also absorb the long-wave (infrared) radiation emitted by Earth's surface, similar to the effects of atmospheric greenhouse gases. But clouds do not cause climate change; they are a feedback to climate change caused by humans.[39] The question is, "Are clouds a positive feedback or a negative feedback to anthropogenic global warming?"

Getting the balance of cooling and warming effects right, and attributing these effects accurately to various cloud types at different altitudes, has been troubling for climate models.[40] Typical modeling experiments consist of a researcher running several global circulation model scenarios and finding that they do not agree on how clouds of various types respond to a warming atmosphere. Another example is to compare observations of clouds (by satellite, for instance) to model predictions and identify failures of the models to depict true cloud conditions.

In a warmer world, will clouds (Figure 4.9) provide a positive or negative feedback? That is, will there be fewer clouds or more, at what elevations, and how will this affect the balance of cooling and warming caused by clouds?[41] Fine-tuning answers to these questions is still the target of active research; however, scientists are increasingly concluding that clouds are not the cause of surface temperature changes, they are instead a feedback in response to those temperature changes because the radiative impact of clouds accounts for little of observed temperature variations.[42]

State-of-the-art climate models disagree on how clouds will respond to warming. Clouds have both warming and cooling effects. Low-level dense clouds tend to reflect sunlight, thus playing a cooling role; high-altitude clouds tend to trap heat, providing amplification to warming caused by other processes. Some models predict that low-level cloud cover will increase in a warmer climate, reflecting more sunlight, and limiting the level of global warming (a negative feedback). Other models predict less cloudiness, thus amplifying global warming (a positive feedback). The way clouds change with warming is of huge importance to global warming predictions. This is the main reason for the differences in warming produced by different climate models.[43]

All of the IPCC climate models[44] reduce low- and middle-altitude cloud cover with warming, a positive feedback. However, there is published research pointing to a negative feedback attributed to clouds. One study[45] of seasonal changes in the tropics

---

[37] See the many learning resources at National Science Foundation, "Clouds: The Wild Card of Climate Change," http://www.nsf.gov/news/special_reports/clouds/downloads.jsp (accessed July 10, 2012).

[38] V. Ramanathan, R. D. Cess, E. F. Harrison, et al., "Cloud Radiative Forcing and Climate: Results from the Earth Radiation Budget Experiment," *Science* 24, no. 4887 (1989): 57–63.

[39] A. E. Dessler, "Cloud variations and the Earth's energy budget." *Geophysical Research Letters* 38 (2011): L19701, doi: 10.1029/2011GL049236

[40] J. E. Kay, B. R. Hillman, S. A. Klein, et al., "Exposing Global Cloud Biases in the Community Atmosphere Model (CAM) Using Satellite Observations and their Corresponding Instrument Simulators," *Journal of Climate* 25, no. 4 (2012), http://journals.ametsoc.org/doi/abs/10.1175/JCLI-D-11-00469.1 (accessed July 10, 2012).

[41] See the animation "From All Sides Now" at the end of the chapter.

[42] A. E. Dessler, "Cloud Variations and the Earth's Energy Budget," *Geophysical Research Letters* 38 (2011): L19701, doi:10.1029/2011GL049236.

[43] K. E. Trenberth and J. T. Fasullo, "Global Warming Due to Increasing Absorbed Solar Radiation," *Geophysical Research Letters* 36 (2009); L07706, doi:10.1029/2009GL037527.

[44] D. A. Randall, R.A. Wood, S. Bony, et al., "Climate Models and Their Evaluation." In S. Solomon, D. Qin, M. Manning, et al. (eds.), *Climate Change 2007: The Physical Science Basis*. Contribution of Working Group I to the Fourth Assessment Report of the Intergovernmental Panel on Climate Change. (Cambridge, UK, Cambridge University Press, 2007).

[45] R. Spencer, W. Braswell, J. Christy, and J. Hnilo, "Cloud and Radiation Budget Changes Associated with Tropical Intraseasonal Oscillations," *Geophysical Research Letters* 34 (2007): L15707, doi:10.1029/2007GL029698.

Figure 4.9. Earth is a cloudier place than many people realize. Low dense clouds reflect sunlight, a cooling action, but high-altitude clouds trap heat coming off Earth's surface, a warming action. How will these opposite effects change in a warmer world and what will be the net effect of clouds on future climate change? Research suggests that clouds tend to produce a positive feedback to global warming, amplifying the effects of greenhouse gases.

Source: Figure from NASA Visible Earth, "The Blue Marble," http://visibleearth.nasa.gov/view_rec. php?id=2429 (accessed July 10, 2012).

Image credit: NASA Goddard Space Flight Center Image by Reto Stöckli (land surface, shallow water, clouds). Enhancements by Robert Simmon (ocean color, compositing, 3D globes, animation). Data and technical support: MODIS Land Group; MODIS Science Data Support Team; MODIS Atmosphere Group; MODIS Ocean Group Additional data: USGS EROS Data Center (topography); USGS Terrestrial Remote Sensing Flagstaff Field Center (Antarctica); Defense Meteorological Satellite Program (city lights).

using satellite data observed a decrease in net radiation (cooling) during the rainy season; this was related to a decrease in ice formation in the atmosphere. Another study[46] used detailed climate modeling to study the behavior of clouds above a warming ocean. Researchers found that low-level clouds thickened (reflecting more sunlight) as the ocean warmed, providing a natural cooling effect in response to the warming.

But there are an equal (or greater) number of papers concluding that clouds amplify warming.[47] In one study,[48] researchers examined measurements from the Clouds and Earth's Radiant Energy System (CERES[49]) instrument onboard NASA's Terra satellite to calculate the amount of energy trapped by clouds as the climate varied over the last decade. The study concluded that warming due to increases in greenhouse gases will cause clouds to trap more heat, which will lead to additional warming, meaning clouds trap more heat, which in turn leads to even more warming—a positive feedback.

[46] P. Caldwell and C. S. Bretherton, "Response of a Subtropical Stratocumulus-Capped Mixed Layer to Climate and Aerosol Changes." *Journal of Climate* 22 (2009): 20–38.

[47] A. C. Clement, R. Burgman, and J. R. Norris, "Observational and Model Evidence for Positive Low-Level Cloud Feedback," *Science* 325, no. 5939 (2009): 460–464.

[48] A. E. Dessler, "A Determination of the Cloud Feedback from Climate Variations over the Past Decade," *Science* 330, no. 6010 (2010): 1523–1527, doi: 10.1126/science.1192546.

[49] See the CERES homepage, http://ceres.larc.nasa.gov/ (accessed July 10, 2012).

Another study[50] focused on a region of the atmosphere over the eastern Pacific Ocean and adjacent land. The clouds here are known to influence present climate, yet most models do poorly in representing them. The model developed by the authors performed well and simulated key features of the modern cloud field, including the response of clouds to El Niño. The improved model was then turned to focus on a warmer climate at the end of the century. The result? The model projected thinner and fewer clouds, and these trends were more pronounced than in other models. The study authors concluded that if their results prove to be representative of the real global climate, then climate is actually more sensitive to greenhouse gases than current global models predict, and even the highest warming predictions for the future would underestimate the real change we could see.[51]

Yet another study[52] examined the change in cloudiness that could occur as storm tracks shift poleward with continued warming. In the first study to document that storm tracks have indeed shifted poleward, researchers also found a related reduction in cloudiness and an increase in the net flux of radiation at the top of Earth's atmosphere in storm track regions. These observations point to a positive feedback: Poleward migration of storms produces a reduction in cloudiness that leads to amplified warming.

Cloud science continues to yield surprises. For instance, a NASA study[53] revealed that Earth's clouds are getting lower. What does this mean? It means that global average cloud height declined by around 1% over a decade, or by around 30 to 40 m (100–130 ft). Most of the lowering was due to fewer clouds occurring at very high altitudes. A consistent reduction in cloud height would allow Earth to cool to space more efficiently, reducing the surface temperature of the planet and potentially slowing the effects of global warming. This might represent a negative feedback, a change resulting from global warming that could counteract its worst effects. Researchers involved with the study state that they don't know exactly what causes the cloud heights to lower, but it must be due to a change in the circulation patterns that give rise to cloud formation at high altitude.

So what does all this discussion about clouds mean? Scientists are still working to nail down the complexities of clouds. A range of instrumentation (satellites, weather balloons, aircraft) is being used by researchers to study a range of cloud types (ice clouds, low-level clouds, high-level clouds, tropical clouds, mid-latitude clouds[54]). The intense work to document cloud processes with direct observations and the constant effort to improve modeling capabilities promise to keep alive the field of clouds and climate change for quite some time.

Solar Radiation

In Chapter 3 we established that the Sun is not responsible for recent climate change.[55,56] Satellites have not detected any increase in solar radiation over the past 35 years, a period when global mean temperature has dramatically risen. Had the Sun been

[50] A. Lauer, K. Hamilton, Y. Wang, V. T. J. Phillips, and R. Bennartz, "The Impact of Global Warming on Marine Boundary Layer Clouds over the Eastern Pacific—A Regional Model Study," *Journal of Climate* 23 (2010): 5844–5863, doi: 10.1175/2010JCLI3666.1.

[51] See the quotation by Dr. Kevin Hamilton: http://www.sciencedaily.com/releases/2010/11/101122172010.htm (accessed July 10, 2012).

[52] F. Bender, V. Ramanathan, and G. Tselioudis, "Changes in Extratropical Storm Track Cloudiness 1983–2008: Observational Support for a Poleward Shift, Climate Dynamics," 2011, doi:10.1007/s0038-011-1065-6.

[53] R. Davies, and M. Molloy, "Global Cloud Height Fluctuations Measured by MISR on Terra from 2000 to 2010," *Geophysical Research Letters* 39, no. 3 (2012), doi: 10.1029/2011GL050506.

[54] See the story at http://www.sciencedaily.com/releases/2011/03/110315142526.htm (accessed July 10, 2012).

[55] M. Lockwood, "Solar Change and Climate: An Update in the Light of the Current Exceptional Solar Minimum," *Proceedings of the Royal Society* A, 2 December 2009, doi 10.1098/rspa.2009.0519. J. Lean, "Cycles and Trends in Solar Irradiance and Climate," *Wiley Interdisciplinary Reviews: Climate Change* 1, January/February (2010): 111–122.

[56] See the animation "Climate Denial Crock of the Week: Solar Schmolar" at the end of the chapter.

responsible for global warming, the entire atmosphere would warm, not just the tropo-sphere, as has been observed; in fact, the stratosphere has cooled over the same period. This is because greenhouse gases are trapping heat in the lower atmosphere. Indeed, none of the following would be true[57] if the Sun were the cause of global warming:

- Warming has been greater at the poles than at the equator.
- Warming has been the same rate at night as during the day.
- Warming has been greatest in the winter, not in the summer.

In each of these cases the opposite would be true had the Sun caused the warming.

However, solar output does vary through time (Figure 4.10), and it is critical that global climate models continue to account for the role that the Sun plays in Earth's climate. Total solar irradiance (TSI)[58] varies in what is known as the solar cycle, the rise and fall (over approximately 11 years) of the number of sunspots on the Sun's surface. Sunspots are dark cool regions, but along the edge of a sunspot solar activity is high; thus when there are a high number of sunspots, TSI is at a maximum. During the 11-year solar cycle, the total energy given off by the Sun varies by 0.1%. The solar cycle also causes a sizeable change in the ultraviolet (UV) radiation produced by the Sun, where most of the impacts are located in the stratosphere (above ~10 km, 6.2 mi).

Figure 4.10. In July 2011, researchers at NASA's Marshall Space Flight Center[59] predicted that the solar maximum of cycle 24 would peak in June 2013 with a relatively low amplitude (or TSI). Solar cycles are numbered beginning with the first confirmed cycle 1755–1766; they average about 10.66 years in length, but cycles as short as 9 years and as long as 14 years have been observed.[60]

SOURCE: http://solarscience.msfc.nasa.gov/predict.shtml

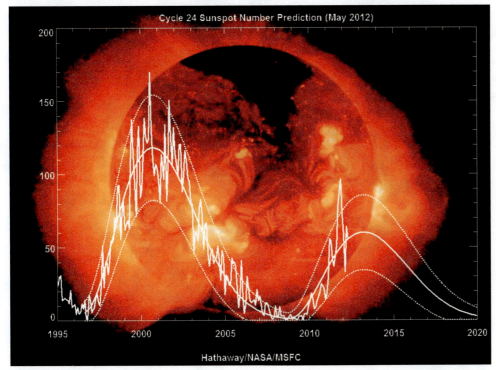

[57] M. Menne, C. Williams, Jr., and M. Palecki, "On the Reliability of the U.S. Surface Temperature Record," *Journal of Geophysical Research – Atmospheres* 115 (2010): D11108, doi:10.1029/2009JD013094. D. D. Parker, "A Demonstration that Large-Scale Warming is not Urban," *Journal of Climate* 19, no. 12 (2006): 2882–2895.

[58] G. Kopp and J. l. Lean, "A New, Lower Value of Total Solar Irradiance: Evidence and Climate Significance," *Geophysical Research Letters*, 38 (2011): L01706.

[59] See http://solarscience.msfc.nasa.gov/predict.shtml (accessed July 10, 2012).

[60] See "Solar Cycle," *Wikipedia*, http://en.wikipedia.org/wiki/Solar_cycle (accessed July 10, 2012).

If TSI only varies by 0.1% and UV radiation affects mainly the stratosphere, are Earth's weather and climate unaffected by the solar cycle? NASA scientists tested[61] this question by simulating 1600 years of varying UV and TSI in climate models. They found[62] that the solar cycle can account for 15% to 20% of rainfall in certain areas. For instance, a solar maximum favors increased precipitation north of the equator (the South Asian monsoon) and decreased precipitation near the equator and at northern mid-latitudes. Complex changes in UV and TSI drive these patterns; increased UV radiation leads to a rise in stratospheric ozone, which warms the tropics and (because of various interactions between the stratosphere and the troposphere) shifts the zone of Hadley Cell circulation to the north, accounting for regional shifts in climate. Increased TSI during the solar cycle causes a rise in sea-surface temperature where cloudiness is low (Northern Hemisphere subtropics), an effect that also favors reduced rainfall near the equator and in the northern mid-latitudes.

Remember, the solar cycle influence on these processes is relatively minor, on the order of 15% to 20%. This influence is likely to change as rising greenhouse gases cause their own changes in climate; stratospheric cooling, increased sea surface temperatures, expanding tropics, accelerating winds, and enhanced Hadley circulation have all been attributed to global warming. How these balance with solar cycle influences adds significant complexity to the challenge of modeling global climate.

This image, taken from a simulation of 20th century climate, depicts several aspects of Earth's climate system. Sea surface temperatures and sea-ice concentrations are shown by the two color scales. The figure also captures sea-level pressure and low-level winds, including warmer air moving north on the eastern side of low-pressure regions and colder air moving south on the western side of the lows. Such simulations, produced by the NCAR-based Community Climate System Model, can also depict additional features of the climate system, such as precipitation. Companion software, released as the Community Earth System Model, will enable scientists to study the climate system in even greater complexity.[63]

SOURCE: University Corporation for Atmospheric Research. http://www.nsf.gov/news/special_reports/clouds/images/photos/large/CCSM4.jpg

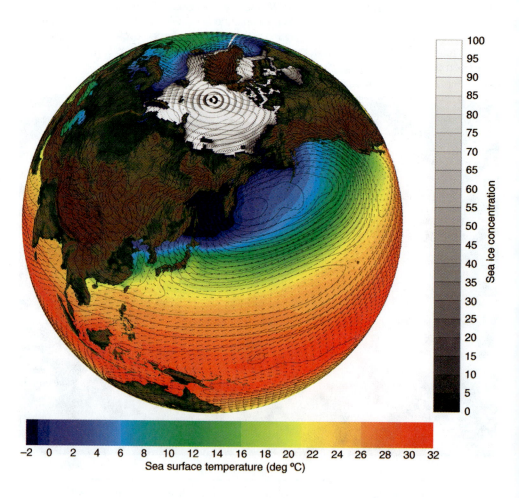

---

[61] D. Rind, J. Lean, J. Lerner, P. Lonergan, and A. Leboissetier, "Exploring the Stratospheric/Tropospheric Response to Solar Forcing," *Journal of Geophysical Research* 113 (2008): D24103, doi:10.1029/2008JD010114.

[62] See the story at http://www.giss.nasa.gov/research/briefs/rind_03/ (accessed July 10, 2012).

[63] http://www.nsf.gov/news/special_reports/clouds/downloads.jsp (accessed July 10, 2012).

GENERAL CIRCULATION MODELS OF CLIMATE

By now you are probably starting to realize that Earth's climate system is very complex. There are many oceanographic, atmospheric, and terrestrial processes (all with some degree of uncertainty) that need to be individually depicted, and their interrelationships depicted, in the form of mathematical calculations. These include frequent but unpredictable processes such as ENSO (located in the Pacific Basin but with global impacts), quasi-predictable processes such as the solar cycle (but each cycle is not identical, and there are long-term trends in solar strength), uncertainties related to clouds,[64] random and unpredictable explosive volcanism, ice-albedo feedback that is not fully understood, and others. These processes all interact with one another over different lengths of time and on different geographic scales and produce various types of feedbacks. It is our awareness of complexity and variability that makes it all the more remarkable that climate models are able to get it right so much of the time. How do we know when they get it right? Because they can reproduce, with amazing fidelity, a range of complex historical observations of climate.

A major test for GCMs is whether they can accurately simulate measured surface temperatures. In this, they succeed well.[65] In Figure 4.11, 100 years of measured temperature changes are plotted as a black line. Two different model results are plotted in red and blue. Blue simulations were produced using only natural factors: solar variation and volcanic activity. They do not match the observed temperature changes very well; in fact they indicate that we would be experiencing global cooling if only natural factors were in control of climate. Red simulations were produced with a combination of natural and human factors, including industrial emission of greenhouse gases and other products of pollution. It is clear that the combination of human and natural factors provides the best match with measured temperatures, leading to the conclusion that human pollution with greenhouse gases is responsible for global warming.

Historical Accuracy; No Guarantee of Future Success

Despite success in reproducing historical climate, it is nonetheless possible that models are achieving the right results for the wrong reasons. That is, assumptions about climate processes represented in a model may be wrong, yet the combined effects of various processes could lead to a model successfully matching historical observations. Scientists tested[66] this possibility by running 11 atmosphere–ocean coupled GCMs. Instead of looking at the models' ability to reproduce 20th century temperatures, the study focused on model skill in recreating global average, Arctic, and tropical climates. Additionally, climate forcings (such as solar activity), feedback systems (like Arctic ice melt or the effects of clouds), and representations of heat storage and transport mechanisms were analyzed.

Of the 11 models tested, eight successfully reproduced global average, Arctic, and tropical temperatures for the past century; most failed to capture warming that occurred in the 1920s and 1930s; three failed to achieve historical accuracy; and two unrealistically depicted either Arctic or tropical temperature change. All the models emphasize climate feedbacks and forcings to different degrees and

[64] T. Andrews, J. Gregory, M. Webb, and K. Taylor, "Forcing, Feedbacks and Climate Sensitivity in CMIP5 Coupled Atmosphere-Ocean Climate Models," *Geophysical Research Letters* 39 (2012): L09712, doi:10.1029/2012GL051607.

[65] See analysis by the IPCC: D. A. Randall, R.A. Wood, S. Bony, et al., "Climate Models and Their Evaluation." In S. Solomon, D. Qin, M. Manning, et al. (eds.), *Climate Change 2007: The Physical Science Basis.* Contribution of Working Group I to the Fourth Assessment Report of the Intergovernmental Panel on Climate Change (Cambridge, UK, Cambridge University Press, 2007).

[66] J. A. Crook and P. M. Forster, "A Balance between Radiative Forcing and Climate Feedback in the Modeled 20th Century Temperature Response," *Journal of Geophysical Research* 116 (2011): D17108, doi:10.1029/2011JD015924.

Figure 4.11. Climate models that can accurately replicate past climate changes build confidence in their ability to predict future changes. The simulations represented by the blue band were produced with only natural factors such as solar variation and volcanic activity. Those shown in red were produced with human greenhouse gas production combined with natural factors. The red band shows that human factors combined with natural factors best account for observed temperature changes (http://www.ipcc.ch/graphics/syr/fig2-5.jpg).

Source: Climate Change 2007: Synthesis Report. Contribution of Working Groups I, II and III to the Fourth Assessment Report of the Intergovernmental Panel on Climate Change, Figure SPM.4. IPCC, Geneva, Switzerland.

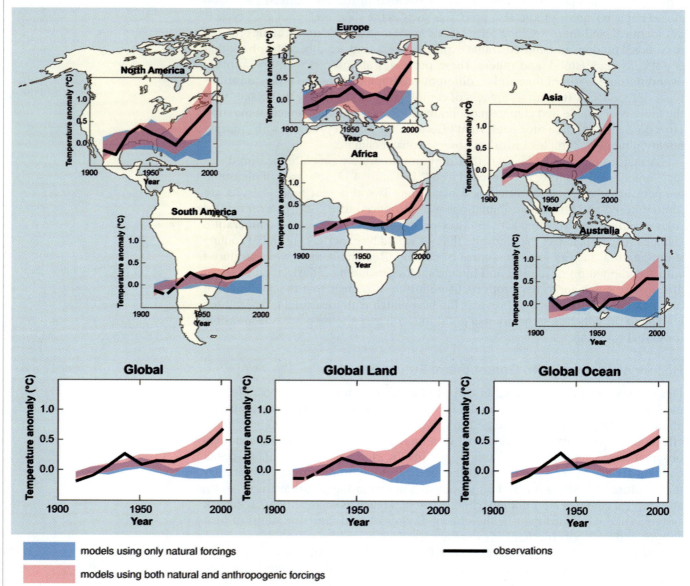

therefore, point out the study authors, model skill in reproducing 20th century climate is not an indication that they will accurately predict future climate. The results suggest that researchers should focus on improving parameterization of the complex relationships among climate forcing and feedback processes that are poorly represented in GCMs.

Modeling Regional Climate

For all their success at simulating climate at the global scale, GCMs are limited in their ability to replicate climate processes at more highly resolved scales (regionally or locally).[67] In regions where the land surface is flat for thousands of kilometers (hundreds of miles), there is no ocean or coastline nearby, land cover is simple, and land use by humans is either absent or uncomplicated, the coarse resolution of a GCM may be enough to accurately simulate weather changes under future climate conditions. However, most land areas are affected by human development and have mountains, coastlines, and changing vegetation characteristics on much smaller scales. In these areas, GCM simulations are not adequate for the practical purposes of planning water resources, changes in storminess, impacts on ecosystems, and individual community planning. In these cases, and others, information is required on a much more detailed scale than GCMs are typically able to provide.

There are basically three approaches to managing this problem:

- Run a GCM at a very fine resolution. The challenge with this approach is that GCM computer time is extremely expensive, and increasing the resolving power of models leads to rapid increases in costs. Typically, funding for these expenses is only obtained through highly competitive government grants such as from the National Science Foundation or other agencies that support scientific research. Additionally, researchers are typically only able to run one climate experiment at a time, each at great cost and commitment of time. This approach also would require a very powerful computer, of which there are relatively few in the world. One of these is the Earth Simulator 2 (ES2), which was developed for three scientific agencies of the Japanese government. Models are also operated by NASA, NOAA, the British Met Office Hadley Center, and others. Another approach is to run a fine-resolution experiment for a short simulation period such as 5 years rather than 50 years. This cuts the computer time needed to complete the simulation, and thus the expense, but it might not answer the key questions desired, such as "What will be the average annual rainfall in my area 50 years in the future?"

- Statistically downscale a coarse resolution simulation. Statistical downscaling assumes that climate variables at a coarse scale (e.g., grid-scale winds, humidity, temperature, rainfall) are related to fine-scale weather (e.g., rainfall at a rain gauge) and that this relationship will hold valid in the future even under changing conditions. By knowing the present statistical relationship between grid-scale climate variables and point-scale measurements, one uses that relationship to define point-scale processes in a future climate. This assumption might or might not be valid.

- Embed a regional-scale (tens of kilometers) climate model (RCM) within a global climate model (hundreds of kilometers). RCMs employ localized models that calculate the equations of climate on a regional scale planted within a GCM. The GCM provides boundary conditions for the calculations in the RCM. Boundary conditions are the set of climate variables (humidity, temperature, wind stress, etc.) calculated by the GCM along the edges of the RCM. These are incorporated by the RCM to make more highly resolved simulations on the smaller grid. RCMs take globally calculated conditions and provide a description of climate that resolves local factors such as topography, coastline, land use, industrialization, and other local parameters. One cannot run an RCM outside of a GCM because the weather in one part of the world is connected to the weather in another part. For instance ENSO, volcanism, and ocean circulation could affect global temperature; it takes a GCM to define these connections.

Increasingly, as decision makers responsible for community sustainability and environmental conservation grow more worried about the long-term impacts of global warming, RCMs are being used to clarify plans for the future. For instance,

---

[67] See discussions at the World Climate Research Program http://www.wcrp-climate.org/ and at Climate Prediction.net http://climateprediction.net/ (accessed July 10, 2012).

GCMs show[68] that in the future, water availability in the western United States will be increasingly tied to extreme events[69]; however, the degree to which this change will affect local communities is poorly defined by the scale of global models. This problem has been improved with the use of regional climate modeling. One study[70] used an ensemble of eight RCMs embedded within the projections of GCMs to estimate future winter average and extreme precipitation in the western United States. Researchers found a consistent and statistically significant increase in the intensity of future extreme winter precipitation events over the western United States. For the years 2038 to 2070, 20-year return-period and 50-year return-period winter storms are modeled to increase across the entire west by 12.6% and 14.4%, respectively. Model results show this increase in storminess will be accompanied by a 7.5% decrease in winter average precipitation in the southwestern United States. For water managers in communities faced with population growth (the southwestern United States is the fastest-growing region in the nation), this type of information is a key element in effective planning for the future.

Intergovernmental Panel on Climate Change

In 2007, the IPCC[71] released its Fourth Assessment Report (AR4) on Climate Change. The AR4 was the product of thousands of scientists, criticizing each other's work, struggling to find common wording to describe results, and ultimately building a description of the impacts of global warming on all sectors of human interest. AR4 was an enormous effort summarizing the current understanding of climate change. The report took six years to produce and involved more than 2,500 scientific expert reviewers and more than 800 authors from more than 130 countries.

Some of their key findings include the following:

- The warming trend over the last 50 years (about 0.13°C or 0.23°F per decade) is nearly twice that for the last 100 years.

- The average amount of water vapor in the atmosphere has increased since at least the 1980s over land and ocean. The increase is broadly consistent with the extra water vapor that warmer air can hold.

- Since 1961, the average temperature of the global ocean down to depths of at least 3 km (1.9 miles) has increased. The ocean has been absorbing more than 90% of the heat added to the climate system, causing seawater to expand and contributing to sea-level rise.

- Global average sea level rose on average by 1.8 mm (0.07 inches) per year from 1961 to 2003. There is high confidence that the rate of observed sea-level rise increased from the 19th to the 20th century.

- Average arctic temperatures increased at almost twice the global average rate in the past 100 years.

- Mountain glaciers and snow cover have declined on average in both hemispheres. Widespread decreases in glaciers and ice caps have contributed to sea-level rise.

- Long-term trends in the amount of precipitation have been observed over many large regions from 1900 to 2005.

[68] G.A. Meehl, C. Tebaldi, H. Teng, and T. C. Peterson, "Current and Future US Weather Extremes and El Niño," *Geophysical Research Letters* 34 (2007): L20704, doi:10.1029/2007GL031027.

[69] S. Emori and S. Brown, "Dynamic and Thermodynamic Changes in Mean and Extreme Precipitation under Changed Climate," *Geophysical Research Letters* 32 (2005): L17706, doi:10.1029/2005GL023272.

[70] F. Dominguez, E. Rivera, D. P. Lettenmaier, and C. L. Castro, "Changes in Winter Precipitation Extremes for the Western United States under a Warmer Climate as Simulated by Regional Climate Models," *Geophysical Research Letters* 39 (2012): L05803, doi:10.1029/2011GL050762.

[71] S. Solomon, D. Qin, M. Manning, et al., (eds.), *Climate Change 2007: The Physical Science Basis*. See http://www.ipcc.ch/pdf/assessment-report/ar4/wg1/ar4-wg1-chapter3.pdf (accessed July 10, 2012).

Global circulation model projections of the future are based on understanding the past and current conditions of global warming. Although new modeling efforts have been published since the IPCC report in 2007, the AR4 remains the benchmark study on which governments are basing development of new policies for counteracting the negative effects of global warming.

In the "Summary for Policymakers," the AR4 presented 12 major conclusions:

- Global atmospheric concentrations of carbon dioxide, methane, and nitrous oxide have increased markedly as a result of human activities since 1750 and now far exceed preindustrial values determined from ice cores spanning many thousands of years.

- Global increases in carbon dioxide concentration are due primarily to fossil fuel use and land-use change, whereas those of methane and nitrous oxide are primarily due to agriculture.

- Understanding of human warming and cooling influences on climate has improved, leading to very high confidence (meaning at least 90% correct) that the globally averaged net effect of human activities since 1750 has been one of warming.

- Warming of the climate system is unequivocal, as is evident from observations of increases in global average air and ocean temperatures, widespread melting of snow and ice, and rising global average sea level.

- At continental, regional, and ocean basin scales, numerous long-term changes in climate have been observed. These include changes in Arctic temperatures and ice, widespread changes in precipitation amounts, ocean salinity, wind patterns, and aspects of extreme weather including droughts, heavy precipitation, heat waves, and the intensity of tropical cyclones.

- Paleoclimate information supports the interpretation that the warmth of the last half century is unusual in at least the previous 1300 years. The last time the polar regions were significantly warmer than present for an extended period (about 125,000 years ago), reductions in polar ice volume led to 4 to 6 m (13 to 20 ft) of sea-level rise.

- Most of the observed increase in globally averaged temperatures since the mid-20th century is very likely (greater than 90% confidence) owing to the observed increase in anthropogenic (human caused) greenhouse gas concentrations. Discernible human influences now extend to other aspects of climate, including ocean warming, continental-average temperatures, temperature extremes, and wind patterns.

- Analysis of climate models together with constraints from observations enables an assessed likely (more than 66% confidence) range to be given for climate sensitivity for the first time and provides increased confidence in the understanding of the climate system response to radiative forcing.

- For the next two decades, a warming of about 0.2°C per decade is projected by climate models for a range of greenhouse gas–emission scenarios. Even if the concentrations of all greenhouse gases and aerosols had been kept constant at year 2000 levels, a further warming of about 0.1°C per decade would be expected.

- Continued greenhouse gas emissions at or above current rates would cause further warming and induce many changes in the global climate system during the 21st century that would very likely (greater than 90% confidence) be larger than those observed during the 20th century.

- There is now higher confidence in projected patterns of warming and other regional-scale features, including changes in wind patterns, precipitation, and some aspects of extremes and of ice.

- Anthropogenic (human-caused) warming and sea-level rise would continue for centuries owing to the time scales associated with climate processes and feedbacks, even if greenhouse gas concentrations were to be stabilized.

To model[72] future global warming and its impacts, researchers must make some assumptions about future greenhouse gas production. The assumptions used in AR4 emerge from considerations of population growth, economic activity, government policies, social patterns, and other complex factors that govern human behavior. To handle these many possibilities, modelers resort to scenario building. In the AR4, four scenarios were defined (with variations):

A1. A future world of very rapid economic growth, global population that peaks in mid-century and declines thereafter, and the rapid introduction of new and more-efficient technologies.

   A1FI. Energy is fossil-fuel intensive.

   A1T. Energy is not fossil fuel intensive.

   A1B. Energy does not rely too heavily on one particular source.

A2. A future world that is characterized by self-reliance, preservation of local identities, and continuously increasing population. Economic growth and technological change are fragmented and slow.

B1. A future world with a global population that peaks in mid-century and declines thereafter. The economy has a focus on service and information, with reductions in material intensity and the introduction of clean and resource efficient technologies aimed at achieving social and environmental sustainability.

B2. A future world with emphasis on local solutions to economic, social, and environmental sustainability. Global population continuously but slowly increases. Economic development and technological change are slower than in other scenarios.

In all of these scenarios it is assumed that no specific climate initiatives are adopted by governments. The results indicate that global temperatures by the end of the century rise between a low (B1 scenario) of 1.1°C (2°F) and a high (A1FI scenario) of 6.4°C (11.5°F).

For each scenario, a sea-level rise assessment is also made. The sea-level rise by the end of the century is modeled to rise between 0.18 to 0.59 m (7 to 23 in). AR4 sea-level estimates do not consider the possibility of ice calving (the physical disintegration of glaciers), and so it is widely perceived that these numbers underestimate the true likely sea-level rise by the end-of-the-century (sea level is discussed further in Chapter 5). Table 4.1 provides specific scenario results for end-of-the-century global

TABLE 4.1 Projected Globally Averaged Surface Warming and Sea-Level Rise at the End of the 21st Century

Scenario	Temperature Rise (°C) in 2090–2099 Relative to 1980–1999 Best Estimate	Likely Range (°C)	Sea-Level Rise (m) in 2090–2099 Relative to 1980–1999
Year 2000 constant emissions	0.6	0.3–0.9	N/A
B1 scenario	1.8	1.1–2.9	0.18–0.38
A1T scenario	2.4	1.4–3.8	0.20–0.45
B2 scenario	2.4	1.4–3.8	0.20–0.43
A1B scenario	2.8	1.7–4.4	0.21–0.48
A2 scenario	3.4	2.0–5.4	0.23–0.51
A1FI scenario	4.0	2.4–6.4	0.26–0.59

[72] See the animation "Global Temperature Model (1885–2100)" at the end of the chapter.

Figure 4.12. Global average temperatures resulting from various greenhouse gas–emission scenarios. Shading denotes the uncertainty in the projection (plus or minus one standard deviation range of individual model annual averages). The orange line is for the scenario where greenhouse gas concentrations were held constant at year 2000 values. The gray bars at right indicate the best estimate (solid line within each bar) and the likely range assessed for the six scenarios. The IPCC AR4 climate models predict that warming will be greatest in the Arctic and over land. These results vary depending on the level and type of future economic activity and the greenhouse gas production that results.

IMAGE CREDIT: Climate Change 2007: The Physical Science Basis. Working Group I Contribution to the Fourth Assessment Report of the Intergovernmental Panel on Climate Change, Figure SPM.5 and Figure SPM.6 (right panel). Cambridge University Press.

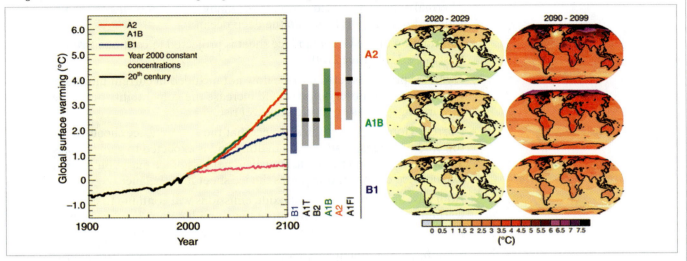

average temperature and sea-level rise. Figure 4.12 shows modeled global temperature changes to the end of the century.

These model scenarios are projected to lead to the following regional-scale patterns:

- Snow cover is projected to contract. Widespread increases in thaw depth are projected over most permafrost regions.

- Sea ice is projected to shrink in both the Arctic and Antarctic under all scenarios. In some projections, Arctic late-summer sea ice disappears almost entirely by the latter part of the 21st century.

- It is very likely that hot extremes, heat waves, and heavy precipitation events will continue to become more frequent.

- It is likely that future tropical cyclones (typhoons and hurricanes) will become more intense, with larger peak wind speeds and more heavy precipitation associated with ongoing increases of tropical sea-surface temperatures. There is less confidence in projections of a global decrease in numbers of tropical cyclones. The apparent increase in the proportion of very intense storms since 1970 in some regions is much larger than simulated by current models for that period.

- Extratropical storm tracks are projected to move poleward, with consequent changes in wind, precipitation, and temperature patterns, continuing the broad pattern of observed trends over the last half-century.

- Increases in the amount of precipitation are very likely in high latitudes, and decreases are likely in most subtropical land regions. This matches observed patterns in recent trends.

- Based on current model simulations, it is very likely that the vertical circulation of the North Atlantic Ocean will slow down during the 21st century. Temperatures in the Atlantic region are projected to increase despite such changes owing to the much larger warming associated with projected increases of greenhouse gases.

- Climate processes are expected to add carbon dioxide to the atmosphere as the climate system warms, but the magnitude of this feedback is uncertain.
- If heating were to be stabilized in 2100 at B1 or A1B levels, a further increase in global average temperature of about 0.5°C would still be expected, mostly by 2200.
- If heating were to be stabilized in 2100 at A1B levels, thermal expansion alone would lead to 0.3 to 0.8 m of sea-level rise by 2300 (relative to 1980–1999). Thermal expansion would continue for many centuries owing to the time required to transport heat into the deep ocean.
- Contraction of the Greenland ice sheet is projected to continue to contribute to sea-level rise after 2100.
- Dynamic processes related to ice flow not included in current models but suggested by recent observations could increase the vulnerability of the ice sheets to warming, increasing future sea-level rise.
- Current global model studies project that the Antarctic ice sheet will remain too cold for widespread surface melting and is expected to gain in mass as a result of increased snowfall. However, net loss of ice mass could occur if dynamic ice discharge dominates the ice sheet mass balance.
- Both past and future carbon dioxide emissions will continue to contribute to warming and sea-level rise for more than a millennium, because of the time scales required for removing this gas from the atmosphere.

How high will global temperature rise by the middle of the century? One study[73] ran almost 10,000 climate simulations on volunteers' home computers to increase the horsepower needed to calculate global climate change using a GCM. The study was the first to run so many simulations using a coupled ocean–atmosphere climate model. Using so many simulations improves definition of some of the uncertainties of previous forecasts that used simpler models or only a few dozen simulations. The modeling experiment found that a global warming of 3°C (5.4°F) by 2050 is equally plausible as a rise of 1.4°C (2.5°F). The results suggest that the world is very likely to cross the "2 degrees barrier" at some point in this century if emissions continue unabated. Thus, those planning for the impacts of climate change need to consider the possibility of warming of up to 3°C (above the 1961–1990 average) by 2050 even on a mid-range emissions scenario. This is a faster rate of warming than most other models predict.

If We Can't Predict Weather, How Can We Predict Climate?

In fact, GCMs do predict climate with accuracy, but at a large scale. Your TV weather forecaster has the harder task of predicting detailed weather, at specific localities, in very short time periods. Predicting climate and predicting weather are very different from each other.

Weather is the short-term (up to a week) state of the atmosphere at a given location. It affects the well-being of humans, plants, and animals and the quality of our food and water supply. Weather is somewhat predictable because of our understanding of Earth's global climate patterns. Climate is the long-term (about 30 years) average weather pattern and is the result of interactions among land, ocean, atmosphere, ice, and the biosphere. Climate is described by many weather elements, such as temperature, precipitation, humidity, sunshine, and wind. Both climate and weather result from processes that accumulate and move heat within and between the atmosphere and the ocean.

---

[73] D. Rowlands, D. Frame, D. Ackerley, et al., "Broad Range of 2050 Warming from an Observationally Constrained Large Climate Model Ensemble," *Nature Geoscience* 5, no 4 (2012): 256, doi: 10.1038/ngeo1430.

Figure 4.13. The change in annual average precipitation projected by the Geophysical Fluid Dynamics Laboratory (NOAA) CM2.1 model for the 21st century. These results are from a model simulation forced according to the IPCC SRES A1B scenario, in which atmospheric carbon dioxide levels increase from 370 to 717 ppm. The plotted precipitation differences were computed as the difference between the 2081 to 2100 20-year average minus the 1951 to 2000 50-year average. Blue areas are projected to see an increase in annual precipitation amounts. Brown areas are projected to receive less precipitation in the future. Note the irregular color bar intervals.

Source: NOAA GFDL Climate Research Highlights Image Gallery: Will the Wet Get Wetter and the Dry Drier? http://www.gfdl.noaa.gov/will-the-wet-get-wetter-and-the-dry-drier (accessed July 10, 2012).

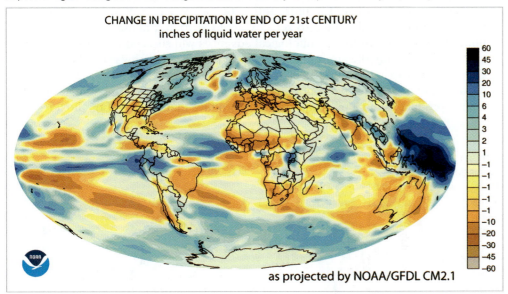

As a general rule, global warming will produce more hot days and fewer cool days in most places.[74] Warming will be greatest over land, and longer, more-intense heat waves will become more common. We will see an increase in the severity of storms, floods, and droughts as rain and snowfall patterns change (Figure 4.13). It has been difficult for the meteorology community to reach agreement on how hurricanes will change with global warming. But there is general agreement that because of warmer ocean surface temperatures hurricanes could decrease in frequency yet increase in intensity[75] as a global average. It is impossible to pin any single unusual weather event on global warming, but emerging evidence suggests that global warming is already influencing the weather.[76] Heat waves, droughts, and intense rain events have increased in frequency during the last 50 years, and human-induced global warming more likely than not contributed to the trend.[77] Climate change is neither proved nor disproved by individual warming or cooling spells. It's the longer-term trends, of a decade or more, that place less emphasis on single-year variability, that count. Nonetheless, unusual bouts of weather, and other weather changes that are a result of naturally occurring patterns, are still consistent with a globally warming world (weather and climate change are discussed further in Chapter 7).[78]

---

[74] G. A. Meehl, C. Tebaldi, G. Walton, D. Easterling, and L. McDaniel, "The Relative Increase of Record High Maximum Temperatures Compared to Record Low Minimum Temperatures in the U.S.," *Geophysical Research Letters* 36 (2009): L23701, doi:10.1029/2009GL040736.

[75] T. R. Knutson, J. L. McBride, J. Chan, et al., "Tropical Cyclones and Climate Change," *Nature Geoscience* 3 (2010): 157–163.

[76] P. Pall, T. Aina, D. A. Stone, et al., "Anthropogenic Greenhouse Gas Contribution to Flood Risk in England and Wales in Autumn 2000," *Nature* 470 (2011): 382–385, doi:10.1038/nature09762.

[77] S. K. Min, X. Zhang, F. W. Zwiers, and G. C. Hegerl, "Human Contribution to More-Intense Precipitation Extremes, *Nature* 470 (2011): 378–381, doi:10.1038/nature09763.

[78] J. Hansen, M. Sato, and R. Ruedy, "Perception of Climate Change," *Proc. Natl. Acad. Sci.* 109 (2012): 14726–14727.

Beyond AR4

Since AR4, climate models have continued to advance.[79] An important development has been the adoption of new scenarios of future climate forcing. Whereas modelers in AR4 used the economic scenarios described B1 through A1FI, for AR5, due in 2013 and 2014, modelers[80] are instead assuming a future characterized by a range of pathways leading to prescribed radiative forcings by the end of the century.[81] Each model pathway is named by the forcing it elicits by the year 2100: 8.5, 6, 4.5, and 2.6 watts per square meter (W/m^2). Researchers chose four trajectories to avoid the mistaken assumption that the middle scenario is the most likely. They also chose pathways considered extreme by some. The most optimistic pathway (2.6 W/m^2) results in greenhouse gas emissions dropping to zero by about 2070 and falling to negative values under the assumption that technological advances will succeed in actually removing carbon dioxide from the atmosphere. The most pessimistic pathway (8.5 W/m^2) pushes CO_2 emissions to a mammoth 1,300 parts per million by the end of the century, a level of forcing so enormous that some consider it inconceivable because humans won't be able to produce enough oil, coal, and gas to produce that much carbon dioxide.

Figure 4.14. a, Observed monthly mean global temperatures (black) and modeled temperature (orange). **b,** Individual contributions of ENSO (purple), volcanic eruption (blue), solar irradiance (green), and greenhouse gas effects (red). Together, the four influences explain 76% of the variance in global temperature observations. Future scenarios are shown as dashed lines. The vertical black dashed lines in **a:** A denotes 2014 with, for example, temperature declines indicating volcanic eruption; B denotes 2019 with, for example, temperature increases due to strong El Niño.

Source: From J. L. Lean and D. H. Rind, "How Will Earth's Surface Temperature Change in Future Decades?" *Geophysical Reearch Letters* 36 (2009): L15708, doi:10.1029/2009GL038932.

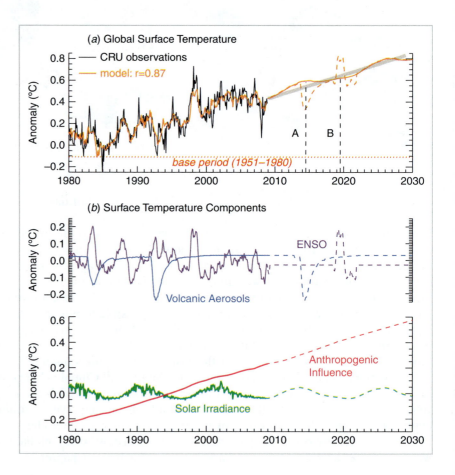

[79] The climate modeling community of researchers has engaged in formal comparisons of modeling results for more than a decade. The latest effort, CMIP5 (Coupled Model Intercomparison Project Phase 5) is designed to provide a context for assessing the mechanisms responsible for model differences in poorly understood feedbacks associated with the carbon cycle and with clouds; examining climate "predictability" and exploring the ability of models to predict climate on decadal time scales; and, more generally, determining why similarly forced models produce a range of responses. See the CMIP5 page http://cmip-pcmdi.llnl.gov/cmip5/ (accessed July 10, 2012).

[80] R. H. Moss, et al., "The Next Generation of Scenarios for Climate Change Research and Assessment," *Nature* 463 (2010): 747.

[81] See discussions at http://www.iiasa.ac.at/Research/ENE/IAMC/rcp.html and http://www.nature.com/nature/journal/v463/n7282/full/nature08823.html (accessed July 10, 2012).

A 2009 study[82] by researchers at the U.S. Naval Research Laboratory and NASA produced high-resolution predictions of climate on the scale of decades. This is especially difficult because natural climate variations can amplify or suppress global warming in ways that GCMs capture poorly. However, by deconstructing recently observed surface temperatures into separate components caused by ENSO, volcanism, solar activity, and human influences, researchers were able to reproduce the past three decades in unusual detail. With this success, the model was used to predict the next 20 years of climate.

From 2009 to 2014, likely increases in greenhouse gases and solar radiation will raise global surface temperature 0.15°C ± 0.03°C (0.25°F ± 0.05°F), at a rate 50% greater than predicted by the modeling performed in AR4. However, because of the 11-year sunspot cycle, solar radiation is modeled to decline in the following five years; thus average temperature in 2019 is projected to be only 0.03°C ± 0.01°C (0.05°F ± 0.02°F) warmer than in 2014. The study (Figure 4.14) concludes that the decade 2010 to 2020 will be comparable to the period from 2002 to 2008, when decreasing solar irradiance also countered much of the global warming. Unable to exactly predict the timing of volcanic activity or ENSO intensity, the study includes scenarios of how a major eruption and a super ENSO would modify temperature projections.

CONCLUDING THOUGHTS

Climate models are not perfect. However, when tested by simulating today's climate, the latest generation of models greatly outperforms their predecessors. This is largely due to more-sophisticated calculations of important climate variables and the growing power of supercomputers.[83]

Figure 4.15 shows the performance of the IPCC AR4 model projections compared to the global mean temperatures since 1980 as measured by various agencies (colored lines).[84] The projections here (black line) are an average of several model outputs (an ensemble), and the white envelope encloses 95% of the model runs. The ensemble projection reproduced observed temperatures from 1980 to 2000 with great skill. Notable misfits occurred at three times: the global cooling caused by Mt. Pinatubo in 1982–1984, the strong El Niño event of 1998, and the extreme solar minimum of 2008. The projections use 2000 as a baseline. That is, the model performed a hindcast (modeled the past) from 1980–2000 and a forecast for the

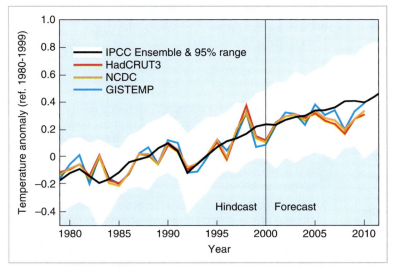

Figure 4.15. Comparison of the annual mean temperature anomalies projected by IPCC AR4 models (IPCC Ensemble) plotted against the surface temperature observations from the British Hadley Climate Center (HadCRUT3), the U.S. National Climate Data Center (NCDC), and the NASA Goddard Institute for Space Studies (GISTEMP). The baseline used is the average global temperature for the period 1980–1999 (as in the 2007 IPCC report), and the envelope in white encloses 95% of the model runs.

Source: After the graph on realclimate.org; see http://www.realclimate.org/index.php/archives/2011/01/2010-updates-to-model-data-comparisons/ (accessed July 10, 2012).

[82] J. L. Lean and D. H. Rind, "How Will Earth's Surface Temperature Change in Future Decades?" *Geophysical Research Letters* 36 (2009): L15708, doi:10.1029/2009GL038932.

[83] T. Reichler and J. Kim, "How Well Do Coupled Models Simulate Today's Climate?" *Bulletin of the American Meteorological Society* 89 (2008): 303–311.

[84] See the discussion at realclimate.org: http://www.realclimate.org/index.php/archives/2011/01/2010-updates-to-model-data-comparisons/ (accessed July 10, 2012).

period 2000–2010. Clearly the model projections demonstrate significant reliability in simulating global temperature.

Climate models[85] are effective tools that grow more powerful with continued research and technological improvement. Model projections warn us that global warming will likely cause serious problems in the future including increased drought, sea-level rise, frequency of warm spells, heat waves, heavy rainfall events, intensity of tropical cyclones (including hurricanes), extreme high tides, sea-ice reductions, and other dangerous physical impacts. Both past and future carbon dioxide emissions will continue to contribute to warming and related effects for more than a millennium as a result of the timescales required for removing this gas from the atmosphere.

ANIMATIONS AND VIDEOS

National Center of Atmospheric Research, "Recipe for a Better Climate Model," http://www.youtube.com/watch?v=TLvCCHNCEdc&NR=1

NASA Center for Climate Simulation: Data Supporting Science, "Supercomputing the Climate," http://www.nasa.gov/topics/earth/features/climate-sim-center.html

El Niño Explained: "El Niño," http://video.nationalgeographic.com/video/player/environment/environment-natural-disasters/landslides-and-more/el-nino.html; "Brian Slocum Explains El Niño," http://www.youtube.com/watch?v=uySu7Zv2cbU

NASA Goddard Institute for Space Science, "Arctic Sea Ice Decline," http://www.nasa.gov/mov/275452main_a010021_H264_640x480.mov

National Center for Atmospheric Research, "From all Sides Now," http://nsfgov.http.internapcdn.net/nsfgov_vitalstream_com/clouds_climate_main.mp4

"Climate Denial Crock of the Week: Solar Schmolar," http://www.youtube.com/watch?v=_Sf_UIQYc20

"Global Temperature Model (1885–2100)" – http://www.youtube.com/watch?v=tBithxUmPiA

NASA Goddard Institute for Space Studies, climate model simulation for IPCC: "Piecing Together the Temperature Puzzle," http://climate.nasa.gov/warmingworld/

COMPREHENSION QUESTIONS

1. Why are climate models important? How are they useful?
2. How are climate observations used in climate models?
3. Describe why scientists are working to increase the resolution of climate model projections.
4. What are some of the impacts of El Niño?
5. What are the implications of declining sea ice in the Arctic?
6. What role do clouds play in climate?
7. Why does explosive volcanism produce global cooling?
8. What is the solar cycle? How does it influence climate?
9. Identify some negative impacts from global warming that are projected by climate models.
10. Describe precipitation changes by the end of the century as projected by climate models.

THINKING CRITICALLY

1. Pick three areas experiencing the worst warming by the end of the century. Describe the major impacts to human civilizations and to natural ecosystems in each area.
2. How would you prepare for climate change in your region?
3. Why are clouds difficult to incorporate into climate models?
4. What is the evidence that climate models are skillful?
5. Describe how modern society might prepare for and adapt to precipitation changes caused by global warming.
6. There is evidence that the solar cycle and TSI will decline over the decade 2011–2021. What additional information would you need to anticipate the impact of this on global climate? Speculate about the potential impacts of a return of TSI to current levels over the decade 2021–2031 if greenhouse gas production does not decrease over the same period.

7. Consider the various economic scenarios used by the IPCC in AR4. Which do you consider most likely to occur over the next decade? Over the next half-century? Why?
8. Study the IPCC AR4 climate impacts projected for the end of the century. Describe the most significant threats to human health.
9. Among most scientists it is a foregone conclusion that significant climate change is unavoidable, especially because greenhouse gas emissions have continued to increase. Pick one area where global warming will affect your life and describe steps you would take to adapt to the change.
10. You have just stepped into an elevator with the mayor of your town; he is a climate skeptic. You have a captive audience for the next 30 seconds. Convince him that climate change is real and that he needs to incorporate this in his leadership.

[85] See animation "Piecing Together the Temperature Puzzle" at the end of the chapter.

CLASS ACTIVITIES (FACE TO FACE OR ONLINE)

ACTIVITIES

1. View the National Center of Atmospheric Research video "Recipe for a Better Climate Model," http://www.youtube.com/watch?v=TLvCCHNCEdc&NR=1, and answer the following questions.

  a. Why are climate models important? How are they useful?

  b. How are climate observations used in climate models?

  c. What is AIRS and why is it useful?

  d. Why is working with people in developing countries important?

2. Study how El Niño works in these videos:

  "El Niño," http://video.nationalgeographic.com/video/player/environment/environment-natural-disasters/landslides-and-more/el-nino.html

  "Brian Slocum Explains El Niño," http://www.youtube.com/watch?v=uySu7Zv2cbU

  Answer the following questions.

  a. What are some of the impacts of El Niño?

  b. What happens in the Pacific Ocean in association with an El Niño?

  c. How does El Niño affect hurricane formation in the Atlantic Ocean?

  d. How do winds change in the Southern Hemisphere when an El Niño occurs?

3. View the NASA Goddard Institute for Space Science video "Arctic Sea Ice Decline," http://www.nasa.gov/mov/275452main_a010021_H264_640x480.mov, and answer the following questions.

  a. Explain the meaning of the yellow lines.

  b. What is the message conveyed by this animation?

  c. What are the implications of declining sea ice in the Arctic?

  d. What can be done to mitigate the problem of declining sea ice?

4. Study the National Center for Atmospheric Research video "From all Sides Now," http://nsfgov.http.internapcdn.net/nsfgov_vitalstream_com/clouds_climate_main.mp4, and answer the following questions.

  a. What instruments do scientists use to study clouds?

  b. Why are clouds of such interest to scientists?

  c. What role do clouds play in climate?

  d. Why are clouds difficult to incorporate in climate models?

5. Watch the NASA video "Piecing Together the Temperature Puzzle" http://climate.nasa.gov/warmingworld/ and answer the following questions.

  a. Describe the pattern of climate change in the climate model simulation.

  b. Which areas of the planet heated fastest in the first part of this century?

  c. Pick three areas experiencing the worst warming by the end of the century. Describe the major impacts to human civilizations and to natural ecosystems in each area.

  d. How would you prepare for climate change in your region?

WHAT IS THE REALITY OF SEA-LEVEL RISE?

Figure 5.0. Devastation of Bolivar Peninsula, Texas, following Hurricane Ike. As sea level continues to rise because of global warming, the damage resulting from coastal hazards such as hurricanes, tsunamis, high waves, and extreme tides will increase in cost, frequency, and magnitude.

IMAGE CREDIT: NOAA.

CHAPTER SUMMARY

Today, rising seas threaten coastal wetlands, estuaries, islands, beaches, reefs, and all types of coastal environments. Human communities living on the coast are subject to flooding by rainstorms that are coincident with high tides, accelerated coastal erosion, groundwater inundation, and saltwater intrusion into streams and aquifers. Sea-level rise threatens cities, ports, and other areas with passive flooding due to rising waters and with damaging flooding that will increase in magnitude when hurricanes and tsunamis strike. Because sea-level rise has enormous economic and environmental consequences, it is important to understand how global warming is creating this threat.

In this chapter you will learn that

1. According to satellite measurements, global mean sea level has risen about 5.7 cm (2.3 in) from 1993 to 2012 at a mean rate of about 3.2 mm/yr (0.13 in/yr). This rise is not uniform across the oceans, however.

2. According to tide gauge measurements, the average global sea level trend from 1962 to 1990 was 1.5 ± 0.5 mm/yr (0.06 in/yr); since 1990 the trend has increased to 3.2 ± 0.4 mm/yr (0.13 in/yr).

3. Ice on Greenland and Antarctica is melting at an accelerating rate that, when combined with thermal expansion of seawater, will likely raise the global mean sea level by 32 ± 5 cm (1 ft) by the year 2050.

4. Global sea level change may correlate to global atmospheric temperature. Using IPCC emission scenarios and temperature projections, semi-empirical modeling indicates that mean sea level may rise 0.75 to 1.90 m (2.5 to 6.2 ft) by 2100.

5. Using physical principles and high end projections of ice sheet decay, coupled global climate models predict global mean sea-level rise of 1.02 m by 2100; low end projections lead to coupled model projections of 0.47 m by 2100.

6. Climate models predict that, because of thermal expansion of deep seawater, sea-level rise will persist for many centuries, even if greenhouse gas emissions cease rising and existing CO_2 is removed from the atmosphere.

7. Globally, sea-level rise will increase coastal erosion, marine inundation from storms and tsunamis, raise the groundwater table and cause drainage problems, cause saltwater intrusion, and threaten coastal communities.

Learning Objective

Sea level is rising today and is projected by climate models to continue rising at an accelerated pace in the decades and centuries ahead. Greenhouse-gas–induced global warming causes ice to melt and ocean water to warm and expand; these two processes are the main causes of global sea-level rise.

Sea level is rising today[1] (Figure 5.1) and will continue to rise in the centuries ahead.[2] Greenhouse-gas-induced global warming leads to the melting of ice in glaciers, and warming of the ocean, which causes ocean water to expand. Melting ice and expanding ocean water are the main causes of global sea-level rise. Climate models[3] predict that, because of thermal expansion of deep ocean water, sea-level rise will persist for many centuries; even if greenhouse gas emissions cease rising and some excess CO_2 is removed from the atmosphere. For this reason, it is appropriate for coastal communities to plan for sea-level rise.[4]

Figure 5.1. Global mean sea-level rise 1992–2011, as measured by satellite detection of the ocean surface.

Source: Cnes/CLS/Legos; http://www.aviso.oceanobs.com/en/news/ocean-indicators/mean-sea-level/ (accessed July 12, 2012).

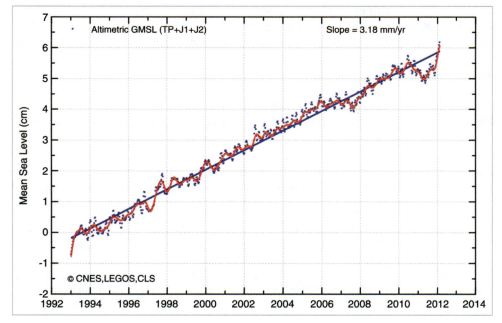

[1] B. Hamlington, R. Leben, S. Nerem, W. Han, and K. Kim, "Reconstructing Sea Level Using Cyclostationary Empirical Orthogonal Functions," *Journal of Geophysical Research* 116 (2011): C12015, doi: 10.1029/2011JC007529.

[2] S. Jevrejeva, J. C. Moore, and A. Grinsted, "Sea Level Projections to AD 2500 with a New Generation of Climate Change Scenarios," *Global and Planetary Change* 80–81 (2011): 14–20 doi: 10.1016/j.gloplacha.2011.09.006.

[3] G. A. Meehl, et al. "Relative Outcomes of Climate Change Mitigation Related to Global Temperature Versus Sea-Level Rise," *Nature Climate Change* (2012), doi: 10.1038/nclimate1529.

[4] See the animation "How Much Will Sea-Level Rise?" at the end of the chapter.

Today, rising seas threaten to forever change coastal wetlands, estuaries, reefs, islands, beaches, and all types of coastal environments. Coastal communities[5] are subject to flooding by rainstorms that are coincident with high tides, accelerated coastal erosion, and saltwater intrusion into streams and aquifers. Sea-level rise threatens cities, ports, coastal communities, and other areas with passive inundation due to rising waters, damaging storm surge associated with hurricanes, and destructive flooding by tsunamis. Because sea-level rise has enormous economic[6] and environmental consequences, it is important to understand how global warming is causing this threat.

RATE OF SEA-LEVEL RISE

Scientists use several types of instruments and geologic proxies to study the behavior of sea level. Geologic records of past sea-level positions include coastal sediments that contain fossil organisms that inhabit ecosystems that are tightly defined by a position between low and high tides. When these fossils are recovered in cores, say from the tidal wetlands lining an estuary, a geologist can interpret their age and depth as a past position of sea level. Physical oceanographers study sea-level behavior using tide gauges located in the calm waters of ports and harbors. Tide gauges measure ocean-level changes through time. Even satellites are used to measure sea-level position. Satellite altimetry is a method of mapping the ocean surface with radar traveling at the speed of light.

Altimeter Studies

Using the time it takes for radar to travel to Earth's surface and back, radar altimeters[7] on satellites can measure the sea surface from space to better than 5 cm (2 in).[8] The TOPEX/Poseidon mission (launched in 1992) and its successors Jason-1 (2001) and Jason-2 (2008) have mapped the sea surface approximately every 10 days for two decades. These missions have led to major advances in physical oceanography and climate studies.[9]

Altimeter measurements indicate that global mean sea level has risen about 6.5 cm (2.5 in) since late 1992 at a mean rate of approximately 3.2 mm/yr (0.13 in/yr; see Figure 5.1).[10] This rise is not uniform across the oceans, however. In some locations

[5] C. Strauss, R. Ziemlinski, J. Weiss, and J. Overpeck, "Tidally Adjusted Estimates of Topographic Vulnerability to Sea-Level Rise and Flooding for the Contiguous United States," *Environmental Research Letters* 7, no. 1 (2012): 014033 doi: 10.1088/1748-9326/7/1/014033.

[6] National Oceanic and Atmospheric Administration, *Adapting to Climate Change: A Planning Guide for State Coastal Managers* (Silver Spring, Md., NOAA Office of Ocean and Coastal Resource Management, 2010). http://coastalmanagement.noaa.gov/climate/adaptation.html (accessed July 12, 2012).

[7] Satellite altimetry measures the time taken by a radar pulse to travel from a satellite to Earth's surface and back to the satellite receiver. Combined with precise satellite location data, altimetry measurements yield sea-surface heights. See "TOPEX/Poseidon," *Wikipedia*, http://en.wikipedia.org/wiki/TOPEX/Poseidon (accessed July 12, 2012).

[8] E. W. Leuliette, R. S. Nerem, and G. T. Mitchum, "Calibration of TOPEX/Poseidon and Jason altimeter data to construct a continuous record of mean sea level change." *Marine Geodesy* 27 (2004): 79–94. See also B. D. Beckley, F. G. Lemoine, S. B. Luthcke, R. D. Ray, and N. P. Zelensky, "A Reassessment of Global and Regional Mean Sea Level Trends from TOPEX and Jason-1 Altimetry Based on Revised Reference Frame and Orbits," *Geophysical Research Letters* 34, no. 14 (2007): L1-4608. See also the NASA entry on "Rising Water: New Map Pinpoints Areas of Sea Level Increase," http://climate.nasa.gov/news/index.cfm?FuseAction=ShowNews&NewsID=16 (accessed July 12, 2012).

[9] Jet Propulsion Laboratory, "Ocean Surface Topography from Space," http://sealevel.jpl.nasa.gov/; "Rising Waters: New Map Pinpoints Areas of Sea-Level Increase," http://climate.nasa.gov/news/?FuseAction=ShowNews&NewsID=16 (accessed July 12, 2012).

[10] See http://sealevel.colorado.edu/ (accessed July 12, 2012).

Figure 5.2. Map of sea-level change 1992–2011. With the Topex/Poseidon, Jason-1 and Jason-2 altimetry missions, the global mean sea level has been calculated on a continual basis since late 1992.

Source: Cnes/CLS/Legos; http://www.aviso.oceanobs.com/en/news/ocean-indicators/mean-sea-level/ (accessed July 12, 2012).

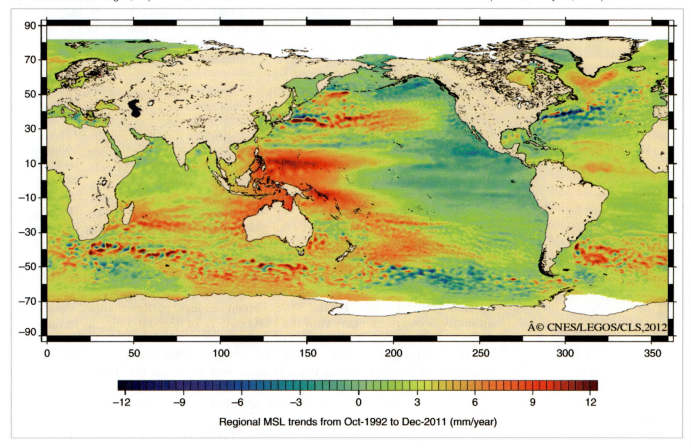

Regional MSL trends from Oct-1992 to Dec-2011 (mm/year)

regional sea level has risen faster than the global average (e.g., the western tropical Pacific[11]), and in other locations regional sea level has risen slower than the global average (e.g., much of the U.S. coastline[12]), and it might have even fallen over the period (e.g., the west coast of the United States[13]).

A map of altimeter measurements reveals the rate of sea-level change since late 1992 on the world's oceans (Figure 5.2). Rates are contoured by color: light blue and green indicate regions where sea level has been relatively stable; green, yellow, and red show areas of sea-level rise; blue indicates areas of sea-level fall. This complex surface pattern largely reflects wind-driven changes in the thickness of the upper layer of the ocean and, to a lesser extent, changes in upper ocean heat content driven by surface circulation.[14] Most noticeable on the map is the dark red area, where sea-level rise in the western Pacific reaches more than 10 mm/yr (0.4 in/yr). This pool

[11] M. A. Merrifield, "A Shift in Western Tropical Pacific Sea Level Trends during the 1990s," *Journal of Climate* 24 (2011) 4126–4138, doi: 10.1175/2011JCLI3932.1.

[12] J.R. Houston and R. G. Dean, "Sea-Level Acceleration Based on U.S. Tide Gauges and Extensions of Previous Global-Gauge Analyses," *Journal of Coastal Research* 27, no. 3 (2011): 409–417.

[13] P. Bromirski, A. J. Miller, R. Flick, and G. Auad, "Dynamical Suppression of Sea-Level Rise Along the Pacific Coast of North America: Indications for Imminent Acceleration," *Journal of Geophysical Research* 116, no. C7 (2011): C07005, doi: 10.1029/2010JC006759.

[14] M. A. Merrifield, "A Shift in Western Tropical Pacific Sea Level Trends during the 1990s."

of rising water has the signature shape of certain phases of quasi-periodic Pacific climate patterns; namely, the La Niña phase of the El Niño Southern Oscillation (ENSO) and the negative phase of the Pacific Decadal Oscillation (PDO). La Niña conditions and the negative phase of the PDO are characterized by pronounced trade winds in the tropical western Pacific. The sea-level buildup in the western Pacific coincides with the absence of strong El Niño events, with the last major El Niño occurring during 1997–1998 and a moderate El Niño in 2010.[15]

Pacific Decadal Oscillation (PDO)

The PDO consists of two phases, each historically lasting 20 to 30 years.[16] In a positive (or warm) phase of the PDO, surface waters in the western Pacific above 20°N latitude tend to be cool, and equatorial waters in the central and eastern Pacific tend to be warm. In a negative (or cool) phase, the opposite pattern develops. Rapid sea-level rise in the western Pacific matches the current negative phase of the PDO.[17] Studies[18] of the history of sea-level change in the western tropical Pacific reveal a strong historical correlation of phases of rapid sea-level rise to negative phases of the PDO. The degree to which this pattern contributes to the global mean rate of sea-level rise observed in satellite altimetry is not known, but the PDO has been recognized as a major factor controlling sea-level change in the Pacific, the world's largest ocean.

Also not known is how long the present negative phase of the PDO will prevail, though it is thought to have begun relatively recently[19] (in 2008; Figure 5.3) and thus might not be responsible for the 20-year mean sea-level pattern mapped by satellite altimetry. Some researchers view the PDO pattern of decadal timing as having broken down in favor of shorter-duration events.[20]

If indeed the Pacific has moved into a negative phase of the PDO, it does not bode well for low-lying islands in the tropical western Pacific that have seen the highest rates of sea-level rise on the planet over the past decade.[21] A negative phase is likely to be accompanied by increased winds, blowing from east to west, that promote higher sea levels in that region.

Although our understanding of the PDO is incomplete, at least one researcher has predicted that the PDO will change sign soon, winds will decrease, and the recent historical stabilization of sea level along the U.S. west coast that results from enhanced trade winds will come to an end and usher in a period of accelerated sea-level rise in the eastern Pacific.[22]

Because winds play an important role in regional sea-level change, it is worthwhile asking, "Are the winds changing as a result of global warming?" Young et al.[23] address this question. They used a 23-year database of satellite altimeter measurements to investigate global changes in oceanic wind speed and wave

[15] For a description of the El Niño Southern Oscillation (ENSO) phenomenon see "El Niño Southern Oscillation," *Wikipedia*, http://en.wikipedia.org/wiki/El_Ni%C3%B1o-Southern_Oscillation (accessed July 12, 2012). See also Chapter 4.

[16] See "Pacific Decadal Oscillation," *Wikipedia*, http://en.wikipedia.org/wiki/Pacific_decadal_oscillation (accessed July 12, 2012).

[17] In April 2008, scientists at NASA's Jet Propulsion Laboratory announced that the Pacific Decadal Oscillation had shifted to its cool (or negative) phase: http://earthobservatory.nasa.gov/IOTD/view.php?id=8703 (accessed July 12, 2012).

[18] Mark A. Merrifield, Philip R. Thompson, and Mark Lander, "Multidecadal Sea Level Anomalies and Trends in the Western Tropical Pacific," *Geophysical Research Letters*, 39, 13, doi:10.1029/2012GL052032, (2012).

[19] See NASA announcement, http://earthobservatory.nasa.gov/IOTD/view.php?id=8703 (accessed July 12, 2012).

[20] See NOAA, http://www.nwfsc.noaa.gov/research/divisions/fed/oeip/ca-pdo.cfm (accessed July 12, 2012).

[21] M. A. Merrifield, "A Shift in Western Tropical Pacific Sea Level Trends during the 1990s."

[22] P. Bromirski, A. J. Miller, R. Flick, and G. Auad, "Dynamical Suppression of Sea-Level Rise along the Pacific Coast of North America: Indications for Imminent Acceleration."

[23] I. Young, S. Zeiger, and A. Babanin, "Global Trends in Wind Speed and Wave Height," *Science* 332 (2012): 451–455.

Figure 5.3. The Pacific Decadal Oscillation (PDO) is thought to have switched to a negative (cool, blue) phase in 2008.[24] Cool phases tend to enhance La Niña conditions and suppress El Niño–like conditions. Thus the switch to a PDO cool phase suggests that high water in the western tropical Pacific could continue to persist owing to strengthened trade winds, at least until the next warm phase is entered.

Source: "Pacific Decadal Oscillation," *Wikipedia*, http://en.wikipedia.org/wiki/Pacific_decadal_oscillation (accessed July 12, 2012).

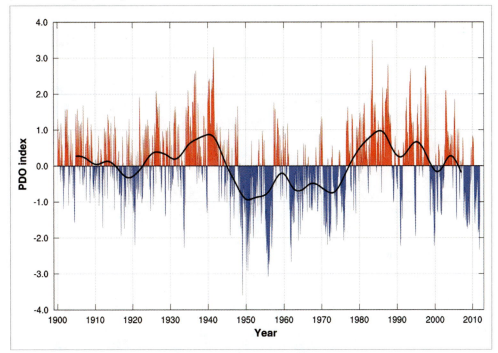

height (Figure 5.4). They discovered a general global trend of increasing wind speed and, to a lesser degree, wave height. The rate of wind speed increase is greater for extreme events compared to the mean condition and indicates the intensity of extreme events is increasing at a faster rate than that of the mean conditions. At the mean and 90th%ile, wind speeds over the majority of the world's oceans have increased by at least 0.25% to 0.5% per year (a 5% to 10% net increase over the past 20 years). The trend is stronger in the Southern Hemisphere than in the Northern Hemisphere. The only significant exception to this positive trend is the central north Pacific, where there are smaller localized increases in wind speed of approximately 0.25% per year and some areas where there is a weak negative trend.

Climate is changing throughout the Pacific, and studies[25] indicate that winds exert an important control on sea-level behavior in the Pacific basin. But how the El Niño Southern Oscillation (ENSO), the PDO, and global warming are linked and how they will continue to interact in the future is largely unknown. For instance, models show that the tropics have expanded, and this has been verified by observations.[26] Presumably related to this is widening of the Hadley cell,[27] the convective system that governs tropical winds and, as seen, it is winds that are currently influencing rates of sea-level change in the low-latitude Pacific. But it is unknown how any of these processes are likely to change in a warmer future.

[24] See NASA, http://earthobservatory.nasa.gov/IOTD/view.php?id=8703 (accessed July 12, 2012).

[25] A. Timmermann, S. McGregor, and F.-F. Jin, "Wind Effects on Past and Future Regional Sea Level Trends in the Southern Indo-Pacific," *Journal of Climate* 23 (2010): 4429–4437, doi: 10.1175/2010JCLI3519.1.

[26] J. Lu, C. Deser, and T. Reichler, "Cause of the Widening of the Tropical Belt Since 1958," *Geophysical Research Letters* 36 (2009): L03803, doi: 10.1029/2008GL036076.

[27] C. M. Johanson and Q. Fu, "Hadley Cell Widening: Model Simulations versus Observations," *Journal of Climate* 22 (2009): 2713–2725.

Figure 5.4. Global contour plots of mean trend (percent per year); wind speed (top) and wave height (bottom). Points that are statistically significant are shown with dots. Researchers have found that since 1990 there is a general global trend of increasing wind speed and, to a lesser degree, wave height. The rate of increase is greater for extreme events as compared to the mean condition.
Source: I. Young, S. Zeiger, and A. Babanin, "Global Trends in Wind Speed and Wave Height," *Science* 332 (2012): 451–455. Reprinted with permission from AAAs.

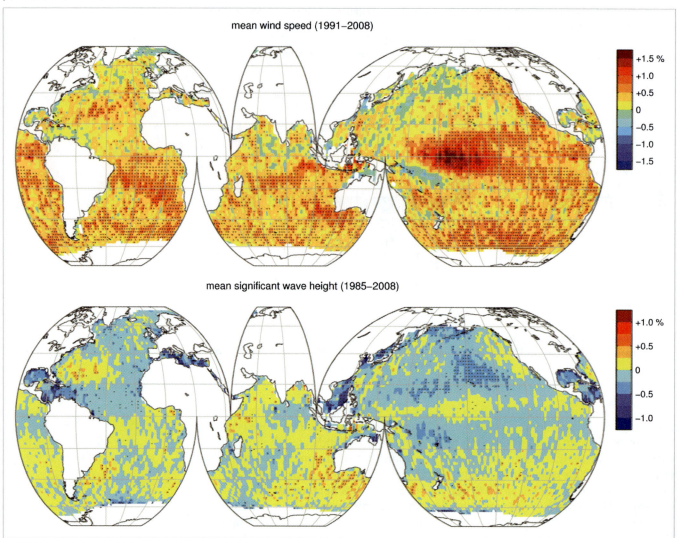

Tide Gauge Studies

In addition to satellite altimetry, sea level is measured around the world using tide gauges.[28] Tide gauges are water-surface-measurement devices mounted on piers, seawalls, and other coastal infrastructure to monitor the rise and fall of the tides and other changes in the ocean surface. Over time the long-term record of changing water level provides information on the relative rate of change between the land the gauge is attached to and the ocean surface it measures. To isolate the tide gauge so that the influence of rising or sinking land does not control the long-term history of water level, modern Global Positioning Systems (GPS) monitor the movement of the gauge. This information is used to resolve a true water level history. Networks of tide gauges provide information on sea-level rise and fall at localities around the world.[29]

---

[28] See the National Atmospheric and Aeronautical Administration website that explains the operation of tide gauges: http://oceanservice.noaa.gov/education/kits/tides/tides10_oldmeasure.html (accessed July 12, 2012).

[29] See the NOAA page for sea-level trends: http://tidesandcurrents.noaa.gov/sltrends/sltrends.shtml (accessed July 12, 2012).

Figure 5.5. A study of tide gauges identified acceleration in the rate of global sea-level rise from approximately 1.56 mm/yr (0.06 in/yr) over the period 1962–1990 to 3.2 mm/yr (0.13 in/yr) between 1990 and 2000. The timing of the acceleration corresponds to similar trend changes in upper ocean heat content and ice melt; it also matches measurements made with satellites over the time period.

Source: Figure from M.A. Merrifield, S.T. Merrifield, and G.T. Mitchum, "An Anomalous Recent Acceleration of Global Sea-Level Rise." *Journal of Climate* 22, (2009): 5772–5781.

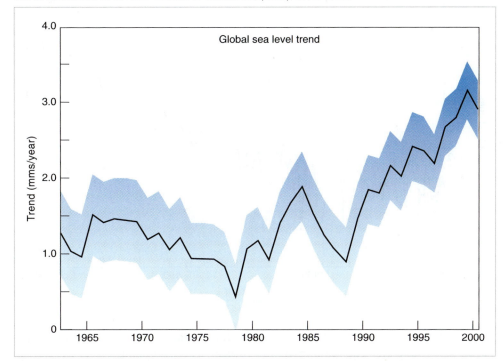

Using the global network of tide gauges, one study[30] identified acceleration in the global rate of sea-level rise that occurred in approximately 1990 (Figure 5.5). The study recognized an average global sea-level trend over the period 1962–1990 of 1.56 mm/yr (0.06 in/yr); however, after 1990 the global trend increased to a rate of 3.2 mm/yr (0.13 in/yr), matching estimates obtained from satellite altimetry. Increased rates in the tropical and southern oceans primarily account for the acceleration. The timing of the global acceleration corresponds to similar trend changes in upper ocean heat content and ice melt.

Another study[31] used the global network of tide gauges in combination with satellite data to establish that global mean sea level rose 19.5 cm (7.7 in) between 1870 and 2004 at an average rate of about 1.44 mm/yr (0.05 in/yr). Over the 20th-century portion of the record, sea level averaged 1.7 mm/yr (0.07 in/yr). This acceleration provided important confirmation of climate models predicting that the rate of sea-level rise will accelerate in response to global warming. If the same acceleration continues, then the amount of rise from 1990 to 2100 will range 28 to 34 cm (11–13 in), which is consistent with IPCC AR4 projections of 18 to 59 cm (7–23 in) of sea-level rise by 2100.

Using a combination of tide gauge and altimeter data, Hamlington[32] used statistical techniques to define the primary components of sea-level change in the

[30] M.A. Merrifield, S.T. Merrifield, and G.T. Mitchum, "An Anomalous Recent Acceleration of Global Sea-Level Rise," *Journal of Climate* 22 (2009): 5772–5781.

[31] J. A. Church and N. J. White, "20th Century Acceleration in Global Sea-Level Rise," *Geophysical Research Letters* 33, no. 1 (2006): L01602.

[32] B. Hamlington, R. Leben, R. Nerem, W. Han, and K. Kim, "Reconstructing Sea Level Using Cyclostationary Empirical Orthogonal Functions," *Journal of Geophysical Research* 116 (2011), C12015, doi: 10.1029/2011JC007529.

Figure 5.6. Using long tide gauge records, researchers reconstructed global sea level since 1700. The shaded portion represents the uncertainties of the reconstruction.

Source: Figure from S. Jevrejeva, J.C. Moore, A. Grinsted, and P. Woodworth, "Recent Global Sea Level Acceleration Started over 200 Years Ago?" *Geophysical Research Letters* 35 (2008): LO8715, doi: 10.1029/2008GL033611; http://www.psmsl.org/products/reconstructions/jevrejevaetal2008.php.

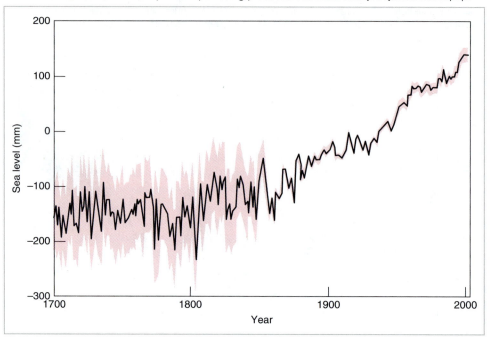

satellite data set and applied these to unravel the characteristics of sea-level change in the tide gauge era. The combined data capture sea-level change over the period 1950–2009. The computed rate of global mean sea-level rise from the reconstructed time series is 1.97 mm/yr (0.077 in/yr) from 1950 to 2009 and 3.22 mm/yr (0.126 in/yr) from 1993 to 2009.

Work[33] with long tide gauge records (Figure 5.6) reveals that sea-level acceleration might have started earlier, more than 200 years ago. By reconstructing global mean sea level since 1700 from long tide-gauge records, researchers concluded that sea-level acceleration began at the end of the 18th century. Sea level rose by 6 cm (2.4 in) during the 19th century and 19 cm (7.5 in) in the 20th century. On the basis of this analysis, they conclude that if the conditions that established the acceleration continue, then sea level will rise 34 cm (13.4 in) over the 21st century.

Tide gauges also provide information on the regional behavior of sea level. The National Research Council[34] studied sea level rise for the coasts of California, Oregon, and Washington. They found that because of vertical land motions resulting from plate tectonics and the ongoing response of Earth's surface to disappearance of North American ice sheets, future sea-level rise is likely to vary along the U.S. west coast. The study projected that, relative to 2000 levels, global sea level will reach 8–23 cm (3–9 in) by 2030, 18–48 cm (7–19 in) by 2050, and 50–140 cm (20–55 in) by 2100. South of Cape Mendocino on the California coast, sea levels are projected to rise an amount similar to global calculations. But north of the cape, along the coasts of northern California, Oregon, and Washington, future sea-level projections are lower than those to the south.

[33] S. Jevrejeva, J.C. Moore, A. Grinsted, and P. Woodworth, "Recent Global Sea Level Acceleration Started over 200 Years Ago?" *Geophysical Research Letters* 35 (2008): LO8715, doi: 10.1029/2008GL033611.

[34] Committee on Sea Level Rise in California, Oregon, and Washington; Board on Earth Sciences and Resources; Ocean Studies Board; Division on Earth and Life Studies; National Research Council, *Sea Level Rise for the Coasts of California, Oregon, and Washington: Past, Present, and Future*, (2012), National Academies Press, Washington, D.C.

On the U.S. east coast, a study[35] of tide gauge records revealed a spatial variation in regional sea-level rise responsible for recent acceleration along a 1000-km-long "hotspot" on the highly populated coast between Cape Hatteras (North Carolina) and New England. Between 1950–1979 and 1980–2009, sea-level rise rate increases were approximately 3 to 4 times higher than the global average and were consistent with a slow-down of North Atlantic circulation.

Coastal Sediment Studies

To extend the record of sea-level changes beyond the era of tide gauges and satellite altimetry, coastal geologists use natural archives of shoreline sediment to reconstruct the past history of sea level. Certain types of sediment can serve as sea-level proxies, including environmental features that grow and collect at the edge of the sea, such as beach sands, shallow-water corals, mud deposited on tidal flats, mangrove roots, and others. One of the most precise sea-level proxies is a type of plankton (foraminifera, a tiny protist) that collects on salt marshes that are only flooded by the highest tides. Because different species of foraminifera live at different levels of the tide, a survey of the types of remains buried in marsh mud tell researchers where the level of the tide was in that particular spot at the time the sediment layer was laid down.[36] As sea level rises (and the land subsides) these microscopic animals are buried by mud that collects in the salt marsh as it maintains its position between high and low tides.

Geologists take cores of salt marshes, analyze the entombed plankton (and plant fragments and other types of remains) that indicate the position of sea level through time, and use radiocarbon to date the sediments and assess the age of the samples. Radiocarbon dating is a method that permits age-dating of organic samples up to an age of about 50,000 years old. For instance, a cored sample from 1.5 m (5 ft) below the marsh surface might contain plankton and plant fossils known to grow only in mud inundated by the full-moon high tide; these are very good indicators of sea level. The same sample might provide a radiocarbon date of 2100 years (or so) before present. Hence, by building a record of changing tide level out of these geologic materials, a sea-level history can be assembled that predates the instrumental record.

It was just such a research effort that produced a reconstructed sea-level history of the North Carolina coast extending 2,100 years into the past[37] (Figure 5.7). Analyzing cores of tide marsh mud, researchers found four phases of persistent sea-level change, a history that applies only to the North Carolina coastal plain because of the unique behavior of Earth's crust from one region to another. (The crust rises in some places and subsides in others, making detailed sea-level records only representative of their home region.) The four phases of sea-level behavior consisted of stable sea level from at least 100 B.C. until A.D. 950; rising sea level from A.D. 950 to A.D. 550 at a rate of 0.6 mm/yr (0.023 in/yr); a period of stable or slightly falling sea level from A.D. 550 until the late 19th century; and finally rising sea level to the present at an average rate of 2.1 mm/yr (0.082 in/yr), representing the steepest century-scale increase of the past two millennia. This rate was initiated between 1865 and 1892, toward the end of the Little Ice Age in the North Atlantic region and the beginning of a clear signal of human-induced global warming.

[35] Asbury H. Sallenger Jr., K.S. Doran, and P.A. Howd, "Hotspot of Accelerated Sea-Level Rise on the Atlantic Coast of North America," *Nature Climate Change*, advance on-line publication, (2012), 24 June, DOI: 10.1038/NCLIMATE1597.

[36] A. Kemp, B. Horton, S. Culver, et al., "Timing and Magnitude of Recent Accelerated Sea-Level Rise (North Carolina, United States)," *Geology* 37 (2009): 1035–1038, doi: 10.1130/G30352A.1.

[37] A. Kemp, B. Horton, J. Donnelly, et al., "Climate Related Sea-Level Variations over the Past Two Millennia," *Proceedings of the National Academy of Sciences* 108, no. 27 (2011): 11017–11022, www.pnas.org/cgi/doi/10.1073/pnas.1015619108.

Figure 5.7. Sea-level history along the North Carolina coast. Blue indicates tide marsh sediment proxy of sea level and uncertainty. Green indicates modern tide gauge sea-level observations. Red indicates model relating sea level and global surface temperature. The rate of sea-level rise along the U.S. Atlantic coast is greater now than at any time in the past 2,000 years and has shown a consistent link between changes in global mean surface temperature and sea level for the past 1,000 years.

SOURCE: Figure after A. Kemp, B. Horton, J. Donnelly, et al, "Climate Related Sea-Level Variations over the Past Two Millennia," Proceedings of the National Academy of Sciences 108, no. 27 (2011): 11017–11022, www.pnas.org/cgi/doi/10.1073/pnas.101561910.

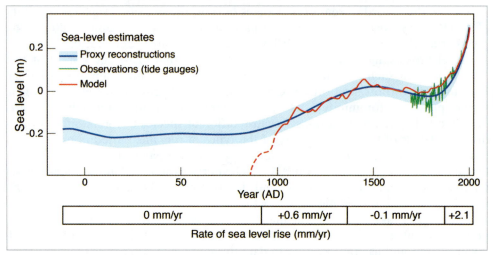

In summary, research shows that today's rate of sea-level rise is the most rapid of the past 2,000 years and that the rate of global mean sea-level rise has accelerated (approximately doubling) over the 20th and 21st centuries[38] and reached more than 3 mm/yr (0.13 in/yr).

SEA-LEVEL COMPONENTS

Global sea-level rise results from a combination of factors. As the oceans absorb heat the water molecules tend to separate and produce thermal expansion; having no other direction to go but upward, they contribute to a rise in sea level. Thermohaline circulation in the North Atlantic, and in the Southern Ocean near Antarctica, cycles warm water downward, leading to thermal expansion in the deep ocean as well (a process that may play out over several centuries). The melting of three forms of ice also contributes to sea-level rise: alpine glaciers in the valleys of mountain systems, ice caps that cover larger surface area than alpine glaciers, and continental ice sheets, of which there are only two, Greenland and Antarctica left over from the last ice age (about 21,000 years ago).[39]

One study[40] considered the various components that go into determining the rate of global mean sea-level change. Using tide gauge data only, researchers calculated that global mean sea level rose between 1972 and 2008 at an average rate of 1.8 ± 0.2 mm/yr (0.071 ± 0.008 in/yr). Using a combination of tide gauges and altimeter observations, they calculated a rate of 2.1 ± 0.2 mm/yr (0.083 ± 0.008 in/yr). The largest contributors to sea-level rise over the period include ocean thermal expansion (0.8 mm/yr; 0.031 in/yr) and melting of various ice forms (0.7 mm/yr; 0.027 in/yr), with Greenland and Antarctica contributing about 0.4 mm/yr (0.016 in/yr). Contributions from melting ice increase throughout the period, as do contributions from thermal expansion, although less rapidly.

---

[38] J. A. Church and N. J. White, "Sea-Level Rise from the Late 19th to the Early 21st Century," *Surveys in Geophysics* 32 (2011): 585–602, doi: 10.1007/s10712-011-9119-1.

[39] See animation "Melting Ice Rrising Seas" at the end of the chapter.

[40] J. Church, N. J. White, L. F. Konikow, et al., "Revisiting the Earth's Sea-Level and Energy Budgets from 1961 to 2008," *Geophysical Research Letters*, 38 (2011): L18601, doi: 10.1029/2011GL048794.

Ocean Warming

The world ocean is so immense that it dominates atmospheric climate, storing more than 90% of the heat in Earth's climate system. (The upper 2.5 m [8.2 ft] of ocean water stores as much heat as the entire atmosphere![41]) Although an increase in the average temperature of the ocean of only 0.01°C (0.018°F) seems small, it is a very large amount of heat. In fact, if this energy were released all at once, the average temperature of the atmosphere would increase by about 10°C (18°F).[42] Thus, a small change in the mean temperature of the ocean represents a very large change in the total heat content of the climate system. It also contributes to sea-level rise, because warming water expands.

The trend of ocean heating has been shown to be quite strong over the longer term,[43] but research indicates that the period 2003 to 2010 showed no net warming of the ocean, and investigators wanted to know why. Using an ensemble of global climate models, one study[44] concluded that an eight-year period without upper ocean warming is not unusual and occurs as a normal event in the model scenarios. Another study,[45] looking at the same problem, concluded that when uncertainties in measurement systems were considered there was, in fact, no missing heat and that Earth has been accumulating heat in the ocean at a rate of 0.5 W/m² (10.8 ft²), with no sign of a decline. This extra energy, they inferred, will eventually find its way back into the atmosphere and increase temperatures on Earth.

In 2012, scientists provided estimates[46] of global ocean warming and its influence on sea level. Over the period 1955–2010, the heat content of the world ocean from 0 to 2000 m (0–6560 ft) depth increased by 0.09°C (0.16°F) and from 0 to 700 m (0–2300 ft) depth it increased by 0.18°C (0.32°F). On this basis, the global ocean accounts for approximately 90% of the warming of the entire Earth climate system that has occurred since 1955. The ocean-warming component of the sea-level trend is 0.54 mm/yr (0.019 in/yr) for depths 0 to 2000 m (0–6500 ft) and 0.41 mm/yr (0.016 in/yr) for depths 0 to 700 m (0–2300 ft).

Understanding how ocean warming (Figure 5.8), and the resulting thermal expansion, contributes to sea-level rise is important to forecast future sea level impacts. Researchers[47] found that from 1961 to 2003, ocean temperatures to a depth of about 700 m (2300 ft) contributed to an average rise in sea level of about 0.5 mm/yr (0.02 in/yr). Although recent warming is greatest in the upper ocean, observations[48] also indicate that the deep ocean below 700 m is warming.

[41] N.L. Bindoff, J. Willebrand, V. Artale, et al., "Observations: Oceanic Climate Change and Sea Level." In S. Solomon, D. Qin, M. Manning, et al. (eds.), *Contribution of Working Group I to the Fourth Assessment Report (AR4) of the Intergovernmental Panel on Climate Change* (Cambridge, U.K., Cambridge University Press, 2007).

[42] S. Levitus, T. Boyer, J. Antonov, H. Garcia, and R. Locarnini, "Ocean Warming 1955–2003." Poster presented at the U.S. Climate Change Science Program Workshop, November 14–16, 2005, Arlington Va., Climate Science in Support of Decision-Making.

[43] J. M. Lyman, S. A. Good, V. V. Gouretski, et al., "Robust Warming of the Global Upper Ocean," *Nature* 465 no. 7296 (2010): 334–337, doi: 10.1038/nature09043.

[44] C. A. Katsman and G. J. van Oldenborgh, "Tracing the Upper Ocean's 'Missing Heat,'" *Geophysical Research Letters* 38 (2011): L14610, doi: 10.1029/2011GL048417.

[45] N. G. Loeb, J. M. Lyman, G. C. Johnson, et al., "Observed Changes in Top-of-the-Atmosphere Radiation and Upper-Ocean Heating Consistent within Uncertainty," *Nature Geoscience* (2012), doi: 10.1038/ngeo1375.

[46] S. Levitus, J. Antonov, T. Boyer, et al, "World Ocean Heat Content and Thermosteric Sea Level Change (0–2000), 1955–2010," *Geophysical Research Letters* 39 (2012): L10603 doi: 10.1029/2012GL051106.

[47] C. M. Domingues, J. A. Church N. J. White, et al., "Improved Estimates of Upper-Ocean Warming and Multi-decadal Sea-Level Rise," *Nature* 453 (2008): 1090–1093, doi: 10.1038/nature07080.

[48] G. C. Johnson and S. C. Doney, "Recent Western South Atlantic Bottom Water Warming," *Geophysical Research Letters* 33 (2006): L14614, doi: 10.1029/2006GL026769. See also G. C. Johnson, S. Mecking, B. M. Sloyan, and S. E. Wijffels, "Recent Bottom Water Warming in the Pacific Ocean," *Journal of Climate*, 13 (2007): 2987–3002; Y. T. Song and F. Colberg, "Deep Ocean Warming Assessed from Altimeters, Gravity Recovery and Climate Experiment, *in situ* Measurements, and a non-Boussinesq Ocean General Circulation Model," *Journal of Geophysical Research* 116 (2011): C02020, doi:10.1029/2010JC006601.

Figure 5.8. This sea-surface temperature map was produced using data from MODIS (Moderate Resolution Imaging Spectroradiometer, a satellite operated by NASA). The data were acquired daily over the whole globe. The red pixels show warmer surface temperatures, yellows and greens are intermediate values, and blue pixels show cold water.
Source: Figure from NASA: http://visibleearth.nasa.gov/view.php?id=54229.

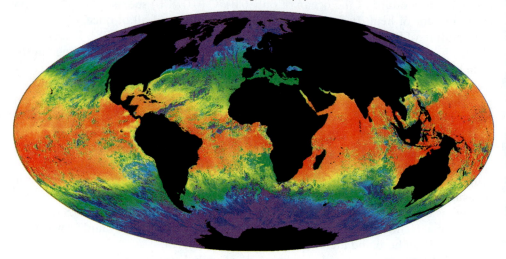

One study[49] combined observations and modeling to conclude that deep ocean warming might have contributed 1.1 mm/yr to the global mean sea-level rise, or one third of the altimeter-observed rate of 3.11 ± 0.6 mm/yr (0.122 ± 0.02 in/yr) over the period 1993–2008. In the IPCC AR4,[50] researchers calculated that thermal expansion of ocean water is responsible for an average 5 mm per decade (0.2 in per decade) of sea-level rise over the 20th century, compared to 18 mm per decade (0.7 in per decade) in the first decade of the 21st century.

According to AR4, global ocean temperature has increased by 0.1°C (0.18°F) from 1961 to 2003 from the surface to a depth of 700 m (2300 ft). Ocean heat content has increased over the upper 3000 m (9842 ft) over the same period, equivalent to absorbing a heating of 0.21 ± 0.04 W/m². During the course of global warming, the oceans have absorbed 90% of the extra heat added to the climate system.

Studies indicate that the temperature of the world's oceans have been trending upward for more than 100 years. In an innovative analysis of temperature records, researchers[51] compared the modern temperature of the ocean as measured by a global deployment of 3000 free-floating probes (the Argo Array[52]) with 300 measurements taken during the historic global voyage of the HMS Challenger (1872–1876), the first systematic exploration of the seas. The study shows a mean warming of the ocean surface of 0.59°C ± 0.12°C (1.06°F ± 0.22°F) over the past century. Below the surface, the mean warming decreases to 0.39°C ± 0.18°C (0.7°F ± 0.32°F) at 366 m (1200 ft) and 0.12°C ± 0.07°C (0.22°F ± 0.13°F) at 914 m (3000 ft). The 0.33°C ± 0.14°C (0.59°F ± 0.25°F) average temperature difference from 0 to 700 m (2300 ft) is twice the value that has been observed globally in that depth

[49] Y. T. Song and F. Colberg, "Deep Ocean Warming Assessed from Altimeters, Gravity Recovery and Climate Experiment, in situ Measurements, and a non-Boussinesq Ocean General Circulation Model."

[50] AR4 (2007), http://www.wmo.int/pages/partners/ipcc/index_en.html (accessed July 12, 2012).

[51] D. Roemmich, W. Gould, and J. Gilson, "135 Years of Global Ocean Warming between the Challenger Expedition and the Argo Programme," *Nature Climate Change* 2, no. 6 (2012): 425–428, doi: 10.1038/nclimate1461.

[52] See the ARGO home page http://www.argo.ucsd.edu/index.html (accessed July 12, 2012).

range over the past 50 years by previous studies,[53] implying a centennial timescale for the present rate of global warming. Warming in the Atlantic Ocean is stronger than in the Pacific.

It is interesting to compare sea-level rise today to conditions during the last interglacial period 120,000 to 130,000 years ago. Although average sea-surface temperatures were only about 0.7°C (1.3°F) above those of the present, researchers believe that global average sea level at that time was several meters higher than today. By analyzing[54] geologic proxies of global sea-surface temperature during the last interglacial, and comparing the data to results of global climate models simulating ocean temperatures over a 200-year period, investigators were able to calculate the contributions to sea-level rise from thermal expansion of seawater and from the melting of Greenland and Antarctica. The study revealed that the thermal expansion component of last interglacial sea-level rise was small, contributing no more than 40 cm (15.7 inches) to global sea level during the two-century period; Antarctic ice sheets must have contributed 2.8 to 4.5 m (9.2 to 14.7 ft) of sea-level rise; and polar ice sheets may be sensitive to small changes in global temperature.

These results have implications for what we can expect in our own warmer world. The study suggests that even small amounts of warming today might have committed us to more ice sheet melting than we previously thought. The ocean temperature during the last interglacial wasn't that much warmer than it is today, yet sea level at the time peaked several meters higher than present. If the research is correct, it indicates that even if we stopped greenhouse gas emissions right now, the troposphere would keep warming, the oceans would keep warming, the ice sheets would keep shrinking, and global sea level would keep rising for a long time. Researchers have concluded that the climate system must experience a series of time lags; greenhouse gas buildup leads to atmospheric warming, ocean warming lags behind the atmosphere, ice melting lags further still, and last is thermal expansion. Authors of the study argue[55] that their work makes the case that humans, by warming the atmosphere and oceans, are pushing Earth's climate toward a threshold where we will be committed to at least 4 to 6 m (13–20 ft) of sea-level rise in coming centuries, with the bulk of the water coming from the melting of the great polar ice sheets.

Melting Ice

Excess heat in Earth's climate system produces thermal expansion of seawater, and it leads to melting of glaciers and sea ice, a decrease in the extent of snow cover, and shifts between snowfall and rainfall. Glacier and snowmelt contribute to sea-level rise, especially from Greenland and Antarctica, the two largest ice-covered regions on the planet. Both of these locations are experiencing accelerating melting (Figure 5.9).[56]

If current rates of ice loss from Greenland and Antarctica continue to mid-century,[57] their combined loss could raise global mean sea level by 15 cm (5.9 in). When added to the predicted sea-level rise due to thermal expansion of seawater (9 cm [3.5 in]) and melting of glacial ice caps (8 cm [3.1 in]), total sea-level rise could reach about 32 cm (12.6 in) by 2050.

[53] S. Levitus, J. I. Antonov, T. P. Boyer, et al., "Global Ocean Heat Content 1955–2008 in Light of Recently Revealed Instrumentation Problems," *Geophysical Research Letters* 36 (2009): L07608.

[54] N. P. McKay, J. T. Overpeck, and B. L. Otto-Bliesner, "The Role of Ocean Thermal Expansion in Last Interglacial Sea-Level Rise," *Geophysical Research Letters* 38 (2011): L14605, doi: 10.1029/2011GL048280.

[55] See http://uanews.org/node/40694 (accessed July 12, 2012).

[56] I. Velicogna, "Increasing Rates of Ice Mass Loss from the Greenland and Antarctic Ice Sheets Revealed by GRACE," *Geophysical Research Letters* 36, (2009): L19503, doi: 10.1029/2009GL040222.

[57] E. Rignot, I. Velicogna, M. R. van den Broeke, A. Monaghan, and J. Lenaerts, "Acceleration of the Contribution of the Greenland and Antarctic Ice Sheets to Sea-Level Rise."

Figure 5.9. Total ice sheet mass balance between 1992 and 2010 for Greenland **(a),** Antarctica **(b),** and the sum of Greenland and Antarctica **(c)** in gigatons per year.[58] Two data types are graphed: The black line and uncertainty (grey) shows the mass balance method,[59] a calculation of glacier discharge, meltwater production, snowfall, and other aspects of changing annual ice mass; the red line and uncertainty (pink) show the GRACE satellite[60] gravity measurement of total ice mass. The acceleration rate in ice sheet mass balance, in gigatons per year squared, is determined from a linear fit of the mass balance method over 18 years (black line) and GRACE data over 8 years (red line). The rate of loss of both ice sheets, plus the rate of thermal expansion of seawater, could produce approximately 32 cm of sea-level rise by mid-century.

SOURCE: Figure after E. Rignot, I. Velicogna, M. R. van den Broeke, A. Monaghan, and J. Lenaerts, "Acceleration of the Contribution of the Greenland and Antarctic Ice Sheets to Sea-Level Rise," *Geophysical Research Letters* 38 (2011): L05503, doi: 10.1029/ 2011GL046583.

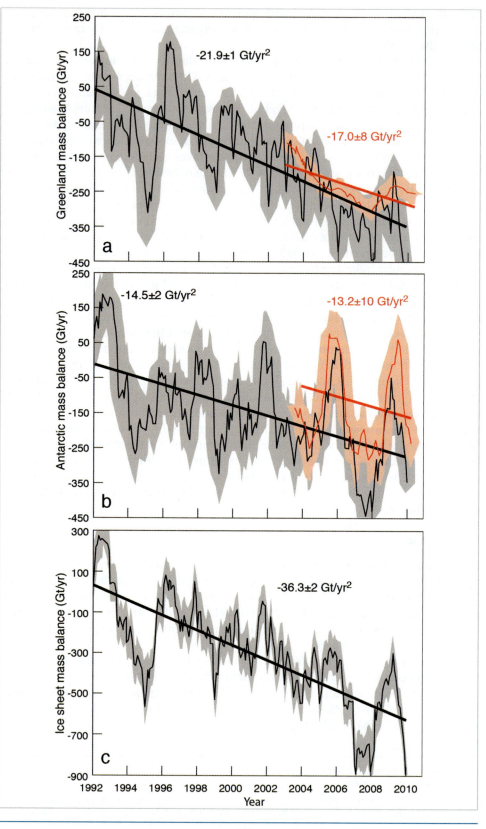

[58] E. Rignot, I. Velicogna, M. R. van den Broeke, A. Monaghan, and J. Lenaerts, "Acceleration of the Contribution of the Greenland and Antarctic Ice Sheets to Sea-Level Rise," *Geophysical Research Letters* 38 (2011): L05503, doi: 10.1029/2011GL046583.

[59] M. J. van den Broeke, J. Bamber, J. Ettema, et al., "Partitioning Recent Greenland Mass Loss," *Science* 326, no. 5955 (2009): 984–986, doi: 10.1126/science.1178176.

[60] See the GRACE homepage at http://www.csr.utexas.edu/grace/ (accessed July 12, 2012).

Antarctica consists of three main geographic regions: the Antarctic Peninsula, West Antarctica, and East Antarctica. In West Antarctica, which has warmed 0.17°C per decade (0.3°F per decade) at the same time that global warming was about 0.11°C per decade (0.2°F per decade), ice loss has increased by 59% in the early 21st century to more than 145 billion tons per year.[61] The yearly loss along the Antarctic Peninsula has increased by 140% to more than 66 billion tons. The East Antarctic ice sheet, by far the largest region of the continent, is also melting. It is losing mass at the rate of approximately 57 Gt (gigatons) per year, apparently caused by increased ice loss since 2006.[62] The East Antarctic ice sheet is experiencing melting along the coastal margin in warming seas and snow accumulation in the hinterlands. Overall, the entire continent of Antarctica is experiencing net melting[63]; all three regions of Antarctica are warming, and the overall rate of ice loss in Antarctica increased by 75% in the past 10 years.[64, 65]

Glaciers that flow into the sea around Antarctica, Greenland, and Canada can form thick (100 to 1000 m; 330 to 3300 ft) floating platforms of ice called ice shelves. Ice shelves constitute the seaward extension of grounded glacier ice. Glaciologists have hypothesized that ice shelves slow the advance of glaciers into the ocean and that when ice shelves melt or fracture, there is a possibility that the adjoining glacier can accelerate. This hypothesis has now been proved correct with direct measurements[66] made by laser altimetry on NASA's ICESat satellite. Additionally, researchers have learned that 20 of the 54 ice shelves studied around Antarctica are experiencing melting by warm ocean currents on their undersides, and as a result adjoining grounded glaciers have accelerated their rate of flow. Data indicate that this form of melting is the dominant cause of recent ice loss from the continent (Figure 5.10). Melting is dramatic in some cases, with some shelves thinning by a few meters per year leading to billions of tons of ice draining into the sea.

The contribution of the Greenland (Figure 5.11) continental glacier to sea-level rise has also been measured. Increased melting of the Greenland ice sheet has been observed, and it is known that the glacier is getting smaller.[67] The balance between annual ice gained and lost is in deficit, and the deficiency tripled between 1996 and 2007.[68] In Greenland, 2007 marked a rise to record levels of the summertime melting trend over the highest altitudes of the ice sheet. Melting in areas above 2000 m (6,560 ft) rose 150% above the long-term average, with melting occurring on 25 to 30 more days in 2007 than the average in the previous 19 years.[69] Crevasses, fractures in the ice that promote sliding, melting, and faster movement, have been seen to grow in some areas over the past two decades, suggesting that mechanical processes are speeding up ice movement into the ocean.[70] Scientists have found that glaciers in

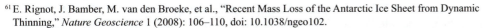

[61] E. Rignot, J. Bamber, M. van den Broeke, et al., "Recent Mass Loss of the Antarctic Ice Sheet from Dynamic Thinning," *Nature Geoscience* 1 (2008): 106–110, doi: 10.1038/ngeo102.

[62] J. L. Chen, C. R. Wilson, D. Blankenship, and B. D. Tapley, "Accelerated Antarctic Ice Loss from Satellite Gravity Measurements," *Nature Geoscience* 2 (2009): 859–862, doi: 10.1038/NGEO694.

[63] E. J. Steig, D. P. Schneider, D. R. Scott, et al., "Warming of the Antarctic Ice-Sheet Surface Since the 1957 International Geophysical Year," *Nature* 457 (2009): 459–462.

[64] E. Rignot, J. Bamber L. Van Den Broeke, et al., "Recent Antarctic Ice Mass Loss from Radar Interferometry and Regional Climate Modeling," *Nature Geoscience* 1, no. 2 (2008): 106–110. doi: 10.1038/ngeo102.

[65] See the animation "Antarctic Ice Flows: A Complete Picture" at the end of the chapter.

[66] H. Pritchard, S. Ligtenberg, H. Fricker, et al., "Antarctic Ice-Sheet Loss Driven by Basal Melting of Ice Shelves," *Nature* 484, no. 7395 (2012): 502, doi: 10.1038/nature10968.

[67] E. J. O. Schrama and B. Wouters, "Revisiting Greenland Ice Sheet Mass Loss Observed by GRACE," *Journal of Geophysical Research* 116 (2011): B02407, doi: 10.1029/2009JB006847.

[68] E. Rignot, J. E. Box, E. Burgess, and E. Hanna, "Mass Balance of the Greenland Ice Sheet from 1958 to 2007," *Geophysical Research Letters* 35 (2008): L20502, doi: 10.1029/2008GL035417.

[69] NASA, "Earth Observatory, Melting Anomalies in Greenland in 2007," http://earthobservatory.nasa.gov/Newsroom/NewImages/images.php3?img_id=17846 (accessed July 12, 2012).

[70] W. Colgan, K. Steffen, W. S. McLamb, et al., "An Increase in Crevasse Extent, West Greenland: Hydrologic Implications," *Geophysical Research Letters* 38 (2011): L18502, doi: 10.1029/2011GL048491.

Figure 5.10. a, Melting of Antarctic ice shelves (rainbow color) is dominated by warm ocean currents sweeping along their undersides[71]; red is thicker ice (greater than 550 m [1,800 ft]), and blue is thinner ice (less than 200 m [656 ft]). **b,** Ice shelves not only melt on their undersides, they can fracture and release large icebergs into the sea. In mid-October 2011, NASA scientists working in Antarctica discovered a massive crack across the Pine Island Glacier,[72] a major ice stream that drains the West Antarctic Ice Sheet. Extending for 30 km (19 mi), the crack was 80 m (260 ft) wide and 60 m (195 ft) deep. Eventually the crack will extend all the way across the glacier and calve a giant iceberg that will cover about 900 km² (350 mi²).

IMAGE CREDIT: Fig. 5.10a: NASA/Goddard CGI Lab. Fig. 5.10b: NASA Earth Observatory image created by Jesse Allen, using data provided courtesy of NASA/GSFC/METI/ERSD AC/JAROS, and U.S./Japan ASTER Science.

(a)

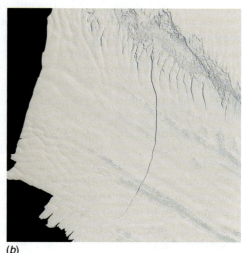

(b)

southern Greenland are melting faster than they were 10 years ago, and the overall amount of ice discharged into the sea has increased from 20 km³ (5 mi³) in 1996 to 54 km³ (13 mi³) in 2005, an increase of 25%.

In 2010 the melting of Greenland ice broke previous records,[73] with melting in some regions extending up to 50 days longer than average. Ice melting on Greenland also spread to previously stable portions of the northwest coast.[74] Researchers reported[75] that melting in 2010 started exceptionally early at the end of April and ended quite late in mid-September; summer temperatures up to 3°C (5.4°F) above average were combined with reduced snowfall; and Nuuk, the capital of Greenland, had the warmest spring and summer since records began in 1873.

How vulnerable is the Greenland ice sheet to melting? Research suggests it may be more susceptible than previously assumed, and melting, at a certain point, becomes irreversible. A model simulation of Greenland ice in a warmer world[76] reveals that the temperature threshold for melting the ice sheet completely is in the range of 0.8°C to 3.2°C (1.4°F to 5.7°F) of global warming, with a best estimate of 1.6°C (2.8°F) above preindustrial levels. Warming has already reached the minimum of this range (0.8°C [1.4°F]), and the world is on track to double this amount sometime in this century. The time it takes before most of the ice in Greenland is lost strongly depends on the level of warming. The more we exceed the threshold, the faster it melts. According to the study, in a business-as-usual scenario of greenhouse-gas emissions, in the long

[71] See discussion, http://www.nasa.gov/topics/earth/features/currents-ice-loss.html (accessed July 12, 2012).

[72] See discussion, http://earthobservatory.nasa.gov/NaturalHazards/view.php?id=76408 (accessed July 12, 2012).

[73] M. Tedesco, X. Fettweis, M. R. van den Broeke, et al., "The Role of Albedo and Accumulation in the 2010 Melting Record in Greenland," *Environmental Research Letters* 6 (2011): 014005, doi: 10.1088/1748-9326/6/1/014005.

[74] S. A. Khan, J. Wahr, M. Bevis, I. Velicogna, and E. Kendrick, "Spread of Ice Mass Loss into Northwest Greenland Observed by GRACE and GPS," *Geophysical Research Letters* 37 (2010): L06501, doi: 10.1029/2010GL042460.

[75] See article, http://www.sciencedaily.com/releases/2011/01/110121144011.htm (accessed July 12, 2012).

[76] A. Robinson, R. Calov, and A. Ganopolski, "Multistability and Critical Thresholds of the Greenland Ice Sheet," *Nature Climate Change* 2 (2012): 429–432, doi: 10.1038/NCLIMATE1449.

Figure 5.11. a, Analysis of satellite and weather station data has shown that Antarctica[77] has warmed at a rate of about 0.12°C (0.22°F) per decade since 1957, for a total average temperature rise of 0.5°C (1°F). **b,** Greenland ice sheet 2011. The map shows where surface melt in 2011 was detected by satellites on more (orange) or fewer (blue) days than the 1979–2010 average. White indicates no difference from average or changes too small to be detected by the satellite. Depending on the data analysis approach, 2011 was either the third most extensive or the sixth most extensive melting year since satellite records began in 1979. Melting was exceptionally high over the western mid-elevations, and the map shows the area swathed in orange. In some places, the melt season lasted up to 30 days longer than average, and it affected 31% of the ice sheet surface, making 2011 one of just three years since 1979 where melt area exceeded 30%.

Source: a from http://earthobservatory.nasa.gov/IOTD/view.php?id=36736; **b** from http://earthobservatory.nasa.gov/IOTD/view.php?id=76596 (accessed July 12, 2012).

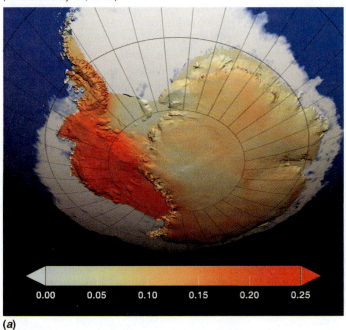

(a)

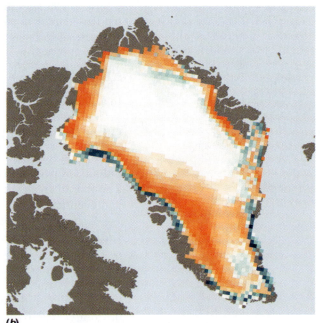

(b)

run humanity might be aiming at 8°C (14.4°F) of global warming. This would result in one fifth of the ice sheet melting within 500 years and a complete loss in 2000 years. Under certain conditions the melting of the ice sheet becomes irreversible because of climate feedbacks. For instance, Greenland's ice is more than 3,000 m (9,842 ft) thick, and much of its surface is located at cooler high altitudes. Prolonged melting will lower the surface of the ice to warmer altitudes, preventing it from rebuilding again. Also, the loss of sunlight-reflecting ice cover and its replacement with heat-absorbing seawater and dark rock will prevent the ice sheet from future growth, even if climate returned to its preindustrial state. Melting the total ice on Greenland would result in a sea-level rise of about 6.5 m[78] (21 ft) and affect many of the world's major cities, which are located on coastlines because of historical ties to shipping.

Greenland's contribution to average sea-level rise is accelerating: Ice losses quickened in 2006–2008 to the equivalent of 0.75 mm/yr (0.03 in/yr) of sea-level rise, from an average 0.46 mm/yr (0.018 in/yr) for 2000–2008.[79] Icebergs breaking away and meltwater runoff are equally to blame for the shrinking ice sheet. Greenland contributions account for between 20% and 38% of the observed yearly global sea-level rise.[80] As glacier acceleration continues to spread northward from its current

[77] E. J. Steig, D. P. Schneider, D. R. Scott, et al., "Warming of the Antarctic Ice-Sheet Surface since the 1957 International Geophysical Year."

[78] R. Z. Poore, R. S. Williams, Jr., and C. Tracey, "Sea Level and Climate: U.S. Geological Survey Fact Sheet 002–00, (2000), http://pubs.usgs.gov/fs/fs2-00/.

[79] M. van den Broeke, J. Bamber, J. Ettema, et al., "Partitioning Recent Greenland Mass Loss," *Science* 326, no. 5955 (2009): 984, doi: 10.1126/science.1178176.

[80] E. Rignot and P. Kanagaratnam, "Changes in the Velocity Structure of the Greenland Ice Sheet," *Science* 311 (2006): 986–989.

focus in southern Greenland, the global sea-level rise contribution from the world's largest island will continue to increase.

Many glaciers draining the Greenland ice sheet end in the ocean. These are especially vulnerable to melting as ocean water warms and causes them to retreat. Although several marine-terminating glaciers are known to have retreated over the past decade, the extent and magnitude of retreat relative to past history was unknown until a study[81] in mid-2011. Scientists used changes in the front positions of 210 marine-terminating glaciers using Landsat imagery spanning nearly four decades, and they compared rates of change with earlier observations to the early 20th century. They found that 90% of the observed glaciers retreated between 2000 and 2010, approaching 100% in the northwest, with rapid retreat observed throughout the entire ice sheet. The retreat today is accelerating and likely began between 1992 and 2000, which was the onset of warming in the region. Previously, during the middle of the 20th century, glaciers were largely stable, and they even advanced, coincident with a cooling period. The early 20th century was warm, and although there was extensive glacier retreat at that time, the current retreat is more widespread.

Other Components of Sea-Level Rise

Retreating mountain glaciers are also contributing to sea-level rise. In fact for the millions of people in communities that depend on seasonal snow and ice melting as a source of freshwater (Los Angeles, San Francisco, and many others), the retreat and eventual loss of these ice centers delivers a fundamental blow to sustainability.

Using a satellite mission that detects changes in the gravitational field on Earth's surface, researchers have completed the first comprehensive study[82] of the contribution of the world's melting glaciers and ice caps to global sea-level rise. The study concluded that Earth's glaciers and ice caps outside of the regions of Greenland and Antarctica are shedding roughly 150 billion tons (or 163 km^3; 39 mi^3) of ice annually, and this is sufficient to raise global mean sea level approximately 0.4 mm/yr (0.016 in/yr).

The data (Figure 5.12) were collected by a satellite named Gravity Recovery and Climate Experiment, or GRACE, a joint effort of space programs in NASA and Germany. The GRACE mission consists of two satellites (launched in 2002) that orbit Earth together 16 times a day at an altitude of about 482 km (300 mi). Traveling together, the two satellites sense subtle variations in Earth's mass and gravitational pull caused by regional changes in the planet's mass, including ice sheets, oceans, and water stored in the soil and in underground aquifers. According to the GRACE data, total sea-level rise from all land-based ice on Earth including Greenland and Antarctica was roughly 1.5 mm/yr (0.06 in/yr) annually from 2003 to 2010.

Warming temperatures lead to the melting of alpine glaciers, and the total volume of glaciers on Earth is declining sharply.[83] Glaciers have been retreating worldwide for at least the last century, and the rate of retreat has increased in the past decade. Only a few glaciers are actually advancing (in locations that are well below freezing and where increased precipitation has outpaced melting). The progressive disappearance of glaciers has implications not only for a rising global sea level but also for water supplies in many communities, both large and small.

Mountain glaciers are retreating and thinning in nearly every mountainous region of the planet. For instance, over the period 2007–2009 a sharp increase in the rate of ice mass loss due to melting made the Canadian Arctic Archipelago the single largest contributor to global sea-level rise outside of Greenland and

---

[81] I. M. Howat and A. Eddy, "Multi-decadal Retreat of Greenland's Marine-Terminating Glaciers," *Journal of Glaciology* 57, no 203 (2011): 389–396.

[82] T. Jacob, J. Wahr, W. T. Pfeffer, and S. Swenson, "Recent Contributions of Glaciers and Ice Caps to Sea-Level Rise," *Nature* 482, no. 7386 (2012): 514–518, doi: 10.1038/nature10847.

[83] See NOAA website at http://www.ncdc.noaa.gov/indicators/ (accessed July 12, 2012).

Figure 5.12. As atmospheric temperatures have risen, the total volume of Earth's glacier ice has declined sharply.[84] This map shows changes in ice thickness (cm/yr) during 2003–2010 as measured by NASA's Gravity Recovery and Climate Experiment (GRACE) satellites, averaged over each of the world's ice caps and glacier systems outside of Greenland and Antarctica. Blue represents ice mass loss, red represents ice mass gain.

Source: http://www.nasa.gov/topics/earth/features/grace20120208.html (accessed July 12, 2012).

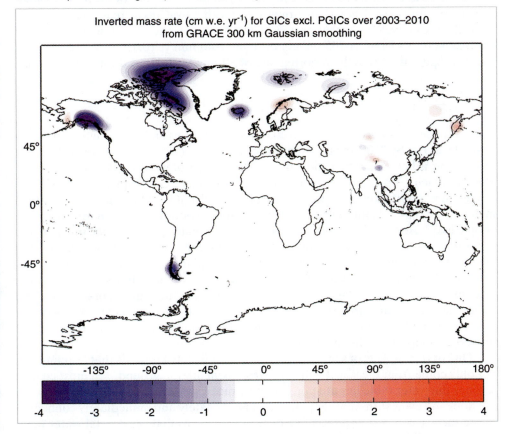

Antarctica.[85] Researchers have documented that melting there was due largely to warmer summertime temperatures, to which rates of ice loss are highly sensitive.

The cumulative mean thinning of the world's mountain glaciers has accelerated from about −1.8 to −4 m (−6 to −13 ft) between 1965 and 1970 to about −12 to −14 m (−40 to −46 ft) of change in the first decade of the 21st century.[86] Over the period 1961–2003, mountain glaciers contributed an estimated 0.5 mm/yr (0.02 in/yr) to global sea-level rise, increasing to 0.8 mm/yr (0.03 in/yr) for the period 1993–2003.[87]

Another important source of sea-level rise is groundwater extraction. As human population increases and the use of water for manufacturing, agriculture, and all types of industrial and domestic purposes continues, the natural renewal rate of groundwater stores has not been able to keep pace with the rate of human use. The vast majority of water drawn from the ground is ultimately released to become

[84] M. F. Meier, M. B. Dyurgerov, U. K. Rick, et al., "Glaciers Dominate Eustatic Sea-Level Rise in the 21st Century," *Science* 317, no. 5841 (2007): 1064–1067.

[85] A. S. Gardner, G. Mohodt, B. Wouters, et al., "Sharply Increased Mass Loss from Glaciers and Ice Caps in the Canadian Arctic Archipelago," *Nature* 473 (2011): 357–360.

[86] M. F. Meier, M. B. Dyurgerov, U. K. Rick, et al., "Glaciers Dominate Eustatic Sea-Level Rise in the 21st Century." See also WWF Nepal Program, *An Overview of Glaciers, Glacier Retreat, and Subsequent Impacts in Nepal, India and China* (Katmandu, World Wildlife Fund, 2005), assets.panda.org/downloads/himalayaglaciersreport2005.pdf (accessed July 12, 2012).

[87] M. B. Dyurgerov and M. F. Meier, *Glaciers and the Changing Earth System: A 2004 Snapshot* (Occasional Paper 58) (Boulder, Colo., Institute of Arctic and Alpine Research, University of Colorado, 2005), http://instaar.colorado.edu/other/occ_papers.html (accessed July 12, 2012).

runoff to the sea. Hence, human withdrawal of groundwater has become a small but measureable component of global sea-level rise.

Using calibrated hydraulic models, analysis of observational data, and inferences about human use, groundwater depletion during the period 1900–2008 has been estimated[88] to equal approximately 4,500 km³ (1,080 mi³). Researchers calculate that this is equivalent to a sea-level rise of 12.6 mm (0.5 in), or more than 6% of the total sea-level rise of the entire period; however, groundwater withdrawal has increased substantially since 1950, and over the recent period (2000–2008) the rate of withdrawal averaged approximately 145 km³/yr (35 mi³/yr) or about 0.40 mm/yr (0.016 in/yr) of sea-level rise. This figure is 13% of the reported rate of 3.1 mm/yr (0.12 in) during the period.

SEA LEVEL BY THE END OF THE CENTURY

Sea-level rise presents challenges to coastal communities and ecosystems. Estimates of sea-level rise by 2100, for a temperature rise of 4°C (7.2°F) or more over the same time frame, is between 0.5 m and 2 m (1.64 to 6.6 ft); potentially a devastating result globally placing up to 187 million people at risk of forced displacement.[89] In the United States approximately 32,000 km² (12,355 mi²) of land lies within one vertical meter of the high tide line, encompassing 2.1 million housing units where 3.7 million people live.[90] Hence it is important that community managers, resource officials, and other decision makers and community groups concerned with natural hazards and environmental conservation begin the process of planning for the impacts of sea-level rise.[91] Because sea level is rising now, and it is very likely to rise at accelerated rates in the future,[92] coastal communities cannot avoid the impacts of sea-level rise by mitigating global warming. Some amount of sea-level rise has already been set into unstoppable motion, and it is thus incumbent upon communities to begin the process of adapting to the inevitable impacts.

One of the impacts of sea-level rise that is not widely known is that flooding will not only come from the sea, it will come from beneath the ground.[93] In most coastal settings the water table lies at approximately mean sea level and rises and falls with the tides and even with weather patterns; it is intimately and immediately connected to the surface of the sea. Thus, a rise in sea level means that the coastal water table will rise, possibly to the point of breaking through the ground surface and creating new wetlands, not a desirable feature in an urban setting at the foundation of a building or road, or in an ecosystem that serves as a refuge for endangered species that are unaccustomed to saturated soil and free-standing water bodies.

Figure 5.13 is a model of sea-level vulnerability of the central urban core of Honolulu, Hawaii. Mapped in blue is the extent of seawater at high tide toward the end of the century when sea level is projected to be in the vicinity of 0.9 m (3 ft)

[88] L. F. Konikow, "Contribution of Global Groundwater Depletion since 1900 to Sea-Level Rise," *Geophysical Research Letters* 38 (2011): L17401, doi: 10.1029/2011GL048604.

[89] R. Nicholls, N. Marinova, J. Lowe, et al., "Sea-Level Rise and its Possible Impacts Given a 'Beyond 4C World' in the Twenty-First Century," *Philosophical Transactions of the Royal Society A* 369 (2011): 161–181 doi: 10.1098/rsta.2010.0291.

[90] C. Tebaldi, B. Strauss, and C. Zervas, "Modeling Sea-Level Rise Impacts on Storm Surges Along US Coasts," *Environmental Research Letters* 7 (2012) 014032, doi: 10.1088/1748-9326/7/1/014032. See also B. Strauss, R. Ziemlinski, J. Weiss, J. Overpeck, "Tidally Adjusted Estimates of Topographic Vulnerability to Sea-Level Rise and Flooding for the Contiguous United States," *Environmental Research Letters* 7 (2012): 014033, doi: 10.1088/1748-9326/7/1/014033.

[91] S. Rahmstorf, "Sea-Level Rise: Towards Understanding Local Vulnerability," *Environmental Research Letters* 7 (2012) 021001, doi: 10.1088/1748-9326/7/021001.

[92] J. E. Hansen, "Scientific Reticence and Sea-Level Rise," *Environmental Research Letters* 2 (2007), 024002, doi: 10.1088/1748-9326/2/2/024002.

[93] D. M. Bjerklie, J.R. Mullaney, J.R. Stone, B.J. Skinner, and M.A. Ramlow, *Preliminary Investigation of the Effects of Sea-Level Rise on Groundwater Levels in New Haven, Connecticut*, U.S. Geological Survey Open-File Report 2012–1025, (2012), at http://pubs.usgs.gov/of/2012/1025/ (accessed July 12, 2012).

Figure 5.13. a, Vulnerability of the Honolulu region from sea-level rise of 0.9 m (3 ft). Blue indicates sea level 0.9 m above present high tide; red indicates groundwater inundation when sea level is 0.9 m above present high tide. **b,** Sea-level rise will likely cause increased coastal erosion, saltwater intrusion, drainage problems, marine inundation, and flooding when it rains. Coastal communities can begin adapting[94] to these problems now if they want to avoid the worst and costliest damage.

(a)

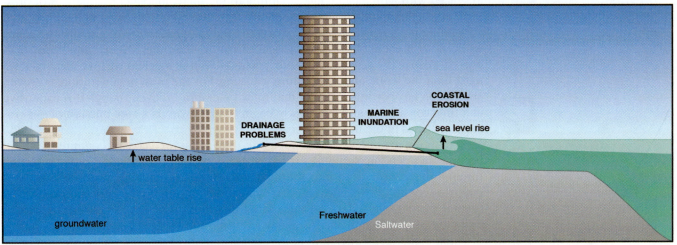

(b)

above present. Mapped in red are lands that will be flooded by the water table under that scenario. These lands in red have an elevation that is below high tide when local sea level is 90 cm (3 ft) above present. They are not connected to the ocean and their inundation is a result of seawater flowing out of storm drainage pipes along road sides, the water table breaking through the land surface to create wetlands, and the collection of runoff into pools because it cannot drain into the ground or the ocean. In non-storm conditions, these lands are more vulnerable to the rise of the water table as a main cause of flooding than direct marine inundation. An aspect of a high water table is that rainfall and runoff of all types will experience restricted drainage due to the saturation of the ground surface by groundwater and the inundation of the storm drainage system by seawater.

Adapting to Sea-Level Rise

Following an analysis[95] and calculation of global sea level projections to the end of the century, study leader Dr. Anthony Dalrymple stated "There will be about 1 meter of sea level rise by 2100." As you might imagine, for a human community located on the edge of the sea, the prospect of 1 m (3.3 ft) or more of sea-level rise represents

---

[94] D. Marcy, A. Allen, W. Sweet, et al., "Incorporating Sea Level Change Scenarios at the Local Level," NOAA Coastal Services Center (2012), http://www.csc.noaa.gov/digitalcoast/_/pdf/slcscenarios.pdf (accessed July 12, 2012).

[95] Committee on Sea Level Rise in California, Oregon, and Washington; Board on Earth Sciences and Resources; Ocean Studies Board; Division on Earth and Life Studies; National Research Council, *Sea Level Rise for the Coasts of California, Oregon, and Washington: Past, Present, and Future* (2012), National Academies Press, Washington, D.C.

a significant challenge, possibly affecting as many as 3.7 million people.[96] As a result of sea-level rise, marine inundation from storm surge and tsunamis will increase in severity and frequency, coastal erosion and flooding will increase, and ecosystems near intertidal elevations will be affected. Studies[97] estimate that by 2050, one third of coastal communities in the United States will see an increase in the frequency of extreme high water levels that currently only happen once per century. Small storms that previously had little impact on the coast will begin, over time, to cause greater damage. Buildings and roads located at the water's edge will experience wave-related flooding and structural damage, placing demands on emergency workers, public works crews, and utility companies. Tourism and private land ownership will be threatened as beach erosion spreads and increases in severity. Ecosystems, human communities, infrastructure, ports and harbors, and other coastal assets will all come under increasing attack during the course of the 21st century. Lands previously dry throughout a tidal cycle will experience flooding at high tide, even in the absence of storms.[98]

Because it is too late to stop the global warming happening today, communities are faced with a dual challenge: adapting to the unavoidable consequences of present warming and mitigating the worst of future warming. Engaging in the process of adaptation is no reason to cease efforts to limit the production of greenhouse gases; working to prevent the worst impacts of future warming is more important than ever. If we love the generation of our grandchildren as much as we love ourselves, we must limit the worst effects of future warming. For the present generation, however, adaptation has become a reality that must be embraced even though it is expensive and time consuming and will slowly begin to take over as a core mission of government agencies in the next few decades. [99]

Adapting to sea-level rise is not a one-time event for a coastal community; it will become an unending series of expenses, decisions, and construction projects lasting through this century and into the next. Adaptation is, in other words, a process[100] requiring community participation; technical skill in geographic information systems (digital cartography); data on the spatial distribution of developed infrastructure, ecosystems, and the built environment; scientific knowledge to build future scenarios of climate change; leadership; and decision making. Fundamentally, a community must engage in a planning process leading to judgments about what assets are at risk and about what assets are to be protected, moved, or sacrificed.

The National Oceanographic and Atmospheric Administration (NOAA) has developed guidance[101] on this issue on behalf of coastal communities around the United States and elsewhere. They articulate a planning process that begins with identifying a planning team, scoping the level of effort, and educating and involving stakeholders. This early phase in sea-level adaptation requires leadership from a small number of knowledgeable individuals, an agency office, or an elected official. NOAA recommends that a new state law or executive order authorizing the climate change adaptation planning process would help ensure it has adequate resources, support, and legitimacy. This could require educating elected officials, which should be done early in the planning process.

[96] C. Tebaldi, B. Strauss, C. Zervas, "Modeling Sea-Level Rise Impacts on Storm Surges along US Coasts," *Environmental Research Letters* 7 (2012): 014032, doi: 10.1088/1748-9326/7/1/014032. See also B. Strauss, R. Ziemlinski, J. Weiss, J. Overpeck, "Tidally Adjusted Estimates of Topographic Vulnerability to Sea-Level Rise and Flooding for the Contiguous United States."

[97] C. Tebaldi, B. Strauss, and C. Zervas, "Modeling Sea-Level Rise Impacts on storm Surges along US Coasts."

[98] See the video "See How Climate Change and Rising Sea Levels Plague American Cities Right Now" at the end of the chapter.

[99] See the animation "Sea-Level Rise" at the end of the chapter.

[100] D. Marcy, A. Allen, W. Sweet, et al., "Incorporating Sea Level Change Scenarios at the Local Level," NOAA Coastal Services Center (2012), http://www.csc.noaa.gov/digitalcoast/_/pdf/slcscenarios.pdf (accessed July 12, 2012).

[101] National Oceanic and Atmospheric Administration, "Adapting to Climate Change: A Planning Guide for State Coastal Managers," http://coastalmanagement.noaa.gov/climate/adaptation.html (accessed July 12, 2012).

Once a planning team is assembled, stakeholders are involved, and responsibilities are established, it is important to assess community vulnerability to sea-level rise. This begins with a technical effort to determine the impacts of sea-level rise, such as accelerated coastal erosion, increased inundation due to tsunami and hurricane storm surge, and drainage problems related to tidal flooding of runoff infrastructure such as suburban storm sewer systems. A focused analysis involves modeling inundation, mapping high-resolution topography in the coastal zone, determining erosion patterns under high rates of sea-level rise, and other technical steps. Planners typically use a geographic information system of map layers depicting community assets, including vulnerable ecosystems, social phenomena, transportation assets, and other key features of the coastal zone. By this stage, the planning team will have identified the sea level–related physical processes likely to affect the coast, examined the associated impacts, and assessed what it is about the coast that could affect its vulnerability to climate change. Armed with this knowledge, the planning team can develop scenarios that illustrate (Figure 5.14) potential impacts and consequences of sea-level rise.

Figure 5.14. Under 1.2 m (4 ft) of sea-level rise, the tourist mecca of Waikiki, Hawaii, would be severely affected by seawater inundation at high tide, and the negative impacts of this would ripple throughout the tourism-based economy of the entire state of Hawaii. This image shows an analysis of sea-level rise impact on the built environment. Buildings are color-coded by vulnerability to higher sea level based on their elevation. Red indicates buildings located at modern high tide. Orange indicates buildings vulnerable to 1 ft (0.3 m) of sea-level rise. Yellow indicates buildings vulnerable to 2 ft (0.6 m) sea-level rise. Green indicates buildings vulnerable to 3 ft (0.9 m) sea-level rise. Purple indicates buildings vulnerable to 4 ft (1.2 m) of sea-level rise. The orange surface shows the uncertainty of flooding on the digital elevation model (topographic surface; uncertainty=0.69 ft at 95%).

IMAGE CREDIT: Figure by Perspective Cartographics, M. Barbee, Honolulu, Hawaii.

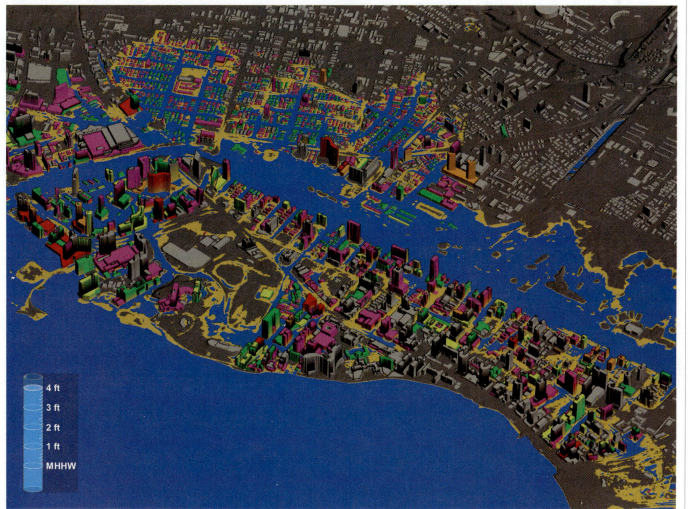

After assessing the vulnerability of community and ecosystem assets to sea-level rise, planners should define the steps needed to adapt to and mitigate negative impacts. This phase requires developing an adaptation strategy characterized by features such as reducing the vulnerability of the built environment to sea-level rise, monitoring and maintaining healthy coastal ecosystems, reducing the expense and building the capacity of disaster response and recovery, protecting critical infrastructure, minimizing economic losses attributable to the impacts of climate change, adapting to sea-level rise in a manner that minimizes harm to the natural environment and loss of public access, and others. Prioritizing these many steps, funding them, and executing the many projects that embody these goals require the development of an action plan that needs implementation and ongoing management and revision throughout the 21st century.

Some planners have begun the arduous process of assessing management options and developing new policies. For instance, state agencies in California have mapped the impact zone of a 1.4 m (4.6 ft) rise in sea level after having calculated that by the end of the century lands at this elevation and lower are vulnerable to negative impacts. They have identified the land and development that is vulnerable to inundation,[102] including 480,000 people, $100 billion in property, 140 schools, 34 police and fire stations, 55 health care facilities, 330 EPA hazardous waste sites, 3,500 miles of roads and highways, 280 miles of railroads, 30 power plants, 28 wastewater treatment plants, and more.[103] For each one of these assets a decision-making process must be engaged to decide on the specific adaptation strategy to be employed (even if it is to abandon the asset), a budget scoped and funded, and a construction project detailed and completed. Adaptation is, in short, a massive operation.

How High?

To properly design community adaptation strategies, it is desirable to have an estimate of sea-level rise this century. An approximation of sea-level rise by certain benchmarks during the course of this century will allow estimates of vulnerability to coastal hazards (such as increased risk from tsunamis, storm surge, and coastal erosion); assessments of flooding and drainage threats to coastal assets such as the built environment, coastal ecosystems, and others; development of climate risk-management policies; and development of urban planning and ecosystem conservation modeling scenarios.

Geologic observations shed light on the natural rate and magnitude of sea-level change. Researchers[104] have reconstructed sea-level fluctuations over the past 22,000 years, spanning the period from the last glacial maximum to the present interglacial warm phase (from about 10,000 years ago to the present). From this work it is apparent that changing climate, in the form of shifts in ice volume and global temperature (the kind of processes operating today), is responsible for driving sea-level changes over this period. Researchers who reconstructed the relationship between climate and sea level predict 4 to 24 cm of sea-level rise over the 20th century, in agreement with AR4 and other reports. When used to forecast sea-level heights over the 21st century on the basis of modeled temperature projections (1.1°C to 6.4°C) in the AR4, the reconstruction predicts 7 to 82 cm (2.8 to 32.3 in) of sea-level rise by the end of this century. Although this range overlaps with and exceeds the AR4 estimate of 18 to 59 cm (7.1–23.2 in),[105] researchers conclude it is sufficiently similar to increase confidence in the projections.

[102] See California Executive Order S-13-08 2008 at http://gov.ca.gov/executive-order/11036/ (accessed July 12, 2012).

[103] See the report by the Pacific Institute at: http://www.pacinst.org/reports/sea_level_rise/index.htm (accessed July 12, 2012).

[104] M. Siddall, T. F. Stocker, and P. U. Clark, "Constraints on Future Sea-Level Rise from Past Sea-Level Change," *Nature Geoscience*, 2 (2009): 571–575, doi: 10.1038/ngeo587.

[105] J. A. Church, J. M. Gregory, N. J. White, S. M. Platten, and J. X. Mitrovica, "Understanding and Projecting Sea Level Change," *Oceanography* 24, no. 2 (2011): 130–143, doi: 10.5670/oceanog.201133.

Another study[106] used a similar approach, developing an equation that estimates the relationship of climate to sea-level change over the past 2,000 years. The researchers found that sea-level rise by the end of the century will be roughly three times higher than predictions in AR4. They also conclude that even if temperature rise were stopped today, sea level will still rise another 20 to 40 cm (7.9–15.7 in) and that actual cooling would be needed to stop the ongoing rise. According to this model, the most optimistic emissions scenario in AR4 (the B1 scenario, which produces a best-estimate temperature rise of 1.8°C [3.24°F] by 2100) will result in a sea-level rise of 80 cm (31.5 in). The most pessimistic scenario in AR4 (the A1FI scenario, which produces a best-estimate temperature rise of 4°C [7.2°F] by 2100), results in a sea-level rise of 1.35 m (4.4 ft). The study estimates that sea level will rise 0.9 to 1.3 m by 2100.

These estimates are consistent with a study published in the Proceedings of the National Academy of Sciences in late 2009[107] and subsequently tested and verified using multiple modeling and statistical tests.[108] This work concludes that a simple relationship links global sea-level variations on time scales of decades to centuries to global mean temperature, namely about 1 m (3.3 ft) of global sea-level rise will result from warming of 1.8°C. This provides a basis for predicting sea level by the end of the century using the IPCC AR4 emission scenarios. The model projects a sea-level rise of 0.75 to 1.90 m (2.5 to 6.2 ft) for the period 1990–2100 (Figure 5.15). The study authors note that to limit the amount of global sea-level rise to a maximum of 1 m (3.3 ft) in the long run, reductions in greenhouse-gas emissions would likely have to be deeper than those needed to limit global warming to 2°C (3.6°F), which is the policy goal now supported by many countries.[109]

Several other researchers have published estimates of sea-level rise by 2100: 0.5 to 1.4 m[110] (1.6–4.6 ft), 0.8 to 2.0 m[111] (2.6–6.6 ft), 1.6 m[112] (5.2 ft); and there are others. These models all employ what is known as semi-empirical modeling; that is, they develop a forecast based on past relationships between temperature and sea level.

MODELING ISSUES[113]   There are two problems with semi-empirical modeling. This approach relies on relatively simple relationships (e.g., global temperature change and global sea-level change) that might not hold true in the warming future, rather than attempting to reproduce the actual physical processes and feedbacks that result in sea-level changes. Semi-empirical modeling also does not account for the high regional variability that characterizes real sea-level change. Regional variability[114] is a term describing the fact that the ocean surface (that is, sea level) is not flat. As shown in Figure 5.2, sea level can simultaneously rise and fall in different parts of the world owing to forcing by winds, currents, differences in heating, vertical

[106] A. Grinsted, J. C. Moore, and S. Jevrejeva, "Reconstructing Sea Level from Paleo and Projected Temperatures 200 to 2100 AD," *Climate Dynamics* 34, no. 4 (2010): 461–472, doi: 10.1007/s00382-008-0507-2.

[107] M. Vermeer, and S. Rahmstorf, "Global Sea Level Linked to Global Temperature," *Proceedings of the National Academy of Sciences* 106, no. 51 (2009): 21527–21532, http://www.pnas.org/content/106/51/21527.full (accessed July 12, 2012).

[108] S. Rahmstorf, M. Perrette, and M. Vermeer, "Testing the Robustness of Semi-Empirical Sea Level Projections." *Climate Dynamics* (2011): doi: 10.1007/s00382-011-1226-7.

[109] See comments from the International Energy Agency: http://in.reuters.com/article/2012/05/16/energy-summit-iea-idINDEE84F0F120120516 (accessed July 12, 2012).

[110] S. Rahmstorf, "A Semi-Empirical Approach to Projecting Sea-Level Rise," *Science* 315, no. 5810 (2007): 368–370.

[111] W. T. Pfeffer, J. T. Harper, and S. O'Neel, "Kinematic Constraints on Glacier Contributions to 21st Century Sea-Level Rise," *Science* 321, no. 5894 (2008): 1340–1343.

[112] E. J. Rohling, K. Grant, C. H. Hemleben, et al., "High Rates of Sea-Level Rise during the Last Interglacial Period," *Nature Geoscience* 1 (2008): 38–42.

[113] Willis and Church (2012) call for sea-level researchers to resolve discrepancies between semi-empirical modeling and fully coupled Earth system modeling projections of future sea-level rise. J.K. Willis and J.A. Church, "Regional Sea-Level Projections," *Science* 336 (2012): 550–551.

[114] For a thorough discussion of regional variability in the context of media reports of sea-level vulnerability by oceanic islands, see S. Donner, "Sea-Level Rise and the Ongoing Battle of Tarawa," *Eos*, 93, no. 17 (2012): 169.

Figure 5.15. Sea-level rise from 1990 to 2100 based on IPCC AR4 temperature projections for three different emission scenarios (labeled on right). The sea-level range projected in the IPCC AR4[115] for these scenarios is shown for comparison in the bars on the right. Also shown is the observations-based annual global sea-level data[116] (red) including artificial reservoir correction.[117]

Source: M. Vermeer, and S. Rahmstorf, "Global Sea Level Linked to Global Temperature," *Proceedings of the National Academy of Sciences* 106, no. 51 (2009): 21527–21532, http://www.pnas.org/content/106/51/21527.full (accessed July 12, 2012).

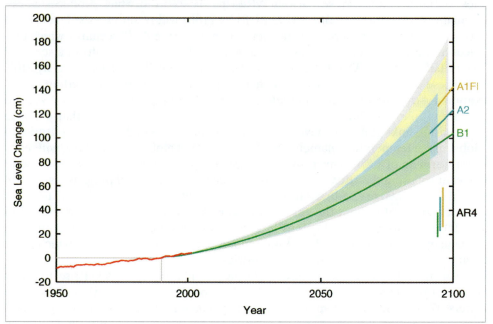

changes in the seafloor and land, and shifts in Earth's gravity field due to melting ice and the weight of water on the sea floor.

Coupled global climate models (GCMs), such as used in AR4, produce estimates of future global mean sea-level rise that are typically significantly less than the results of semi-empirical modeling. This is because GCMs do not account for the dynamics of surging continental-scale ice sheets on Greenland and Antarctica. For instance, AR4 projected a global mean sea-level rise of 18 to 59 cm (7.1–23.2 in) by 2100, but it lacked an assessment of ice dynamics (such as ice acceleration, fracturing, and collapse) in Greenland and Antarctica because these processes are poorly understood.

More-recent global climate modeling using AR4 economic scenarios has been geared toward improving understanding of regional variability in sea-level rise, but it has not advanced understanding of ice sheet dynamics. To account for this gap, researchers can insert a fixed value for ice sheet decay based on satellite observations and expert judgment.[118] This is especially important given that melting ice sheets are now the largest contributor[119] to sea-level rise and the rate at which they are melting is accelerating.[120] An 18-year satellite study found that from one year to

---

[115] S. Solomon, D. Qin, M. Manning, et al. (eds.), *Contribution of Working Group I to the Fourth Assessment Report (AR4) of the Intergovernmental Panel on Climate Change.*

[116] J. A. Church and N. J. White, "A 20th-Century Acceleration in Global Sea-Level Rise," *Geophysical Research Letters* 33 (2006): L01602.

[117] B. F. Chao, Y. H. Wu, and Y. Li, "Impact of Artificial Reservoir Water Impoundment on Global Sea Level," *Science* 320 (2008): 212–214.

[118] C. A. Katsman, A. Sterl, J. J. Beersma, et al., "Exploring High-End Scenarios for Local Sea-Level Rise to Develop Flood Protection Strategies for a Low-Lying Delta—the Netherlands as an Example," *Climatic Change* 109, nos. 3–4 (2010): 617–645, doi: 10.1007/s10584-011-0037.

[119] See http://www.sciencedaily.com/releases/2011/03/110308150228.htm (accessed July 12, 2012).

[120] E. Rignot, I. Velicogna, M. R. van den Broeke, A. Monaghan, and J. Lenaerts, "Acceleration of the Contribution of the Greenland and Antarctic Ice Sheets to Sea-Level Rise."

the next the Greenland ice sheet lost mass faster than it did the year before, by an average of 21.9 Gt per year. In Antarctica, the year-over-year speedup in ice mass lost averaged 14.5 Gt (see Figure 5.9).

In one study of sea level, researchers[121] used a coupled model that included regional estimates of thermal expansion, glacioisostatic land-level changes,[122] and local effects of small glaciers and ice caps to project regional variations in sea level by the end of the century. For the AR4 A1B scenario (moderate economic growth), regional variability in sea level was found to range from −3.91 m (−12.8 ft, sea level fall) to +0.79 m (2.6 ft), with a global mean of +0.47 m (1.5 ft), which is significantly less than the projections of the semi-empirical models; however, this projection lacked an assessment of ice dynamics in Greenland and Antarctica. When a fixed value for ice dynamics is included in the projection (0.41 m [1.3 ft], sea-level equivalent for the Antarctic Ice Sheet and 0.22 m [0.7 ft] sea-level equivalent for the Greenland Ice Sheet; based on Katsman et al., 2010), the global mean sea-level rise by 2100 increased to 1.02 m (3.3 ft).

Because GCM does not (yet) account for the physics of sliding ice sheets, some researchers are relying on expert judgment. One researcher[123] asked experts for their estimate of the ice contribution to sea-level rise by 2100. A survey of 28 colleagues (half of whom responded) produced a best estimate of 32 cm (1 ft) of sea-level rise resulting from ice sheet losses. That results in a total rise of 61 to 73 cm (2 to 2.4 ft) from all sources by the end of the current century. Any calculation of global mean such as this must, however, take into account local and regional variability of sea level when planning for the impacts (an area of research where the GCMs excel).

An important outcome of GCM sea-level projections is a picture of the regional variability of future sea level. As Greenland continues to lose ice during the 21st century, the land rebounds upward because there is not as much weight pushing down on it. As a result, the North Atlantic seafloor flexes downward around the Greenland coast in a seesaw-like action called lithospheric flexure, a form of glacioisostatic land level change. This (ironically) results in falling sea level near Greenland (and other coastal areas experiencing melting and flexing upward). Another important phenomenon is regional differences in thermal expansion. For instance, as Arctic sea ice continues to retreat during the 21st century, the Arctic Ocean—more so than other parts of the ocean—will absorb heat and experience thermal expansion.

How do these various regional-scale processes affect sea level by the end of the century[124]? Figure 5.16[125] shows projected sea level by 2100 under the AR4 A1B economic scenario, with a fixed high-end estimate of ice-sheet decay. What is clear from this projection is that some areas of the globe may actually experience net sea-level fall (North Atlantic, Southern Ocean), and other areas may experience net sea-level rise (Pacific and Indian oceans). If Greenland and Antarctica continue their high rate of decay, the mean global sea level change is projected to be approximately 1.02 m (3.3 ft), although there will be a high degree of regional variability with regard to the ocean surface.

[121] A. Slangen, C. Katsman, R. Van de Wal, L. Vermeersen, and R. Riva, "Towards Regional Projections of Twenty-First Century Sea-Level Change Based on IPCC SRES Scenarios," *Climate Dynamics* 38, nos. 5–6 (2011): 1191–1209, doi: 10.1007/s00382-011-1057-6.

[122] Glacio-isostatic land level changes occur when the land rebounds upward after a glacier melts and its weight is removed from the crust. Much of the land throughout northern Europe and Canada is uplifting in this way because the weight of the continental glaciers during the last ice age has been removed.

[123] Meeting Briefs, "Climate Outlook Looking Much the Same or Even Worse," *Science* 334 (2011): 1616.

[124] C. C. Hay, E. Morrow, R. E. Kopp, and J. X. Mitrovica, "Fostering Advance in Interdisciplinary Climate Science Sackler Colloquium: Estimating the Sources of Global Sea-Level Rise with Data Assimilation Techniques," *Proceedings of the National Academy of Sciences* 2012, doi: 10.1073/pnas.1117683109.

[125] A. Slangen, C. Katsman, R. van de Wal, L. Vermeersen, and R. Riva, "Towards Regional Projections of Twenty-First Century Sea-Level Change Based on IPCC SRES Scenarios," *Climate Dynamics* 38, no. 5–6 (2012): 1191–1209, doi 10.1007/s00382-011-1057-6.

Figure 5.16. Coupled global climate model mean sea-level anomaly (in meters) with regard to global mean sea-level change (1.02 m) for scenario A1B between 1980–1999 and 2090–2099, for a scenario with adapted ice sheet contributions of 0.22 m for the Greenland Ice Sheet and 0.41 m for the Antarctic Ice Sheet.[126]

Source: A. Slangen, C. Katsman, R. van de Wal, L. Vermeersen, R. Riva, "Towards Regional Projections of Twenty-First Century Sea-Level Change Based on IPCC SRES Scenarios," Climate Dynamics 38, no. 5–6 (2012): 1191–1209, doi 10.1007/s00382-011-1057-6. Climate Dynamics, permission needed.

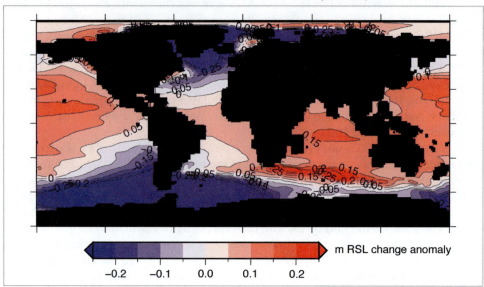

m RSL change anomaly

CONSENSUS If Greenland and Antarctica continue to undergo the observed high rate of decay,[127] a global sea-level rise of approximately 1 m (3.3 ft) by the end of the 21st century is emerging as a consensus[128] of the scientific community.[129] This elevation, as a planning target for the sea-level rise adaptation groups around the world, makes a robust guideline for identifying risk and vulnerability of coastal assets.

If greenhouse-gas concentrations were stabilized today, sea level would nonetheless continue to rise for hundreds of years.[130] After 500 years, sea-level rise from thermal expansion alone would have reached only half of its eventual level, which models suggest could lie within ranges of 0.5 to 2 m (1.6 to 6.6 ft). Glacier retreat will continue and the loss of a substantial fraction of Earth's total glacier mass is likely. Areas that are currently marginally glaciated are likely to become ice-free. But it is unlikely that greenhouse gases will be stabilized soon, so we can probably count on additional atmospheric heating and sea-level rise.

In one study,[131] researchers found that if the current warming trends continue, by 2100 Earth will likely be at least 4°C (7.2°F) warmer than present, with the Arctic at least as warm as it was nearly 130,000 years ago, when the Greenland ice sheet was a mere fragment of its present size. Study leader Jonathan T. Overpeck of the University

[126] A. Slangen, C. Katsman, R. van de Wal, L. Vermeersen, and R. Riva, "Towards Regional Projections of Twenty-First Century Sea-Level Change Based on IPCC SRES Scenarios."

[127] E. Rignot, I. Velicogna, M. R. van den Broeke, A. Monaghan, and J. Lenaerts, "Acceleration of the Contribution of the Greenland and Antarctic Ice Sheets to Sea-Level Rise."

[128] Committee on Sea Level Rise in California, Oregon, and Washington; Board on Earth Sciences and Resources; Ocean Studies Board; Division on Earth and Life Studies; National Research Council, *Sea Level Rise for the Coasts of California, Oregon, and Washington: Past, Present, and Future* (2012), National Academies Press, Washington, D.C.

[129] P. Linwood P. King, C. Mohn, et al., "Estimating the Potential Economic Impacts of Climate Change on Southern California Beaches," *Climatic Change* 109, no. S1 (2011): 277, doi: 10.1007/s10584-011-0309-0.

[130] S. Solomon, G. K. Plattner, R. Knutti, and P. Friedlingstein, "Irreversible Climate Change due to Carbon Dioxide Emissions," *Proceedings of the National Academy of Sciences* 106, no. 6 (2009): 1704–1709.

[131] J. Overpeck, B.L. Otto-Bliesner, G.H. Miller, D.R. et al., "Paleoclimate Evidence for Future Ice Sheet Instability and Rapid Sea-Level Rise," *Science* 311, no. 5768, (2006): 1698–1701.

of Arizona in Tucson says, "The last time the Arctic was significantly warmer than present day, the Greenland Ice Sheet melted back the equivalent of about 2 to 3 meters (6.6 to 9.8 ft) of sea level." The research also suggests the Antarctic ice sheet melted substantially, contributing another 2 to 3 m (6.6–9.8 ft) of sea-level rise. The ice sheets are melting already. The new research suggests melting could accelerate, thereby raising sea level as fast as, or faster than, 1 m (3.3 ft) per century. [132]

Global sea-level rise has accelerated in response to warming of the atmosphere and the ocean and melting of the world's ice environment. Projections indicate that a 1 m (3.3 ft) rise by the end of this century is plausible. It has been pointed out[133] that observed sea-level rise has exceeded the best case projections thus far (Figure 5.17). This has been independently confirmed by other studies as well; observed global average sea level rose at a rate near the upper end of projections of the IPCC Third and Fourth assessment reports.[134]

The map of sea-level change produced by satellite altimetry (Figure 5.2) suggests that sea-level rise will have significant local variability (rising in some areas, falling in others, but as a global mean, rising overall) that is worthy of continued research to improve our understanding of localized impacts and adaptation needs. Planners should consider this variability, as impacts to coastal assets will be scaled to localized sea-level change. The scientific community is converging on a consensus that it is appropriate to plan for a 1 m (3.3 ft) rise in mean sea level by the end of the century.[135] As stated in an update to the IPCC in late 2009,[136]

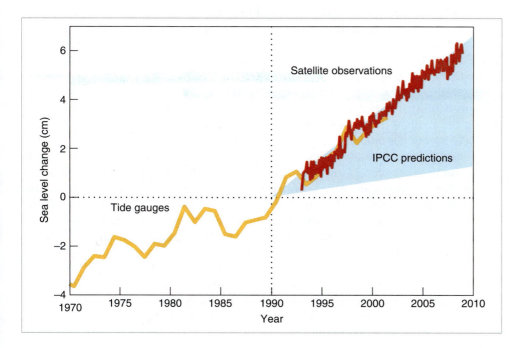

Figure 5.17. Sea-level change during 1970–2010. The tide-gauge data[137] are indicated in yellow and satellite data[138] in red. The blue band shows the high and low projections of the IPCC Third Assessment report for comparison.[139]

SOURCE: The Copenhagen Diagnosis, 2009: Updating the world on the Latest Climate Science (1st Edition). I. Allison, N. L. Bindoff, R.A. Bindschadler, P.M. Cox, N. de Noblet, M.H. England, J.E. Francis, N. Gruber, A.M. Haywood, D.J. Karoly, G. Kaser, C. Le Quéré, T.M. Lenton, M.E. Mann, B.I. McNeil, A.J. Pitman, S. Rahmstorf, E. Rignot, H.J. Schellnhuber, S.H. Schneider, S.C. Sherwood, R.C.J. Somerville, K.Steffen, E.J. Steig, M. Visbeck, A.J. Weaver. The University of New South Wales Climate Change Research Centre (CCRC), Sydney, Australia, 60pp.

[132] See the animation "Climate Denial Crock of the Week: Sea-Level Rise Accelerating" at the end of the chapter.

[133] R.A. Pielke, "Climate Predictions and Observations," *Nature Geoscience* 1 (2008): 206.

[134] J. A. Church and N. J. White, "Sea-Level Rise from the Late 19th to the Early 21st Century," *Surveys of Geophysics* (2011): doi 10.1007/s10712-011-9119-1.

[135] S. Rahmstorf, "Sea-Level Rise: Towards Understanding Local Vulnerability," *Environmental Research Letters* 7 (2012): 021001, doi: 10.1088/1748-9326/7/021001.

[136] I. Allison, N. L. Bindoff, R.A. Bindoff, et al., *The Copenhagen Diagnosis*, 2009.

[137] J. A. Church and N. J. White, "A 20th Century Acceleration in Global Sea-Level Rise, *Geophysical Research Letters* 33 (2006): L01602.

[138] A. Cazenave, K. Dominh, S. Guinehut, et al., "Sea Level Budget over 2003–2008: A Reevaluation from GRACE Space Gravimetry, Satellite Altimetry and ARGO," *Global and Planetary Change* 65 (2009): 83–88.

[139] Figure from I. Allison, N. L. Bindoff, R.A. Bindoff, et al., *The Copenhagen Diagnosis, 2009: Updating the World on the Latest Climate Science.* (University of New South Wales Climate Change Research Centre, Sydney).

By 2100, global sea-level is likely to rise at least twice as much as projected by AR4, for unmitigated emissions it may well exceed 1 meter (3.3 feet). The upper limit has been estimated as ~2 meters (6.6 feet) sea-level rise by 2100. Sea-level will continue to rise for centuries after global temperatures have been stabilized and several meters of sea-level rise must be expected over the next few centuries.

As mentioned earlier in this chapter, the National Research Council[140] found that, relative to 2000 levels, global sea level will reach 8–23 cm (3–9 in) by 2030, 18–48 cm (7–19 in) by 2050, and 50–140 cm (20–55 in) by 2100. The rate of global mean sea-level rise will have to increase if these predictions are to be realized. Nonetheless, when planning for the risk and vulnerability to sea-level rise, these estimates provide a reasonable guideline.

THE EEMIAN ANALOGUE

As the science of sea-level analysis develops, projections of how high global mean sea level will rise by 2100 continue to change. Thus far, there is general agreement that sea level could rise as much as 1 m (3.3 ft) by the end of the century, with some researchers suggesting that a 2 m (6.6 ft) rise is not out of the question.[141] One source of information for improving our understanding of how high and how fast sea-level rise could continue is to look to the natural analogue of the last interglacial climate period, known by its European name, the Eemian (see discussion in Chapter 3).

The Eemian interglacial, approximately 120,000 to 130,000 years ago, was apparently warmer than the preindustrial climate of the past 10,000 years by 1.5°C to 2°C (2.7°F to 3.6°F), and polar temperatures at the time might have reached 3°C to 5°C (5.4°F to 9°F) above present.[142] Researchers have used this episode as an analogue for the ice volume and sea-level position of a warmer world (Figure 5.18).[143] They found a 95% probability that global sea level peaked at least 6.6 m (21.6 ft) higher than today during the Eemian. They found that it is likely (67% probability) to have exceeded 8.0 m (26 ft), but it is unlikely (33% probability) to have exceeded 9.4 m (30.9 ft). They also found that the rate of sea-level rise to peak heights ranged between 5.6 mm/yr and 9.3 mm/yr (between 22 inches per century and 36.6 inches/ per century). It is well known that ice sheets are very sensitive to climate, and because global ice volume takes some time to equilibrate with climate, ice sheets have a long-term vulnerability to even relatively low levels of sustained global warming. Hence, excess heat in the climate system today has the potential to drive ice sheet melting far into the future.

As shown by Figure 5.18, temperature during the Eemian (diamonds and green polynomial fit), and in the late 21st century as projected by the optimistic B1 scenario of AR4 (blue with 1 standard deviation, dashed envelope) are nearly identical, which suggests that the climate of the last interglacial might provide a reasonable analogue for establishing the response of ice sheets to global warming. If it is accurate that Eemian sea levels were approximately 7 to 9 m (23–29.5 ft) above present[144] and that

---

[140] Committee on Sea Level Rise in California, Oregon, and Washington; Board on Earth Sciences and Resources; Ocean Studies Board; Division on Earth and Life Studies; National Research Council, *Sea Level Rise for the Coasts of California, Oregon, and Washington: Past, Present, and Future* (2012), National Academies Press, Washington, D.C.

[141] S. Rahmstorf, "A New View of Sea-Level Rise," *Nature Reports Climate Change* 4, no. 1004 (2010): 44–45, doi: 10.1038/climate.2010.29.

[142] P.U. Clark and P. Huybers, "Interglacial and Future Sea Level," *Nature* 462 (2009): 856–857.

[143] R.E. Kopp, F.J. Simons, J.X. Mitrovica, A.C. Maloof, and M. Oppenheimer, "Probabilistic Assessment of Sea Level during the Last Interglacial Stage," *Nature* 462 (2009): 863–868.

[144] R.E. Kopp, F.J. Simons, J.X. Mitrovica, A.C. Maloof, and M. Oppenheimer, "Probabilistic Assessment of Sea Level during the Last Interglacial Stage." *Nature* 462 (2009): 863–868.

Figure 5.18. Global temperatures during the Eemian interglacial 125,000 years are nearly identical to projected global temperatures at the end of the 21st century under the B1 (the most optimistic) economic scenario of the AR4. Purple diamonds depict global paleoclimate temperature data and uncertainty for the Eeemian, and the green line is a polynomial model of these data. The B1 scenario is shown as a blue line, and one standard deviation is shown as a dashed envelope.

Source: Reprinted with permission from Macmillan Publishers Ltd: P.U. Clark and P. Huybers, "Interglacial and Future Sea Level," *Nature* 462 (2009): 856–857, Copyright 2009.

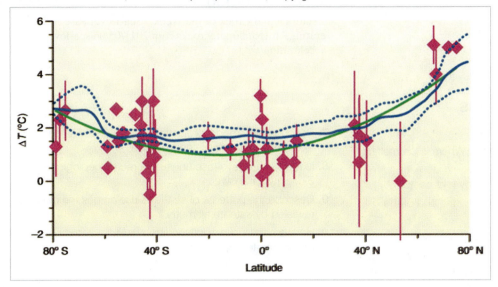

higher sea level resulted from higher temperatures, "the disconcerting message is that the equilibrium response of sea level to 1.5°C to 2°C (2.7°F to 3.6°F) of global warming could be an increase of 7–9 m (23–29.5 ft)."[145] A rise of this magnitude would, of course, be devastating to coastal communities, and the hundreds of millions of individuals who live in them, around the world.

CONCLUSION

Sea-level rise could affect coastal communities in a number of ways. Coastal erosion will increase, marine inundation will worsen, coastal ecosystems will evolve or go extinct, and the coastal water table will rise,[146] causing severe drainage problems. Although these problems have not yet become endemic in much of the developed world, the tropical western Pacific and the islands of Micronesia are flooding at today's high tides, and age-old communities are suffering impacts to their food and drinking water.[147] Many are considering migrating to new lands.

Although coastal communities in developed nations are still awakening to the inevitabilities of sea-level rise and planning for adaptation is not widespread, as impacts grow (and expenses with them), sea-level rise is likely to become the leading issue in coastal community planning in coming years.[148]

---

[145] 145 P.U. Clark and P. Huybers, "Interglacial and Future Sea Level," *Nature* 462 (2009): 856–857.

[146] D. Bjerklie, J. Mullaney, J. Stone, B. Skinner, and M. Ramlow, "Preliminary Investigation of the Effects of Sea-Level Rise on Groundwater Levels in New Haven, Connecticut," *U.S. Geological Survey Open-File Report 2012–1025*, 46 (2012), http://pubs.usgs.gov/of/2012/1025/ (accessed July 12, 2012).

[147] See Executive Summary: Climate Change in the Federated States of Micronesia: Food and Water Security, Climate Risk Management, and Adaptive Strategies, http://icap.seagrant.soest.hawaii.edu/executive-summary-climate-change-federated-states-micronesia-food-and-water-security-climate-risk-ma (accessed July 12, 2012).

[148] J.L. Weiss, J.T. Overpeck, and B. Strauss, "Implications of Recent Sea-Level Rise Science for Low-Elevation Areas in Coastal Cities of the Conterminous U.S.A." *Climatic Change Letters* (2011): doi: 10.1007/s10584-011-0024.

ANIMATIONS AND VIDEOS

"How Much Will Sea-Level Rise?" http://vimeo.com/5188725

"Melting Ice, Rising Seas," http://www.youtube.com/watch?v=gbnW3MK8wgY

"Antarctic Ice Flows: A Complete Picture," http://www.nasa.gov/multimedia/videogallery/index.html?media_id=106877491

See how climate change and rising sea levels plague American cities right now, http://www.treehugger.com/climate-change/see-how-climate-change-rising-sea-levels-plague-american-cities-right-now-video.html

"Sea-level rise," http://www.youtube.com/watch?v=dPOT5TRRL3E

Climate Denial Crock of the Week, "Sea-Level Rise Accelerating," http://climatecrocks.com/2011/05/06/sea-level-rise-accelerating/

COMPREHENSION QUESTIONS

1. What are the sources of sea-level rise?
2. Melting ice is a source of sea-level rise. Where is ice melting occurring?
3. How high is sea level projected to rise by mid-century?
4. How high is sea level projected to rise by the end of the century?
5. Describe the process of planning for community adaptation to sea-level rise.
6. Why do scientists study sea-level history during the Eemian?
7. Describe how model projections of sea-level rise compare to the observations of sea-level rise.
8. Is today's sea level-rise unusual in recent geologic history? Explain your answer.
9. Describe the patterns of sea-level rise across the globe as revealed by satellite altimetry.
10. How is sea-level rise influenced by winds, air temperature, and ocean temperature?

THINKING CRITICALLY

1. Describe how global sea level has changed over the past 2000 years.
2. Compare the maps in Figures 5.2 and 5.3. How are they related or different? Describe any relationships between what is depicted in the two maps.
3. Global mean sea level is rising at about 3.2 mm/yr. Describe the data you would need to create a sea level "budget"; that is, to assign components of this rate to the various processes that cause global sea-level change. How would you get these data?
4. You are asked to appear before a congressional hearing into sea-level rise. Explain how global warming causes sea-level rise.
5. Since satellite altimetry measurements began approximately 20 years ago, the rate of sea-level rise has not accelerated. Rather, it has stayed a fairly consistent 3.2 mm/yr. Yet at the same time, the air temperature has continued to increase. Explain this apparent paradox.
6. Imagine you are part of a city-planning team asked to rank the vulnerability of transportation assets (roads, harbors, rail, airports, bus lines and terminals, maintenance yards, etc.) in the face of sea-level rise. How would you proceed? What data would you need? How would you get the data? In fact this is a real-life issue, and such planning is taking place today in cities in the United States.
7. The Eemian offers an analogue to a future with higher seas. You live in a coastal region where there is evidence of an old Eemian shoreline. How would you use this to improve your understanding of the pattern and impacts of sea-level rise locally?
8. You are building a home on a beach. What design features will you use to mitigate the negative impacts of sea-level rise?
9. You live one block from the ocean. Describe how you are vulnerable to sea-level rise.
10. Describe the negative impacts of sea-level rise on a coastline. How would you design a road to mitigate each impact? A parking lot? A school building?

CLASS ACTIVITIES (FACE TO FACE OR ONLINE)

ACTIVITIES

1. Visit the NASA website "Global Climate Change," http://climate.nasa.gov/evidence/, and answer the following questions.
   a. What is the evidence for rapid climate change?
   b. Describe the rate of global sea-level rise?
   c. Describe some facts about Earth's climate that are not in dispute.
   d. Choose two of the key lines of evidence that Earth's climate is changing and describe their impact where you live.

2. Watch the video "Climate Denial Crock of the Week: Sea-Level Rise Accelerating," http://climatecrocks.com/2011/05/06/sea-level-rise-accelerating/, and answer the following questions.
   a. Do climate deniers on TV accurately portray the sea-level rise problem?
   b. What are the primary causes of global sea-level rise?
   c. What important cause of global sea-level rise is not included in the IPCC AR4? Why is this important?

d. When "rapid dynamical changes in ice flow" are included in future sea-level rise estimates, how high could the sea level rise by the end of the century?
   e. How would a sea-level rise of 0.8 to 2 meters by the end of the century impact the coastlines?

3. Read the article "Rising Sea Levels Set to Have Major Impacts around the World," http://www.sciencedaily.com/releases/2009/03/090310104742.htm.
   a. Describe the major points of the article.
   b. Carefully summarize the wording used by the experts in describing sea level at the end of the century. What do they state in a definitive manner and what do they state in a tentative manner? Why do they use these styles of communication?
   c. What are some of the impacts of sea-level rise as described in the article?
   d. This website provides "Related Stories" on sea-level rise. Explore these and provide a summary of impacts, projections, research methods, and understanding of the causes of sea-level rise.

HOW DOES GLOBAL WARMING AFFECT OUR COMMUNITY?

FIGURE 6.0. On July 11, 2012, the United States Department of Agriculture announced that more than 1,000 counties in 26 states qualified as natural disaster areas—the largest total area ever declared a disaster zone by the agency. The extent of the damage to crops is depicted in this vegetation anomaly map based on data from NASA's Terra satellite. Brown areas show where plant growth was less vigorous than normal; cream colors depict normal levels of growth; and green indicates abnormally lush vegetation. Data were not available in the gray areas due to snow or cloud cover. Climate change is accompanied by an increase in dangerous weather, including drought, flooding, heat waves, and others.

IMAGE CREDIT: NASA Earth Observatory image by Jesse Allen, Earth Observatory, using data provided by Inbal Reshef, Global Agricultural Monitoring Project.

CHAPTER

6

CHAPTER SUMMARY

Climate change impacts to human communities include: stresses to water resources, threats to human health, shifting demand on energy supply, disruptions to transportation and agriculture, and increased vulnerability of society and ecosystems to future climate change. In the United States, extreme weather events have increased in number and magnitude and are likely to do so in the future. Severe heat waves and record-setting temperatures are occurring with greater frequency. Among other impacts are the spread of diseases not historically prevalent in North America, retreat of tundra and northern and arctic ecosystems, increased occurrence of drought and flooding, sea-level rise, decreased snow pack and retreating glaciers, changes in the timing of seasons, and ecological impacts.

In this chapter you will learn that:

- All geographic regions of the U.S. are experiencing negative effects from climate change and are growing increasingly vulnerable to future climate change.
- Coastal areas and islands in the Pacific and Caribbean are at increasing risk from sea-level rise and storm surge.
- Climate change is threatening water resources, transportation, ecosystem, and agriculture sectors, as well as society, energy, and health sectors.
- Over the past 50 years the average temperature in the U.S. has risen more than 1.1°C (2°F) and is projected to continue to rise as greenhouse gas production continues.
- Extreme weather events, such as heat waves, regional droughts, snowstorms, and flooding, have become more frequent and intense during the past half century.
- Rain and snowfall as a whole have increased (about 5%); however, there have been important regional and seasonal differences leading to severe localized impacts, including drought and increased flooding hazards.
- Freshwater resources are changing in most states and adaptation plans should be developed to manage these changes.
- Sea level has risen along most of the coast; the majority of the shoreline is eroding.

Learning Objective

According to the U.S. Global Change Research Program (USGCRP), which reports to Congress and the President on the effects of global climate change, we have already experienced significant impacts from climate change.

The effects of global warming can be identified in every sector of the U.S. economy and every region of the continent. Effects are found in human society and health; ecosystems, water resources, energy and transportation infrastructure; and agriculture. Many of these effects are distributed regionally. For instance, the south and southwest have plunged into severe drought, leading to an increase in wildfires, decreased drinking water, and growing heat stress in cities and within ecosystems. The New England, Midwest, and southern Canadian regions have seen a sharp increase in flooding owing to higher levels of rainfall and more violent rainstorms.[1] Nationwide,[2] extreme summer temperatures are already occurring more frequently and heat waves will become normal by mid-century if the world continues on a business-as-usual schedule of emitting greenhouse gases.[3,4]

Overall there is a new intensity and destructiveness to weather events.[5] Across the continent, air temperature has increased and ecosystems that have evolved under the regular timing of seasonal patterns are experiencing increased disease and infestation. Coastal communities have sustained losses due to shoreline erosion and the landfall of storms. Unusually intense snowstorms have hit the east coast, winters across North America have become warmer, climate is establishing new records of warmth,[6] there are unusually high numbers of tornados in the Midwest, and summers throughout the continent are characterized by record-breaking heat waves.

Scientists have concluded that these trends are consistent with the expected influence of rising air temperature created by global warming.[7] There is 40% more carbon dioxide in the atmosphere now than there was only 19 years ago. As we saw in earlier chapters, this greenhouse gas is effective at trapping heat in the troposphere. Researchers report that the temperature of the ocean surface in summer 2009 was the warmest ever recorded,[8] land-surface temperatures in 2012 were the warmest yet observed, 2005, 2010, and 2012 were, in-turn, the warmest years on record, 2011 was the warmest La Niña year in recorded history, and the decade 2001 to 2011 was the warmest in modern human history.[9] This powerful warming trend has led to numerous direct and indirect impacts throughout North America.

[1] T. R. Karl, J. M. Melillo, and T. C. Peterson, "Global Climate Change Impacts in the United States," 2009, http://www.globalchange.gov/publications/reports/scientific-assessments/us-impacts (accessed July 12, 2012).

[2] P. Duffy and C. Tebaldi, "Increasing Prevalence of Extreme Summer Temperatures in the U.S.," *Climatic Change* 111, no. 2 (2012): 487, doi: 10.1007/s10584-012-0396-6.

[3] The international Energy Agency reports that global CO_2 emissions reached a record high of 31.6 Gt in 2011. This represents an increase of 1.0 Gt over 2010, or 3.2%. Coal accounted for 45% of total energy-related CO_2 emissions in 2011, followed by oil (35%) and natural gas (20%). See: http://www.iea.org/newsroomandevents/news/2012/may/name,27216,en.html (accessed July 12, 2012).

[4] See the video "White House Releases Landmark Climate Report" at the end of the chapter.

[5] S. Rahmstorf and D. Coumou, "Increase of Extreme Events in a Warming World," *Proceedings of the National Academy of Sciences* 108, no. 44 (2011): 17905–17909 www.pnas.org/cgi/doi/10.1073/pnas.1101766108.

[6] See the NASA report, http://www.earthobservatory.nasa.gov/IOTD/view.php?id=77465 (accessed July 12, 2012).

[7] K. Guirguis, A. Gershunov, R. Schwartz, and S. Bennett, "Recent Warm and Cold Daily Winter Temperature Extremes in the Northern Hemisphere," *Geophysical Research Letters* 38 (2011): L17701, doi: 10.1029/2011GL048762, 2011.

[8] See NOAA, "Warmest Global Sea-Surface Temperatures for August and Summer": http://www.noaanews.noaa.gov/stories2009/20090916_globalstats.html (accessed July 12, 2012).

[9] See NASA Goddard Institute for Space Studies Graphs and Plots page, http://data.giss.nasa.gov/gistemp/graphs/ (accessed July 12, 2012).

Figure 6.1. Map showing global temperature anomalies (departures from the average March temperatures of 1951–1980) for March 2012. A massive heat wave hit the North American continent in March 2012. East of the Rocky Mountains, 25 states had their warmest March on record; 15 more states were in their top 10 warmest. More than 15,000 temperature records were broken—evenly split between daytime highs and nighttime highs—and there were 21 instances of nighttime low temperatures that were warmer than the former daytime records.

Source: NASA Earth Observatory, http://earthobservatory.nasa.gov/IOTD/view.php?id=77671 (accessed July 12, 2012).

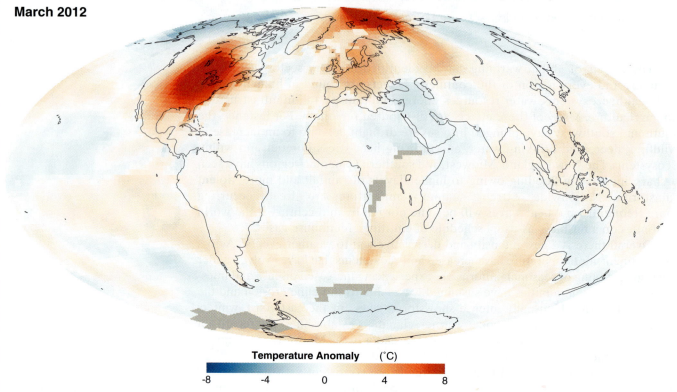

March 2012

Temperature Anomaly   (°C)

-8   -4   0   4   8

IT'S GETTING HOT OUT THERE

As a result of a heat wave that hit North America in July 2011, 9,000 daily temperature records were broken or tied, including 2,755 daytime highs and 6,171 nighttime highs. Another heat wave hit in the spring and summer of 2012 (Figure 6.1), making each month the warmest on record (records go back to 1895) and the 12 month and 6 month periods ending in October 2012 the hottest yet observed.

In June 2009 the United States Global Change Research Program (USGCRP)[10] produced a comprehensive National Climate Assessment[11] on the status of climate change impacts in the U.S.[12] The program is housed under the Executive Office of the President and its role is to coordinate and integrate research among thirteen federal agencies on climate change and its implications.[13] By examining the influence of changing climate on transportation, water resources, ecosystems, human health, and other key economic sectors, National Climate Assessments produced by the USGCRP offer an examination of impacts and play a valuable role in the development of new policy on local, state, and federal levels.

[10]See their website at: http://www.globalchange.gov/ (accessed July 12, 2012).

[11]T. R. Karl, J. M. Melillo and T. C. Peterson, *Global Climate Change Impacts in the United States*. Cambridge University Press (2009). http://www.globalchange.gov/publications/reports/scientific-assessments/us-impacts (accessed July 12, 2012).

[12]Available at: http://www.globalchange.gov/publications/reports/scientific-assessments/us-impacts (accessed July 12, 2012).

[13]See the overview statement for the U.S. Global Change Research Program at: http://www.globalchange.gov/about/overview (accessed July 12, 2012).

TEMPERATURE TRENDS

Temperature changes in North America vary by location (Figure 6.2). According to the U.S. Environmental Protection Agency,[14] average temperatures have risen across the lower 48 states since 1901, with an increased rate of warming over the past 30 years. Eight of the top 10 warmest years on record for the lower 48 states have occurred since 1990, and the last 10 five-year periods have each sequentially been the warmest five-year periods on record. (Average global temperatures show a similar trend, and 2001 to 2011 was the warmest decade on record worldwide.) Temperatures in parts of the North, the West, and Alaska have increased most. U.S. average temperature has risen more than 1.1°C (2°F) over the past 50 years and will rise more in the future; how much more depends on the amount of heat-trapping gas that is emitted globally and the sensitivity of the climate system to further perturbations.

Many types of extreme weather events, such as heat waves, intense rainfall, and regional droughts, have become more frequent and strong during the past 40 to 50 years and are projected to increase as global warming continues.[15] Medical experts are worried that the number of extremely hot days is likely to increase in the future. The elderly, persons with medical disabilities, children, and others with limiting physical conditions (that normally cause little problem) are very vulnerable on extremely hot days,[16] especially when power-generating facilities cease working owing to excess demand (blackout). During a blackout, air conditioning is not available, elevators stop working, and streetlights go out. Home medical equipment might not run, and because telecommunications fail, it may be impossible for emergency vehicles and personnel to respond to problems.

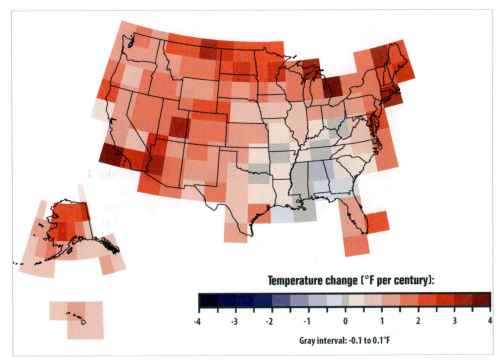

Temperature change [°F per century]:

-4   -3   -2   -1   0   1   2   3   4

Gray interval: -0.1 to 0.1°F

Figure 6.2. Rate of temperature change in the United States, 1901–2008. This figure shows how average air temperatures have changed in different parts of the United States since the early 20th century (since 1901 for the lower 48 states, 1905 for Hawaii, and 1918 for Alaska).

Source: United States Environmental Protection Agency, "Climate Change Indicators in the United States," http://www.epa.gov/climatechange/indicators.html (accessed July 12, 2012).

[14]U.S. Environmental Protection Agency, "Climate Change Indicators in the United States," 2010, http://www.epa.gov/climatechange/indicators.html (accessed July 12, 2012).

[15]S. Rahmstorf and D. Coumou, "Increase of Extreme Events in a Warming World."

[16]K. L. Ebi, J. Balbus, P.L. Kinney, et al., "Effects of Global Change on Human Health." In J.L. Gamble (ed.), K.L. Ebi, F.G. Sussman, and T.J. Wilbanks (authors), *Analyses of the Effects of Global Change on Human Health and Welfare and Human Systems, Synthesis and Assessment Product 4.6* (Washington, D.C., U.S. Environmental Protection Agency, 2008), pp. 39–87.

Figure 6.3. Compared to the period 1961–1979, most areas of the United States have warmed 0.5°C to 1°C (1°F to 2°F), resulting in longer warm seasons and shorter, less-intense cold seasons. By the middle and end of the century, the average U.S. temperature is projected to increase by 4°C to 6°C (7°F to 11°F) under the higher-emissions scenario and by approximately 2°C to 3.6°C (4°F to 6.5°F) under the lower-emissions scenario.

Source: USGCRP, http://www.globalchange.gov/resources/gallery?func=viewcategory&catid=1 (accessed July 12, 2012).

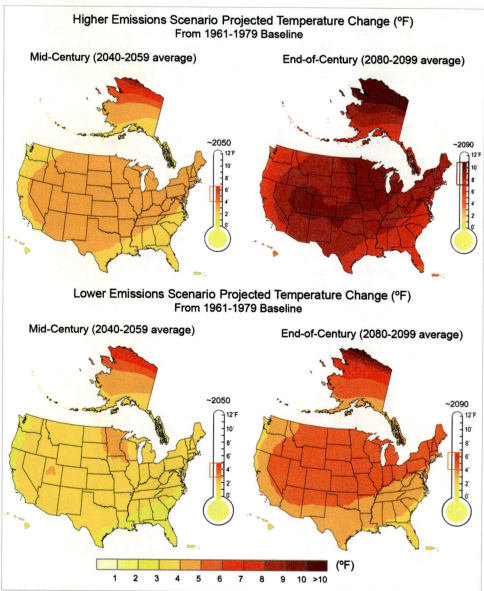

Temperature increases in the next couple of decades are already largely determined by recent emissions. As a result (Figure 6.3), the likely temperature by 2020 is nearly the same regardless of whether there are high or low greenhouse gas emissions. But toward the middle and end of the century, temperature may be significantly different depending on whether we control emissions (low scenario) or do not (high scenario). By the end of the century, the average national temperature is projected to increase by approximately 3.8°C to 6.1°C (7°F to 11°F) under the higher-emissions scenario and by 2.2°C to 3.6°C (4°F to 6.5°F) under the lower-emissions scenario. These ranges are due to differences among climate model results for the same emissions scenarios.

PRECIPITATION TRENDS

A simple general rule among climatologists studying climate change is "wet areas will become wetter and dry areas will become drier." According to the USGCRP, annual precipitation (including rain and snowfall) has increased across the United States an average of about 5% over the past 50 years (Figure 6.4); however, there are important regional and seasonal differences. In the Northeast and throughout the Great Plains and Midwest, precipitation has increased. In the Southeast, decreased precipitation has occurred in winter, spring, and summer (but not fall), affecting the growing seasons and survivability of economically important crops. In the Northwest, decreases have occurred in summer, fall, and winter (but not spring). Precipitation has also generally decreased during the summer and fall in the Southwest, whereas winter and spring, which are the wettest seasons in California and Nevada, have had increased precipitation.

Projections of future precipitation generally indicate that southern areas, particularly in the west, could become drier and that northern areas could become wetter. The amount of rain falling in the heaviest downpours has increased approximately 20% on average in the past century, and this trend is very likely to continue, with the largest increases in the wettest places.

Despite the overall increase in precipitation across North America, drought has grown to be a major worry in many states. A study by the Massachusetts Institute of Technology[17] using climate models predicts that droughts due to changes in precipitation and temperature could increase, with "very substantial and almost universally experienced increases in drought risk" by 2050. Results indicate that drought severity will vary by region, with the Southwest and Rocky Mountain states likely to experience the largest increases in drought frequency. Study authors concluded that climate change could increase the longevity of droughts in many regions, causing droughts that would otherwise be mild to become severe or even extreme.

Historical data point to no clear trend in drought for the nation as a whole. Nonetheless, the USGCRP reports that increasing temperatures over the past five decades have made droughts more severe and widespread. If precipitation had

Figure 6.4. Over the past 50 years the annual average precipitation in the U.S. has increased about 5%, although there have been important regional differences.

SOURCE: USGCRP, http://www.globalchange.gov/resources/gallery?func=viewcategory&catid=1 (accessed July 12, 2012).

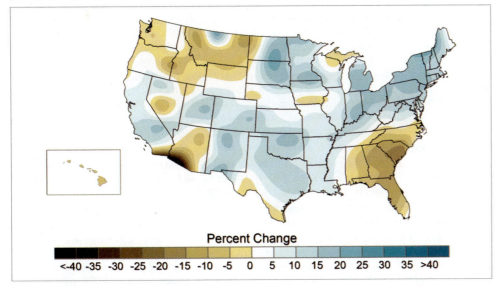

[17]K. Strzepek, G. Yohe, J. Neumann, and B. Boehlert, "Characterizing Changes in Drought Risk for the United States from Climate Change," *Environmental Research Letters* 5, no. 044012 (2010): 1–9.

not also increased, higher temperatures would have extended the area in severe to extreme drought, perhaps by as much as 30%.[18]

Studies indicate that drought is likely to grow in frequency and severity in some regions.[19] For instance, as the dry subtropics expand northward, the Southwest is likely to experience increasing periods of severe drought.[20] Rising air temperatures have also led to earlier melting of the seasonal snowpack in the western region.[21] Because western states depend on runoff from melting snow and ice, changes in the timing and amount of runoff can aggravate existing problems with already limited water supplies in the region.

CLIMATE IMPACTS TO PLANNING SECTORS

When planning for the future, decision-makers typically consider seven planning sectors. These are water resources, energy supply and use, transportation, agriculture, ecosystems, human health, and society. Climate change affects each of these.

Water Resources

The water cycle is powered by solar energy. When the troposphere becomes warmer—as is happening now—the process is accelerated. Evaporation rates increase, which increases the amount of moisture circulating in the atmosphere, leading to an increase in the frequency of intense rainfall and snowfall events, mainly over land areas. More precipitation falls as rain rather than snow, which causes the peak discharge of streams in the spring season to arrive earlier; the danger of summer drought increases; and there is less freshwater available in summer and fall for agriculture, drinking, and other human uses, when demand is highest. In addition, more precipitation comes in the form of heavier rains and snow storms rather than light events.[22]

Over the past 50 years the heaviest 1% of rain events increased by nearly 20%[23]; increases were greatest in the Northeast. For every 0.55°C (1°F) rise in temperature, the capacity of the atmosphere to hold water vapor increases by about 4%.[24] In addition, with the expansion of the tropics and other climate zones, changes in atmospheric circulation shift storm tracks northward. As a result, dry areas experiencing a decrease in storminess can become drier and wet areas experiencing an increase in storminess can become wetter. The arid Southwest and South are projected to

[18]K. E. Kunkel, P.D. Bromirski, H.E. Brooks, et al., "Observed Changes in Weather and Climate Extremes." In T. R. Karl, G. A. Meehl, C. D. Miller, et al. (eds.), *Weather and Climate Extremes in a Changing Climate: Regions of Focus: North America, Hawaii, Caribbean, and U.S. Pacific Islands.* Synthesis and Assessment Product 3.3. (Washington, DC, U.S. Climate Change Science Program, 2008), pp. 35–80.

[19]W. J. Gutowski, G.C. Hegerl, G.J. Holland, et al., "Causes of Observed Changes in Extremes and Projections of Future Changes." In T. R. Karl, G. A. Meehl, C. D. Miller, et al. (eds.), *Weather and Climate Extremes in a Changing Climate: Regions of Focus: North America, Hawaii, Caribbean, and U.S. Pacific Islands.* Synthesis and Assessment Product 3.3. (Washington, D.C., U.S. Climate Change Science Program, 2008), pp. 81–116.

[20]E. R. Cook, P.J. Bartlein, N. Diffenbaugh, et al., "Hydrological Variability and Change." In *Abrupt Climate Change.* Synthesis and Assessment Product 3.4. (Reston, VA, U.S. Geological Survey, 2008), pp. 143–257.

[21]P. Lemke, J. Ren, R. B. Alley, et al., "Observations: Changes in Snow, Ice and Frozen Ground." In S. Solomon, D. Qin, M. Manning, et al. (eds.), *Contribution of Working Group I to the Fourth Assessment Report (AR4) of the Intergovernmental Panel on Climate Change* (Cambridge, U.K., Cambridge University Press, 2007). pp. 337–383.

[22]See the video at the end of the chapter: "Climate Change and the Water Cycle"; http://www.youtube.com/watch?v=BIbys6VQpVk&feature=related (accessed July 12, 2012).

[23]W. J. Gutowski, G. C. Hegerl, G. J. Holland, et al., "Causes of Observed Changes in Extremes and Projections of Future Changes." In T. R. Karl, G. A. Meehl, C. D. Miller, et al. (eds.), *Weather and Climate Extremes in a Changing Climate: Regions of Focus: North America, Hawaii, Caribbean, and U.S. Pacific Islands.* Synthesis and Assessment Product 3.3. (Washington, D.C., U.S. Climate Change Science Program, 2008), pp. 81–116.

[24]G. C. Hegerl, F. W. Zwiers, P. Braconnot, et al., "Understanding and Attributing Climate Change." In S. Solomon, D. Qin, M. Manning, et al. (eds.), *Contribution of Working Group I to the Fourth Assessment Report (AR4) of the Intergovernmental Panel on Climate Change* (Cambridge, U.K., Cambridge University Press, 2007), pp. 663–745.

experience longer and more-severe droughts from the combination of increased evaporation and reductions in precipitation.[25] Precipitation and runoff are likely to increase in the Northeast and Midwest in winter and spring. Regional changes are summarized in Box 6.1.

Are U.S. states preparing for these changes in water resources? The answer is "maybe." One study[26] found that more than one in three counties in the United States could face a "high" or "extreme" risk of water shortages due to climate change by the middle of the 21st century. As climate change affects communities across the continent, some states[27] are preparing for the impacts on water resources (New York, Pennsylvania, California, Oregon, Alaska, and Washington). These states are reducing carbon pollution and planning for climate change effects. Yet many states are not acting and remain unprepared including Texas, Alabama, Ohio, Indiana, Kansas, Montana, North Dakota, and South Dakota, and others.

Transportation

Transportation is a huge daily activity in North America. People and materials are moved by vast fleets of cars, trains, airplanes, trucks, and ships. The great majority of passenger travel occurs by automobile for shorter distances and by airplane for longer distances. In descending order, most types of cargo travel by railroad, truck, pipeline, or boat; air shipping is typically used for perishables and premium express shipments.[28] According to the U.S. Department of Transportation,[29] employment in the national transportation and material movement industry accounts for approximately 7.4% of all employment and more than $1 out of every $10 produced in the U.S. gross domestic product.

Climate change poses definite threats to transportation activities. According to a 2008 study by the National Research Council (NRC),[30] five categories of climate change are of particular concern: increases in very hot days and heat waves, increases in Arctic temperatures, rising sea levels, increases in intense precipitation events, and increases in hurricane intensity. These changes in the environment will have significant effects on transportation, affecting the way professionals design and maintain the system of roads, airports, harbors, bridges, rail lines, and other elements that keep transportation moving. Decisions made today will affect how well the transportation system adapts to climate change in the future.[31]

SEA-LEVEL RISE AND STORM SURGE   Historically, the growth of communities has been tied to the transportation advantages of ports and harbors. But coastal areas are repeatedly assaulted by high waves and winds, storms, and tsunamis, and if sea

[25]J. H. Christensen, B. Hewitson, A. Busuioc, et al., "Regional Climate Projections." In S. Solomon, D. Qin, M. Manning, et al. (eds.), *Contribution of Working Group I to the Fourth Assessment Report (AR4) of the Intergovernmental Panel on Climate Change* (Cambridge, U.K., Cambridge University Press, 2007), pp. 847–940.

[26]S. Roy, L. Chen, E. Girvetz, et al., "Projecting Water Withdrawal and Supply for Future Decades in the U.S. under Climate Change Scenarios," *Environmental Science & Technology* 46, no. 5 (2012): 2545–2556, 120210153558000 doi: 10.1021/es2030774.

[27]See the National Resources Defense Council study at http://www.nrdc.org/water/readiness/ (accessed July 12, 2012).

[28]See "Transportation in the United States," *Wikipedia*, http://en.wikipedia.org/wiki/Transportation_in_the_United_States (accessed July 12, 2012).

[29]See the website http://www.bts.gov/publications/freight_in_america/html/nations_freight.html (accessed July 12, 2012).

[30]Transportation Research Board, *Special Report 290: Potential Impacts of Climate Change on U.S. Transportation*, Committee on Climate Change and U.S. Transportation, Transportation Research Board, Division on Earth and Life Studies (Washington, D.C., National Research Council of the National Academies, 2008).

[31]T. R. Karl, J. M. Melillo, and T. C. Peterson, "Global Climate Change Impacts in the United States," 2009, http://www.globalchange.gov/publications/reports/scientific-assessments/us-impacts (accessed July 12, 2012).

BOX 6.1

Regional Changes in Precipitation and Runoff

Coastal Alaska, Yukon River Basin, Coastal British Columbia

- Increased spring flood risks[1]
- Retreat and disappearance of glaciers, leading to impacts on stream discharge and associated aquatic ecology
- Flooding of coastal wetlands by sea-level rise
- Changes in estuary salinity and ecology

Pacific Coast States: Alaska, Washington, Oregon, California

- More winter rainfall and less snowfall
- Earlier spring peak in stream runoff
- Increased fall and winter flooding
- Decreased summer water supply,[2] which is likely to produce changes in estuary and stream ecologies and negatively affect availability of water for irrigation

Washington, Oregon, Idaho, Montana, Wyoming

- Increasing drought

California, Nevada, Utah, Arizona, New Mexico, parts of Oklahoma and Colorado

- Decreased runoff

Rocky Mountain Region[3]

- Rise in the snow line, switch from snowfall to rainfall earlier in spring and late winter, earlier snowmelt
- More frequent rain or snow
- Earlier peaks in stream discharge and reductions in summer stream discharge and summer soil moisture
- Likely decrease in surface runoff by mid-century
- Rising stream temperatures[4] with impacts on stream ecology and species composition

Southwest[5]

- Changes in snowpack and runoff leading to declines in groundwater recharge (freshwater used for drinking and irrigation)[6]
- Increased stream water temperatures that will change aquatic ecosystems
- Higher frequency of intense precipitation events and risk of flash floods

---

[1]A. Loukas and M. Quick, "The Effect of Climate Change on Floods in British Columbia," *Nordic Hydrology* 30 (1999): 231–256.

[2]J. M. Melack, J. Dozier, C. R. Goldman, et al., "Effects of Climate Change on Inland Waters of the Pacific Coastal Mountains and Western Great Basin of North America," *Hydrological Processes* 11 (1997): 971–992. See also A. F. Hamlet and D. P. Lettenmaier, "Effects of Climate Change on Hydrology and Water Resources in the Columbia River Basin," *Journal of the American Water Resources Association* 35, no. 6, (1999): 1597–1624.

[3]M. W. Williams, M. Losleben, N. Caine, and D. Greenland, "Changes in Climate and Hydrochemical Responses in a High-Elevation Catchment in the Rocky Mountains, USA," *Limnology and Oceanography* 41, no. 5 (1996): 939–946.

[4]F. R. Hauer, J. S. Baron, D. H. Campbell, et al., "Assessment of Climate Change and Freshwater Ecosystems of the Rocky Mountains, USA and Canada," *Hydrological Processes* 11 (1997): 903–924.

[5]B. Hurd, N. Leary, R. Jones, and J. Smith, "Relative Regional Vulnerability of Water Resources to Climate Change," *Journal of the American Water Resources Association* 35, no. 6 (1999): 1399–1410.

[6]U.S. Environmental Protection Agency, *Climate Change and Arizona*. Publication EPA 236-F-98-007c (Washington, D.C., Environmental Protection Agency, 1998).

Midwest and Canadian Prairies[7]

- Annual stream flow could increase in some areas and decrease in others, but a decrease in summer stream discharge is expected
- Severe drought[8] and heat waves are likely to become more frequent
- Semiarid zones may become drier

Arctic and Sub-Arctic Coasts of Alaska and Canada[9]

- Arctic sea ice is retreating and thinning with each year, resulting in a 1- to 3-month extension of the annual ice-free seasons
- Where possible, ecosystems will shift to more northerly and higher elevation conditions as they give way to ecosystems migrating from the south
- Decreased ice cover produces a positive-feedback effect of reduced sunlight reflection, and increased ground heating could lead to acceleration of snow and ice cover loss

Great Lakes Region[10]

- Precipitation increases, with lake level declines due to reduced snow pack melt and resulting runoff
- Reduced ice cover, or some years with no ice cover
- Changes in zooplankton and phytoplankton biomass
- Northward migration of fish species and loss of coldwater species
- Reduced runoff[11] leading to reduced hydro power production and shallow shipping channels

Northeast and East[12]

- Decreased snow cover and large reductions in stream flow
- Rising sea level and decreasing sea ice could produce accelerated coastal erosion, saline intrusion into coastal aquifers, and changes in the magnitude and timing of ice freeze-up and break-up on lakes and cold coastal regions, which can affect spring flooding,[13] eliminate bog ecosystems, and shift the distribution and migration patterns of fish species

Southeast, Gulf, and Mid-Atlantic regions

- Heavily populated coastal floodplains at risk from flooding from extreme precipitation and hurricanes
- Lower base flow to streams, larger peak flows, and longer droughts possible, as well as precipitation increase and changes in runoff and stream discharge
- Prolonged drought of growing severity likely in the Southeast,[14] even while heavy rainfall increases threat of flooding
- Gulf of Mexico hypoxic zone and other impacts to coastal systems likely to continue to grow as a result of polluted runoff,[15] sea-level rise, accelerated coastal erosion, and saltwater intrusion to low-lying coastal plain aquifers
- Changes to estuarine and wetland ecosystems, biotic processes, and species distribution also possible

---

[7]C.A. Woodhouse and J.T. Overpeck, "2000 Years of Drought Variability in the Central United States," *American Meteorological Society Bulletin* 79 (1998): 2693–2714.

[8]D. M. Wolock and G. J. McCabe, "Estimates of Runoff Using Water-Balance and Atmospheric General Circulation Models," *Journal of the American Water Resources Association* 35, no. 6 (1999): 1341–1350.

[9]B. Maxwell, *Responding to Global Climate Change in Canada's Arctic*, Vol. II of the Canada Country Study: Climate Impacts and Adaptation. (Downsview, Ontario: Environment Canada, 1997).

[10]N. L. Hofmann, et al. *Climate Change and Variability: Impacts on Canadian Water*, Vol. II of the Canada Country Study: Climate Impacts and Adaptation. (Downsview, Ontario: Environment Canada, 1998).1–120.

[11]P. Chao, "Great Lakes Water Resources: Climate Change Impact Analysis with Transient GCM Scenarios," *Journal of the American Water Resources Association* 35, no. 6 (1999): 1499–1508.

[12]F. K. Hare, R. B. B. Dickinson, and S. Ismail, "Climatic Variation over the Saint John Basin: An Examination of Regional Behavior," *Climate Change Digest*, CCD 1997. 97-02, Atmospheric Environment Service, Toronto.

[13]M. V. Moore, M. L. Pace, J. R. Mather, et al., "Potential Effects of Climate Change on Freshwater Ecosystems of the New England/Mid-Atlantic Region," *Hydrological Processes* 11 (1997): 925–947.

[14]P. J. Mulholland, G. R. Best, C. C. Coutant, et al., "Effects of Climate Change on Freshwater Ecosystems of the Southeastern United States and the Gulf Coast of Mexico," *Hydrological Processes* 11 (1997): 949–970.

[15]J. F. Cruise, A. S. Limaye, and N. Al-Abed, "Assessment of Impacts of Climate Change on Water Quality in the Southeastern United States," *Journal of the American Water Resources Association* 35, no. 6 (1999): 1539–1550.

level rises, the potential for them to do even worse damage increases. In one study[32] it was calculated that today's 100-year floods could instead occur every decade or two because of the effects of sea-level rise.

Along the Gulf Coast area, an estimated 3,862 km (2,400 mi) of major roadway and 395 km (246 mi) of freight rail lines are at risk of marine inundation within the next 50 to 100 years owing to a combination of sea-level rise and land subsidence.[33] Because the Gulf Coast transportation network is interdependent and relies on minor roads and other low-lying infrastructure, the risks of service disruptions are likely to be even greater.

Sea-level rise causing marine and groundwater inundation of roads, railroads, airports, seaports, and pipelines would potentially affect commercial transportation activity valued in the hundreds of billions of dollars annually. The NRC study concluded that six of the nation's top 10 freight gateways may be threatened by sea-level rise. Seven of the 10 largest ports are located on the Gulf Coast. The region is also home to the U.S. oil and gas industry, with its offshore drilling platforms, refineries, and pipelines. Roughly two thirds of all U.S. oil imports are transported through the region.

Global climate change is viewed as having high potential to change the nature of storms, especially hurricanes, and their impact on the coast. Some studies have identified an increase in storm frequency for some areas and a decrease in others.[34] More-intense storms, especially when coupled with sea-level rise, could result in far-reaching and damaging storm surges. An estimated 96,560 km (60,000 mi) of coastal highway are already exposed to periodic flooding from coastal storms and high waves.[35]

RAINFALL INTENSITY Although total precipitation has increased by only 5%, the heaviest 1% of events increased by 20%.[36] Intense precipitation can cause severe damage to transportation assets. For instance, the Great Flood of 1993[37] caused catastrophic flooding along 500 miles of the Mississippi and Missouri river system, paralyzing rail, truck, and marine traffic and affecting a fourth of all U.S. freight. During the June 2008 Midwest flood, the second record-breaking flood (statistically defined as the 100-year flood) in 15 years, dozens of levees in Iowa, Illinois, and Missouri were breached or overtopped, and the runoff inundated huge populated areas. Although highway and rail bridges largely survived, access and approach roads and rail lines were under water, and rail, roadway, and marine transport was shut down for weeks. Events like these are likely to occur more often in a warming world.

If more precipitation falls as rain rather than snow in winter and spring, the increased runoff raises the risk of landslides, mudflows, stream floods, and rockfalls, which can prompt road closures, more road repair, and reconstruction of rail lines and roadways. More-frequent heavy precipitation also causes increases in weather-related accidents, traffic and rail delays, and disruptions in a network

[32]N. Lin, K. Emanuel, M. Oppenheimer, and E. Vanmarcke, "Physically Based Assessment of Hurricane Surge Threat under Climate Change," *Nature Climate Change* 2 (2012): 462–467, doi: 10.1038/nclimate1389.

[33]R. S. Kafalenos, K .J. Leonard and D. M. Beagan, "What Are the Implications of Climate Change and Variability for Gulf Coast Transportation?" In M. J. Savonis, V. R. Burkett, and J. R. Potter (eds.), *Impacts of Climate Change and Variability on Transportation Systems and Infrastructure: Gulf Coast Study, Phase I.* Synthesis and Assessment Product 4.7 (Washington, D.C., U.S. Department of Transportation, (2008), pp. 4–1 to 4F-27.

[34]T. Li, M. H. Kwon, M. Zhao, et al., "Global Warming Shifts Pacific Tropical Cyclone Location," *Geophysical Research Letters* 37 (2010): L21804, doi:10.1029/2010GL045124.

[35]Transportation Research Board Special Report 290 (2008).

[36]K. E. Kunkel, P. D. Bromirski, H. E. Brooks, et al., "Observed Changes in Weather and Climate Extremes."

[37]Transportation Research Board Special Report 290 (2008).

already challenged by increasing congestion.[38] Local governments[39] also need to anticipate and budget for the impact of increased flooding on evacuation routes, construction activities, urban infrastructure, and congested traffic locations (e.g., commuter choke points).

EXTREME HEAT As the world warms, the frequency and duration of days characterized by extreme heat will increase. Extreme heat, especially when 32.2°C (90°F) and above for sustained periods (Figure 6.5), affects the transportation sector in several costly and potentially dangerous ways: Asphalt softens and can develop ruts from heavy traffic, affecting the safe operation of cars and trucks; railroad tracks can warp and deform, leading to speed restrictions and derailment in the worst cases; transportation vehicles of all types can overheat; and tires can deteriorate, leading to concerns about safe operation and raising maintenance costs.[40]

Extreme heat also raises the possibility of health and safety problems for highway workers, construction crews, and vehicle operators. The U.S. Occupational Safety and Health Administration states that concern over heat stress for moderate to heavy work begins at about 26.6°C (80°F) and varies from place to place depending on humidity levels, urban heat island (the tendency of cities to be hotter than surrounding countryside) effects, and winds.[41]

DROUGHT Drought occurs when rising air temperatures, especially when accompanied by decreasing precipitation, increase evaporation and create dry conditions. In a warmer world, the impact of increasing drought will be even greater. In North America, even in those parts where total annual precipitation might not decrease, the frequency of rainfall and snowfall events are projected to drop.[42]

Drought causes significant problems for transportation activities. For example, because of drought, wildfires are projected to grow in duration, frequency, and intensity, especially in the southwest. These catastrophic events threaten communities and infrastructure and cause road and rail closures in affected areas. Increased susceptibility to wildfires during droughts threatens roads and other transportation infrastructure directly, or it causes road closures because of fire threat or reduced visibility, such as has occurred in Texas, Oklahoma, New Mexico, Florida, California, and other states. Areas deforested by wildfires are also at increased susceptibility to mudslides. River transport is seriously affected by drought, with reductions in the routes available, shipping season, and cargo-carrying capacity.

[38]J. R. Potter, V.R. Burkett, and M.J. Savonis, "Executive Summary." In M. J. Savonis, V. R. Burkett, and J. R. Potter (eds.), *Impacts of Climate Change and Variability on Transportation Systems and Infrastructure: Gulf Coast Study, Phase I*. Synthesis and Assessment Product 4.7. (Washington, D.C., U.S. Department of Transportation, (2008), pp. ES-1 to ES-10.

[39]R. S. Kafalenos, K. J. Leonard, D. M. Beagan, et al., "What Are the Implications of Climate Change and Variability for Gulf Coast Transportation?" In M. J. Savonis, V. R. Burkett, and J. R. Potter (eds.), *Impacts of Climate Change and Variability on Transportation Systems and Infrastructure: Gulf Coast Study, Phase I*. Synthesis and Assessment Product 4.7 (Washington, D.C., U.S. Department of Transportation, (2008), pp. 4–1 to 4F-27.

[40]C. B. Field, L. D. Mortsch, M. Brklacich, et al., "North America." In M.L. Parry, O.F. Canziani, J.P. Palutikof, P.J. van der Linden, and C.E. Hanson (eds.), *Climate Change 2007: Impacts, Adaptation and Vulnerability. Contribution of Working Group II to the Fourth Assessment Report of the Intergovernmental Panel on Climate Change* (Cambridge, U.K., Cambridge University Press, 2007), pp. 617–652.

[41]Occupational Safety and Health Administration, "Heat Stress." In OSHA Technical Manual, Section III: Chapter 4. (Washington, D.C., Occupational Safety and Health Administration, 2008), http://www.osha.gov/dts/osta/otm/otm_iii/otm_iii_4.html (accessed July 12, 2012).

[42]W. J. Gutowski, G. C. Hegerl, G. J. Holland, et al., "Causes of Observed Changes in Extremes and Projections of Future Changes." In T. R. Karl, G. A. Meehl, C. D. Miller, et al. (eds.), *Weather and Climate Extremes in a Changing Climate: Regions of Focus: North America, Hawaii, Caribbean, and U.S. Pacific Islands*. Synthesis and Assessment Product 3.3 (Washington, D.C., U.S. Climate Change Science Program, 2008), pp. 81–116.

Figure 6.5. **a,** The average number of days per year when the maximum temperature exceeded 32.2°C (90°F) from 1961 to 1979. **b,** The projected number of days per year above 32.2°C (90°F) by the 2080s and 2090s for lower emissions of greenhouse gases. **c,** The projected number of days per year above 32.2°C (90°F) by the 2080s and 2090s for higher emissions of greenhouse gases. Much of the southern United States is projected to have more than twice as many days per year above 32.2°C (90°F) by the end of this century.

Source: USGCRP, http://downloads.globalchange.gov/usimpacts/pdfs/Global.pdf (accessed July 12, 2012).

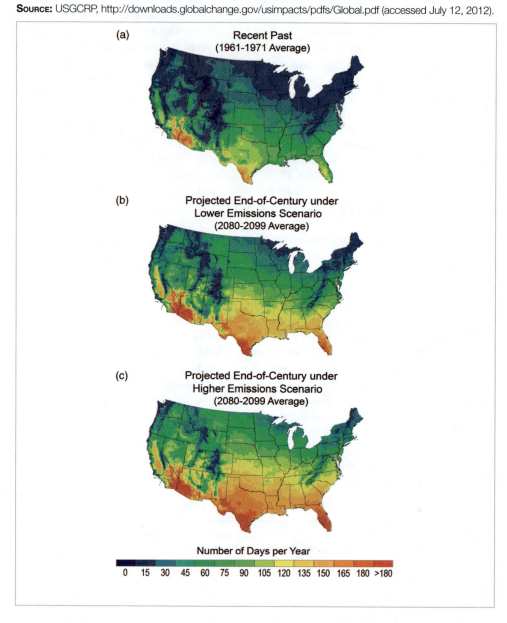

STORMINESS Future projections[43] of tropical cyclones indicate that greenhouse warming will cause the globally averaged intensity of tropical cyclones to shift toward stronger storms, with intensity increases of 2% to 11% by 2100. Studies also project the globally averaged frequency of tropical cyclones to decrease by 6% to 34%. Modeling studies project substantial increases in the frequency of the most intense cyclones and increases in the precipitation rate on the order of 20% within 100 km of the storm center. As a result, the transportation sector would experience precipitation impacts, wind impacts, and storm surge impacts. Stronger hurricanes have longer periods of intense precipitation, and the amount of rainfall is expected to be higher.

[43]T. Knutson, J. McBride, J. Chan, et al., "Tropical Cyclones and Climate Change," *Nature Geoscience* 3 (2010): 157–163, doi:10.1038/ngeo779.

Higher wind speeds lead to greater damage, and damage increases exponentially with wind speed.[44] Higher wind speeds and low air pressures in a storm produce higher storm surges and waves. These increases in transportation vulnerability require new methods of planning for future impacts, because statistics based on historical patterns are less likely to provide accurate projections of future conditions.[45]

Storms have costly results: a higher probability of infrastructure failures, such as damaged decking on bridges, washed away roads and rail lines, debris left on roads and rail lines, emergency evacuations, damage to signs and lighting fixtures, and reduction in the useful life of highways and rail lines exposed to flooding. On the Gulf Coast, more than one third of the railways are likely to flood when hit by a storm surge of 5.5 m (18 ft).[46]

Shipping is especially vulnerable to major storms. Freighters have to be diverted around storms, and their sailing schedules are delayed. Work on offshore drilling platforms, coastal pumping facilities, sewage-treatment plants, and other marine-dependent activities comes to a halt, and costly evacuations are needed. Infrastructure associated with these activities is heavily damaged by high winds, waves, and storm surge. Harbor infrastructure, such as cranes, docks, and other terminal facilities, damaged during severe storms costs billions of dollars to replace.

Ecosystems

An ecosystem is an interdependent system of plants, animals, and microorganisms. The natural resources that humans depend on are largely made possible by the healthy state of ecosystems around the world. The air we breathe, clean water, lumber, food, and even our safety from a number of natural hazards (such as landslides, flooding, and others) are due in part or in whole to a healthy planet made up of healthy ecosystems. The key to a healthy ecosystem is the interdependence of its components, and if one or more of these components is negatively affected by global warming, the entire system is less robust and less resistant to stress. Without the support of the other organisms within their own ecosystem, life forms would not survive, much less thrive.[47]

High-altitude and high-latitude ecosystems across the world have already been negatively affected by changes in climate. The Intergovernmental Panel on Climate Change (IPCC) reviewed studies of biological systems and concluded[48] that 20% to 30% of species assessed may be at risk for extinction from climate change impacts in this century if global mean temperatures exceed 2°C to 3°C (3.6°F to 5.4°F) relative to preindustrial levels. The USGCRP[49] concludes that ecosystem processes have been affected by climate change, and they document the following developments:

- Large-scale shifts have occurred in the ranges of species and the timing of the seasons and animal migration.
- Fires, insect pests, disease pathogens, and invasive weed species have increased.

[44]C. W. Landsea, "A Climatology of Intense (or Major) Atlantic Hurricanes," *Monthly Weather Review* 121, no. 6 (1993): 1710–1713.

[45]J.E. Hay, R. Warrick, C. Cheatham, et al., "*Climate Proofing: A Risk-Based Approach to Adaptation*" (Manila, Asian Development Bank, 2005), http://www.adb.org/Documents/Reports/Climate-Proofing/default.asp (accessed July 12, 2012).

[46]R. S. Kafalenos, K. J. Leonard, D. M. Beagan, et al., "What Are the Implications of Climate Change and Variability for Gulf Coast Transportation?" In M. J. Savonis, V. R. Burkett, and J. R. Potter (eds.), *Impacts of Climate Change and Variability on Transportation Systems and Infrastructure: Gulf Coast Study, Phase I.* Synthesis and Assessment Product 4.7. (Washington, D.C., U.S. Department of Transportation, (2008), pp. 4-1 to 4F-27.

[47]U.S. Environmental Protection Agency, *Climate Change-Health and Environmental Effects, Ecosystems and Biodiversity*, http://www.epa.gov/climatechange/effects/eco.html (accessed July 12, 2012).

[48]M.L. Parry, O.F. Canziani, J.P. Palutikof, P.J. van der Linden, and C.E. Hanson (eds.), "Climate Change 2007: Impacts, Adaptation and Vulnerability. Contribution of Working Group II to the Fourth Assessment Report of the Intergovernmental Panel on Climate Change" (Cambridge, U.K., Cambridge University Press, 2007).

[49]T. R. Karl, J. M. Melillo, and T.C. Peterson, *Global Climate Change Impacts in the United States* (New York, Cambridge University Press, 2009), http://www.globalchange.gov/publications/reports/scientific-assessments/us-impacts (accessed July 12, 2012).

- Deserts and semiarid lands are likely to become hotter and drier, feeding a self-reinforcing cycle of invasive plants, fire, and erosion.
- Coastal and near-shore ecosystems are already under multiple stresses. Climate change and ocean acidification will exacerbate these stresses.
- Arctic sea ice ecosystems are already being adversely affected by the loss of summer sea ice, and rapidly rising temperatures and further changes are expected.
- The habitats of some mountain species and coldwater fish, such as salmon and trout, are very likely to contract in response to warming.
- Some of the benefits that ecosystems provide to society will be threatened by climate change, and others will be enhanced.

Ecosystems are very sensitive to changing temperatures, shifts in precipitation, variations in seasonal timing, and other processes normally associated with climate change. These shifts in established patterns have a strong influence on the processes that control growth and development in ecosystems. Higher temperatures generally speed up plant growth, rates of decomposition, and how rapidly nutrients are cycled; however, factors, such as extreme temperatures, lack of water, soil desiccation, the spread of hardy weeds, and others also influence these rates.

Researchers[50] have observed that spring now arrives an average of 10 days to two weeks earlier than it did 20 years ago, and the growing season is lengthening over much of North America. Migratory bird species are returning earlier. Northeastern birds that winter in the south now arrive back in the northeast an average of 13 days earlier than they did during the first half of the 20th century. Birds wintering in South America arrive back in the northeast an average of four days earlier. The range boundaries of species have shifted poleward, with a mean velocity of 6 km (3.7 mi) per decade, as well as upward in elevation.[51] Measurements indicate that forest growth has risen over the past several decades owing to young forests reaching maturity sooner, more carbon dioxide in the atmosphere, longer growing seasons, and increased deposition of nitrogen from the atmosphere.[52]

Climate change is also causing the geographic range of species to shift northward and upward in elevation. Trees, flowers, birds and insects face the arrival of spring at ever-earlier dates compared to the past 30 years, providing some of the clearest evidence that nature is responding to climate change. The timing of life-cycle events such as blooming, migration, and insect emergence has changed unevenly, however, prompting concern that further warming could disrupt interactions between species, such as feeding and pollination.[53] For example, the ranges of many butterfly species have expanded northward, contracted at the southern edge, and shifted to higher elevations as warming has continued.

In the future, forest tree species are expected to shift their ranges northward and upslope in response to climate change (Figure 6.6). Some common forest types, such as oak and hickory, are projected to expand; others, such as maple, beech, and birch, are projected to contract. Still others, such as spruce and fir, are likely to disappear from the United States altogether.[54] Although some forests might derive near-term benefits from an extended growing season, longer periods of hot weather could stress trees and make them more susceptible to wildfires, insect damage, and

[50]M. G. Ryan, S. R. Archer, R. Birdsey, et al., "Land Resources." In P. Backlund, A. Janetos, D. Schimel, et al. (eds.), *The Effects of Climate Change on Agriculture, Land Resources, Water Resources, and Biodiversity in the United States*, Synthesis and Assessment Product 4.3 (Washington, D.C., U.S. Department of Agriculture), pp. 75–120.

[51]C. Parmesan, and G. Yohe, "A Globally Coherent Fingerprint of Climate Change Impacts across Natural Systems." *Nature* 421, (2003): 37–42, 2 January, doi:10.1038/nature01286.

[52]M. G. Ryan, S. R. Archer, R. Birdsey, et al., "Land Resources."

[53]R. Wilson and D. Roy, "Ecology: Butterflies Reset the Calendar," *Nature Climate Change* 1 (2011): 101–102, doi:10.1038/nclimate1087.

[54]M. G. Ryan, S. R. Archer, R. Birdsey, et al., "Land Resources."

Figure 6.6. These maps show current and projected future forest types. Major changes are projected for many regions in the United States. For example, in the Northeast, under a mid-range warming scenario, the currently dominant maple–beech–birch forest type is projected to be completely displaced by other forest types in a warmer future.

SOURCE: USGCRP, http://www.globalchange.gov/resources/gallery?func=viewcategory&catid=1 (accessed July 12, 2012).

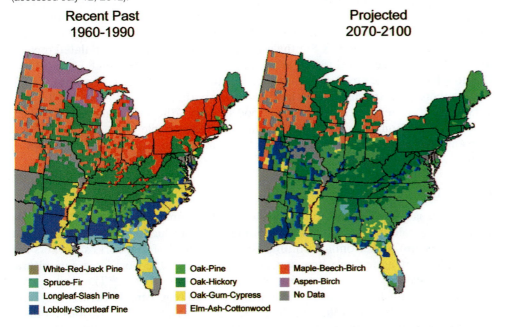

disease. Climate change has likely already increased the size and number of forest fires, insect outbreaks, and tree deaths, particularly in Alaska and the West.[55] The area burned in western U.S. forests from 1987 to 2003 is almost seven times larger than the area burned from 1970 to 1986. In the last 30 years, the length of the wildfire season in the West has increased by 78 days.

Global warming is changing marine ecosystems as well. Oceanic plankton and various species of marine fish are shifting northward into cooler water.[56] The timing of plankton blooms is changing, and coral reefs are experiencing stress related to warmer waters. In 2005, the Caribbean basin experienced high water temperatures that resulted in widespread coral bleaching, with some sites in the U.S. Virgin Islands seeing 90% of the coral bleached. Coral bleaching occurs when symbiotic algae that provide coral polyps with food leave the polyp as a result of stress, such as high water temperatures. Coral might not immediately die, and may recover if the microscopic algae return, but if the algae do not return the coral usually die within a period of months. Some corals began to recover when water temperatures decreased, but later that year disease appeared, striking the previously bleached and weakened coral. To date, a net 50% of the corals in Virgin Islands National Park have died from the bleaching and disease events. In the Florida Keys, summer bleaching in 2005 was also followed by disease in September. What is the likely future of coral reefs? According to speakers at a meeting of coral reef biologists sponsored by the American Association for the Advancement of Science,[57] the reality may be that reefs face a future of

[55]D.B. Fagre, C.W. Charles, C.D. Allen, et al., *Climate Change Science Program, Thresholds of Climate Change in Ecosystems*. A Report by the U.S. Climate Change Science Program and the Subcommittee on Global Change Research. Reston, Va., U.S. Geological Survey, 2009).

[56]A. Janetos, L. Hansen, D. Inouye, et al., "Biodiversity." In P. Backlund, A. Janetos, D. Schimel, et al. (eds.), *The Effects of Climate Change on Agriculture, Land Resources, Water Resources, and Biodiversity in the United States*, Synthesis and Assessment Product 4.3 (Washington, D.C., U.S. Department of Agriculture), pp. 151–181.

[57]Natural Sciences and Engineering Research Council, "Will Coral Reefs Disappear?" *ScienceDaily* (2010), http://www.sciencedaily.com/releases/2010/02/100221200908.htm (accessed July 12, 2012).

declining coral cover and a breakdown of the physical structure of reefs. But no one is predicting that coral reefs will go extinct; they will continue to survive, but only in certain habitats, such as shaded areas and regions bathed in cool waters.

A number of studies[58] indicate that without major changes in the emission of greenhouse gases within the next decade, severe ecosystem effects are likely by the end of the century. If allowed to occur, the following conditions may have a negative impact on the quality of human life, health, and happiness: high temperature rise, especially over land—some 5°C to 6°C (10°F) over much of the United States; Dust Bowl conditions over the U.S. Southwest and many other heavily populated regions around the globe; sea level rise of around 32 cm (1 ft) by 2050, then 80 cm to 1.8 m (2.8 to 6 ft) (or more) by 2100, rising some 15 to 30 cm (6 to 12 in) (or more) each decade thereafter; species loss on land and sea—perhaps 50% or more of all biodiversity; more-extreme weather;[59] loss of food security—the increasingly difficult task of feeding 7 billion, then 8 billion, and then 9 billion people in a world in an ever-worsening climate; myriad direct health effects; and unanticipated effects known as "unknown unknowns."

Agriculture

Agriculture is a key element in the development of human civilization. Prior to the Industrial Revolution, most humans labored for the production of food, animal, and plant goods and fuels from agriculture. Today one third of the world's workers are still employed in agriculture; however, despite the size of the workforce, agricultural production accounts for less than 5% of the gross world product.[60] The major products of agriculture can be broadly categorized as food (e.g., cereals, vegetables, fruits, and meat), fibers (e.g., cotton, wool, hemp, silk, and flax), fuels (e.g., various biofuels, methane, ethanol, and biodiesel), and raw materials (e.g., lumber, bamboo, and plant resins). Food production, among the most important direct activities of agriculture, is an increasingly global concern, especially because of tensions arising from human population growth, global warming impacts to soil and water availability, and international trade.

IMPACTS ON CROPS   Crops respond to changing climate based on the interrelationship of three factors: rising temperatures, changing water resources, and increasing carbon dioxide.[61] Warming air temperatures and increased carbon dioxide generally cause plants that are below their optimum temperature to grow faster, thus producing benefits in the form of increased yields (yield is a measure of agricultural output); however, for some plants, such as cereal crops, faster growth means there is less time for the grain itself to grow and mature, and instead the stalk grows faster. This leads to lower yield. Many weeds, insects, and pathogens also respond positively to increased warmth and CO_2 levels.

Some noncereal crops show positive responses to elevated carbon dioxide and low levels of warming, but higher levels of warming negatively affect the growth of food plants. This is because soil moisture is decreased, competition with invasive weeds increases costs, and the combined effect of these competing factors leads to

[58]A. P., Sokolov, P. H. Stone and C. E. "Forest, Probabilistic Forecast for Twenty-First-Century Climate Based on Uncertainties in Emissions (Without Policy) and Climate Parameters," *Journal of Climate* 22 (2009): 5175–5204, doi: 10.1175/2009JCLI2863.1; International Energy Agency, *World Energy Outlook 2011* (Paris, International Energy Agency, 2011); S. Solomon, G. Kasper R. Plattner, R. Knutti, and P. Friedlingstein, "Irreversible Climate Change due to Carbon Dioxide Emissions," *Proceedings of the National Academy of Sciences* 106 (2009): 1704–1709.

[59]See the video at the end of the chapter: "Research Meteorologists See More Severe Storms Ahead: The Culprit: Global Warming."

[60]See "Agriculture," *Wikipedia*, http://en.wikipedia.org/wiki/Agriculture (accessed July 12, 2012).

[61]J. Hatfield, K. Boote, P. Fay, et al., "Agriculture." In P. Backlund, A. Janetos, D. Schimel, et al. (eds.), *The Effects of Climate Change on Agriculture, Land Resources, Water Resources, and Biodiversity in the United States*, Synthesis and Assessment Product 4.3 (Washington, D.C., U.S. Department of Agriculture), pp. 21–74.

diminished yields. Analysis of crop responses suggests that even a moderate increase in temperature will decrease yields of corn, wheat, sorghum, bean, rice, cotton, and peanut crops in the United States.

As a result of global warming, extreme weather events such as heavy downpours and droughts, extreme temperature days, and an early end to winter are growing in frequency.[62] These reduce crop yields because excesses or deficits of water have negative impacts on plant growth.[63] Rain and snowfall have become less frequent but more intense, a pattern that is projected to continue across the United States.[64] Excessive rainfall delays spring planting, which jeopardizes profits related to early season production of high-value crops, such as melon, sweet corn, and tomatoes. When flooding causes fields to become unusable during the growing season, crop losses occur owing to low oxygen levels in the soil, increased susceptibility to root diseases, and increased soil compaction due to the use of heavy farm equipment on wet soils.

For instance, in spring 2008, heavy rains caused the Mississippi River to rise to about 2.1 m (7 ft) above flood stage. Hundreds of thousands of acres of cropland were inundated just as farmers were preparing to harvest wheat and plant corn, soybeans, and cotton, with net losses estimated at around $8 billion.[65] Some farmers were put out of business and others will be recovering for years to come. The flooding caused severe erosion in some areas and also caused an increase in runoff and leaching of agricultural chemicals into surface water and groundwater.

Weeds, insect pests, and various types of crop and animal diseases benefit from warming, and weeds also benefit from higher carbon dioxide concentration. Some historically aggressive weeds have been confined to the South because they cannot cross certain winter temperature thresholds. For instance, the kudzu vine has invaded 2.5 million acres of the Southeast and is a carrier of the fungal disease soybean rust, which represents a major and expanding threat to U.S. soybean production. Sixty-four percent of the southern soybean crop is lost each year to weeds, whereas only 22% is lost on farms to the north.[66] As winter warming increases (Figure 6.7), these weeds will find a foothold on northern farmland. Stress on crop plants will increase, requiring more attention to pest and weed control. As pesticide and herbicide use increases, so will costs and consumer prices.[67]

EFFECTS ON FARM ANIMALS   High temperature and high humidity stress animals, too. Milk production declines in dairy operations, it takes longer for cows in meat operations to reach their target weight, the conception rate in cattle falls, and swine growth rates decline due to heat. Swine, beef, and milk production are all projected to decline in a warmer world.[68]

Cool night air allows animals stressed by heat to recover. Heat from recent heat waves has, however, not lifted at night and livestock, unable to recover, have died. (Individual states have reported losses of 5,000 head of cattle in a single heat wave.) Warmer winter temperatures, the early arrival of spring, and summer heat also increase the presence of parasites and disease pathogens. The cost of new housing facilities, treatments, food types, medicines, and other animal care logistics necessary to cope with these new stresses is passed on to the consumer.

[62]D. Medvigy, and C. Beaulieu, "Trends in Daily Solar Radiation and Precipitation Coefficients of Variation since 1984," *Journal of Climate* 25 (2011): 1330–1339, doi: 10.1175/2011JCLI4115.1.

[63]B. Kahn, "Innovative Farmers Look to Climate Forecasts for an Edge," NOAA Climate Services, *ClimateWatch Magazine*, http://www.climatewatch.noaa.gov/article/2012/innovative-farmers-look-to-climate-forecasts-for-an-edge (accessed July 12, 2012).

[64]K. E. Kunkel, P.D. Bromirski, H.E. Brooks, et al., "Observed Changes in Weather and Climate Extremes."

[65]National Climatic Data Center, "Climate of 2008: Midwestern U.S. Flood Overview" (2008), http://www.ncdc.noaa.gov/oa/climate/research/2008/flood08.html (accessed July 12, 2012).

[66]D.C. Bridges (ed.), *Crop Losses Due to Weeds in the United States*. (Champaign, Ill., Weed Science Society of America, 1992).

[67]J. Hatfield, K. Boote, P. Fay, et al., "Agriculture."

[68]J. Hatfield, K. Boote, P. Fay, et al., "Agriculture."

Figure 6.7. Winter temperature trends 1975 to 2007. Temperatures are rising fastest in winter, especially in many key agricultural areas. The decrease in the length of the freezing season allows insect pests, weeds, and crop diseases to expand northward and thrive. Over the past 30 years, average winter temperatures in the Midwest and northern Great Plains have increased more than 3.9°C (7°F).

SOURCE: USGCRP, http://www.globalchange.gov/resources/gallery?func=viewcategory&catid=1 (accessed July 12, 2012).

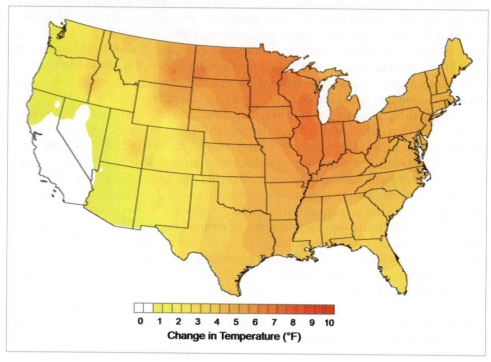

A warming world poses challenges to crop and livestock production. Because some plant species respond positively to warmer conditions and higher CO_2 concentrations, agriculture may be one of the sectors most adaptable to climate change. Even so, increased heat, pests, water stress, diseases, flooding, and weather extremes will require crop and livestock production to adapt.

Society

Human population growth and vulnerability to climate impacts are inextricably linked. In late October 2011, the world population reached 7 billion; in just over 10 years, more than one billion people have been added to the human race, the same number gained over the entire nineteenth century. Although there is evidence that the rate of population growth will slow, it is far from stabilizing: UN predictions indicate that by mid-century, the total number will have reached somewhere between 8 and 10.5 billion.

The consequences of expanding human population look severe for a world with increasingly stressed energy and food supplies. Most of the two to three billion people born between now and 2050 will live in the cities and towns of low-income countries in Africa and Asia, where fertility rates continue to be high. What makes this especially worrisome is that the areas likely to experience most growth are also those likely to be most affected by climate change and least able to cope with the extra demand on resources. At present, 13% of the world's population lives in at-risk coastal areas, and 75% of those people are located in Asia. Just over 50% of the world's inhabitants now live in urban areas; this number will rise to almost 69% by 2050.[69]

[69]"A Scary Statistic," *Nature Climate Change* 1, (2011): doi:10.1038/nclimate1255.

Globally, one billion people are now suffering from food shortages and water scarcity, a figure that could triple within 40 years. Yet the planet is already feeling the strain of its seven billion human inhabitants, documented in lost species, rivers that have run dry, air that is polluted with chemicals, and sprawling development that has degraded ecosystems. If those being born between now and 2050 continue on the development path of the present population, they will experience these effects to an even greater extent. They could also experience a lesser-appreciated outcome of the relationship of extreme climate and overcrowding: conflict. Researchers[70] have found evidence of a link between civil unrest and higher temperatures; one result showed that 21% of 234 conflicts during the period 1950 to 2004 were probably set off as a result of the high temperatures associated with El Niño events. Much of the civil conflict that arises globally is also sparked by resource scarcity and access issues. As the climate warms and populations grow, regions that are hotspots for both are more likely to become conflict zones.

In the United States, more than 80% of the population resides in urban areas. They are among the most rapidly changing environments on Earth[71] and are host to myriad social problems, including neighborhood degradation, traffic congestion, crime, unemployment, poverty, poor air and drinking water quality, and inequities in health and well-being. Urban communities also have unique vulnerabilities to climate change because they are analogous to complex ecosystems consisting of multifaceted and interconnected regional and national economies and infrastructure. The growth in size and complexity compound the impact of increased heat, water shortages, and extreme weather events. The negative influence of these stressors is intensified by the aging infrastructure, buildings, and populations that abound in cities; however, urban settings also present opportunities for adaptation through technology, infrastructure, planning, and design.[72]

Because cities absorb, produce, and retain more heat than the surrounding countryside, they alter local climates through the urban heat island effect. This process has raised average urban air temperatures by 1.1°C to 2.7°C (2°F to 5°F) more than surrounding areas over the past century, and by up to 11°C (20°F) more at night.[73] These temperature increases, on top of warmer air induced by global warming, affect the health, comfort, energy costs, air quality, water quality and availability, and even violent crime rate in urban areas.[74]

Sea-level rise, storm surge, and increased hurricane intensity, projected to grow worse in future decades, are all looming threats for coastal cities. New Orleans, Miami, and New York are particularly at risk, and they would have difficulty coping with the sea-level rise projected by the end of the century under a higher emissions scenario (1 m [3.3 ft] or more of higher sea level).[75] Analyses of population centers in the U.S. Northeast indicate that the potential impacts of climate change are likely to be negative, but that policy changes can reduce vulnerability.[76]

[70] S. M. Hsian, K. C. Meng, and M. A. Cane, "Civil Conflicts Are Associated with Climate Change," *Nature* 476, (2011) 438–441; doi:10.1038/nature10311.

[71] I. van Kamp, K. Leidelmeijer, G. Marsman, and A. de Hollander, "Urban Environmental Quality and Human Well-Being: Towards a Conceptual Framework and Demarcation of Concepts; a Literature Study." *Landscape and Urban Planning*, 65(1–2), (2003): 5–18.

[72] T. J. Wilbanks, P. Kirshen, D. Quattrochi, P. et al., "Effects of Global Change on Human Settlements." In Gamble, J. L. (ed.), K. L. Ebi, F. G. Sussman, and T. J. Wilbanks (authors), *Analyses of the Effects of Global Change on Human Health and Welfare and Human Systems*, Synthesis and Assessment Product 4.6 (Washington, D.C., U.S. Environmental Protection Agency, 2008), pp. 89–109.

[73] S. Grimmond, "Urbanization and Global Environmental Change: Local Effects of Urban Warming," *Geographical Journal* 173, no. 1 (2007): 83–88.

[74] C.A. Anderson, "Heat and Violence," *Current Directions in Psychological Science* 10, no. 1 (2001): 33–38.

[75] M. Vermeer and S. Rahmstorf, "Global Sea Level Linked to Global Temperature," *Proceedings of the National Academy of Sciences* 106, no. 51 (2009): 21527–21532, http://www.pnas.org/content/106/51/21527.full (accessed July 12, 2012).

[76] C. Rosenzweig and W. Solecki (eds.), *Climate Change and a Global City: The Potential Consequences of Climate Variability and Change—Metro East Coast* (New York, Columbia Earth Institute, 2001).

Urban areas concentrate activities that produce heat-trapping emissions and thus afford advantages in managing and limiting these gases.[77] Cities have a large role to play in reducing heat-trapping emissions, and many are pursuing such actions. For example, more than 900 cities have committed to the U.S. Mayors' Climate Protection Agreement to advance emissions-reduction goals by making transportation more efficient and relieving stress on nearby natural settings.[78]

Over the past century, U.S. population growth has been most rapid in the South, near the coasts, and in large urban areas. The four most populous states in 2000—California, Texas, Florida, and New York—account for 38% of the total growth and share significant vulnerability to coastal storms, severe drought, sea-level rise, air pollution, water shortages, and urban heat island effects. But population is shifting toward the Mountain West (Montana, Idaho, Wyoming, Nevada, Utah, Colorado, Arizona, and New Mexico), a region projected to increase by 65% from 2000 to 2030 and representing one third of all U.S. population growth. Simultaneously, populations of southern coastal areas on the Atlantic and on the Gulf of Mexico are projected to continue to grow. As a result, more Americans will be living in the areas most vulnerable to the effects of climate change including flooding, decreased water resources, shifts in weather patterns toward more extreme events, and increased sea-level rise and storm surge.

Heat waves and poor air quality are projected to increase in a warmer world. Research shows[79] that atmospheric conditions that produce heat waves are often accompanied by stagnant air and poor air quality. The simultaneous occurrence of these factors plus drought negatively affects quality of life, especially in cities. Poor air quality resulting from the lack of rainfall, high temperatures, and stagnant conditions during a heat wave can lead to unhealthy air quality days throughout large parts of the country. Climate change is projected to increase the likelihood of such episodes.[80]

Energy Supply and Use

Energy is at the heart of the global warming challenge. The production and use of energy, and the resultant greenhouse gas emissions, are the primary cause of global warming. Climate change, in turn, will affect our production and use of energy. The majority of U.S. greenhouse gas emissions, about 87%, come from energy production and use.[81]

In most American cities the demand for air conditioning will grow while the demand for space heating will decrease. Studies[82] find that the demand for cooling energy increases from 5% to 20% per 1°C (1.8°F) of warming, and the demand for heating energy drops by 3% to 15% per 1°C (1.8°F) of warming. These are only

[77] J.L. Gamble, K.L. Ebi, A. Grambsch, et al., "Introduction." In Gamble, J.L. (ed.), K.L. Ebi, F.G. Sussman, and T.J. Wilbanks (authors), *Analyses of the Effects of Global Change on Human Health and Welfare and Human Systems*, Synthesis and Assessment Product 4.6 (Washington, D.C., U.S. Environmental Protection Agency, 2008), pp. 13–37.

[78] United States Conference of Mayors, "U.S. Conference of Mayors Climate Protection Agreement," as endorsed by the 73rd Annual U.S. Conference of Mayors meeting, Chicago, 2005. http://usmayors.org/climateprotection/agreement.htm (accessed July 12, 2012).

[79] J. X. L. Wang and J. K. Angell, "Air Stagnation Climatology for the United States (1948–1998)." NOAA/Air Resources Laboratory atlas no.1. (Silver Spring, Md., NOAA Air Resources Laboratory, 1999).

[80] L. R. Leung and W. I. Gustafson Jr., "Potential Regional Climate Change and Implications to U.S. Air Quality," *Geophysical Research Letters* 32 (2005): L16711, doi:10.1029/2005GL022911.

[81] T. J. Wilbanks, V. Bhatt, D. E. Bilello, et al., "Effects of Climate Change on Energy Production and Use in the United States." A Report by the U.S. Climate Change Science Program and the Subcommittee on Global Change Research (Washington, D.C., Department of Energy, Office of Biological & Environmental Research, 2007).

[82] M. J. Scott and Y.J. Huang, "Effects of Climate Change on Energy Use in the United States." In T. J. Wilbanks, V. Bhatt, D. E. Bilello, et al. (eds.), "Effects of Climate Change on Energy Production and Use in the United States." Synthesis and Assessment Product 4.5 (Washington, D.C., U.S. Climate Change Science Program, 2007), pp. 8–44.

partially offsetting trends, and the net change will be an increase in demand. Cooling a building is primarily powered by electricity. This can be supplied by renewable energy sources such as hydropower, solar and wind power, geothermal energy, and traditional carbon-based power sources. Heating is supplied primarily by natural gas and fuel oil. Because nearly half of the nation's electricity is currently generated from coal, these factors together have the potential to increase total national carbon dioxide emissions. However, improved energy efficiency, development of noncarbon energy sources, and/or carbon capture and storage technologies can combine to limit and even reduce emissions.

Climate change also places stress on the energy production network of human communities. Generation of electricity in thermal power plants (coal, nuclear, gas, or oil) is water intensive. Power plants rank only slightly behind irrigation in terms of freshwater withdrawals in the United States. There is a high likelihood that water shortages will limit power plant electricity production in many regions. By 2025, water limitations on electricity production in thermal power plants are projected[83] for Arizona, Utah, Texas, Louisiana, Georgia, Alabama, Florida, California, Oregon, and Washington state.

A warmer climate is characterized by more-extreme weather events such as windstorms, ice storms, floods, tornadoes, and hail.[84] As a result, the transmission systems of electric utilities could experience a higher rate of failure. Development of new energy facilities could be restricted by siting concerns related to sea-level rise, exposure to extreme events, and increased costs resulting from a need to provide greater protection from extreme events.

Power plant operations can be affected by extreme heat waves. For example, intake water that is normally used to cool power plants becomes warm enough during extreme heat events that it compromises power plant operations. High demand for cooling can overwhelm electricity production, causing blackouts. In the summer heat wave of 2006, for example, electric power transformers failed in several areas (including St. Louis, Missouri, and Queens, New York) as a result of high temperatures, causing interruptions of electric power supply. During the record-setting heat wave of 2012, a rolling windstorm called a "derecho"[85] passed from west to east across the continental United States and brought down power lines causing blackouts in hundreds of communities just when the demand for cooling was greatest.

If climate change leads to increased cloudiness, solar energy production could be reduced. Wind energy production would be reduced if wind speeds increase above or fall below the acceptable operating range of the technology. Changes in growing conditions could affect biomass production, a transportation and power plant fuel source that is rising in importance.

Demographic trends in the United States are increasing energy use. The population is shifting to the South and the Southwest (Figure 6.8), where air conditioning use is high. There is an increase in the square footage built per person and increased electrical needs in residential and commercial buildings, and air conditioning is being implemented in new places and by persons in income levels who had not previously embraced it.

As changes in precipitation take place in various regions of the country, the hydropower industry may be affected positively or negatively. Increases in hurricane intensity, frequency, and location will likely cause disruptions to oil and gas operations in the Gulf of Mexico (such as occurred in 2005 with Hurricane Katrina and 2008

[83] S. R. Bull, D. E. Bilello, J. Ekmann, M. J. Sale, and D. K. Schmalzer, "Effects of Climate Change on Energy Production and Distribution in the United States." In T. J. Wilbanks, V. Bhatt, D. E. Bilello, et al (eds.), *Effects of Climate Change on Energy Production and Use in the United States*. Synthesis and Assessment Product 4.5 (Washington, D.C., U.S. Climate Change Science Program, 2007), pp. 45–80.

[84] "Research Meteorologists See More Severe Storms Ahead: The Culprit—Global Warming," http://www.sciencedaily.com/videos/2009/0109-global_warming_causes_severe_storms.htm (accessed July 12, 2012).

[85] See the NOAA website on the "Historic Derecho of June 29, 2012": http://www.erh.noaa.gov/rnk/events/2012/Jun29_derecho/summary.php.

Figure 6.8. This map shows the percentage change in county population between 1970 and 2008 and illustrates large increases in places that require air conditioning (red, orange, and maroon). Some places had enormous growth, including influxes of hundreds of thousands of people. For example, counties in the vicinity of South Florida, Atlanta, Los Angeles, Phoenix, Las Vegas, Denver, Dallas, and Houston all had very large increases.

Source: USGCRP, http://www.globalchange.gov/resources/gallery?func=viewcategory&catid=1 (accessed July 12, 2012).

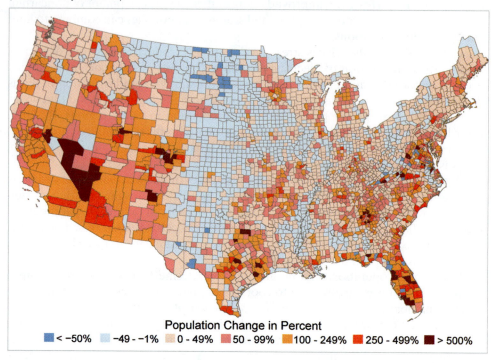

Population Change in Percent

■ < −50%   ■ −49 − −1%   ■ 0 − 49%   ■ 50 − 99%   ■ 100 − 249%   ■ 250 − 499%   ■ > 500%

with Hurricane Ike). Public concerns about global warming will alter perceptions and valuations of energy technology alternatives. These effects will play a role in energy policies in the United States.

Human Health

Human health[86] could suffer impacts (Figure 6.9) from climate change related to heat stress, extreme weather events and flooding, waterborne diseases, poor air quality, and diseases transmitted by insects and rodents. There are direct health effects from ailments caused or exacerbated by air pollution and airborne allergens and many climate-sensitive infectious diseases. In general, warming is likely to make it more challenging to meet air quality standards necessary to protect public health. For instance, rising temperature and carbon dioxide concentration increase pollen production and prolong the pollen season in a number of plants with highly allergenic pollen, presenting a health risk.[87]

As temperatures rise and heat waves occur with greater frequency and intensity,[88] the population of senior citizens (currently 12% and projected to be 21% by 2050;

[86] World Health Organization, *Constitution of the World Health Organization – Basic Documents*, 45th ed, Supplement, October 2006, http://www.who.int/governance/eb/who_constitution_en.pdf. (accessed July 12, 2012).

[87] K. L. Ebi, J. Balbus, P.L. Kinney, et al., "Effects of Global Change on Human Health."

[88] W. J. Gutowski, G. C. Hegerl, G. J. Holland, et al., "Causes of Observed Changes in Extremes and Projections of Future Changes." In T. R. Karl, G. A. Meehl, C. D. Miller, et al. (eds.), *Weather and Climate Extremes in a Changing Climate: Regions of Focus: North America, Hawaii, Caribbean, and U.S. Pacific Islands*. Synthesis and Assessment Product 3.3 (Washington, D.C., U.S. Climate Change Science Program, 2008), pp. 81–116.

Figure 6.9. This chart shows the distribution of deaths for 11 hazard categories as a percentage of the total of 19,958 deaths due to these hazards from 1970 to 2004. Heat/drought ranks the highest, followed by severe weather, which includes events with multiple causes such as lightning, wind, and rain.

SOURCE: USGCRP, http://www.globalchange.gov/resources/gallery?func=viewcategory&catid=1 (accessed July 12, 2012).

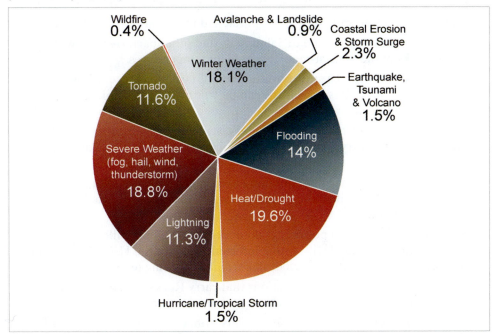

more than 86 million people), those with diabetes, and those with heart disease, is increasing. This population is vulnerable to the stresses associated with heat waves. Heat is already the leading cause of weather-related deaths in the United States. More than 3,400 deaths between 1999 and 2003 were reported as resulting from exposure to excessive heat and humidity.[89] A study of climate change impacts in California[90] projects that by the 2090s, annual heat-related deaths in Los Angeles would increase by two to three times under a lower emissions scenario and by five to seven times under a higher emissions scenario.

Poor air quality, especially in cities, is a serious concern across North America. Half of all Americans, 158 million people, live in counties where air pollution fails to meet national health standards. Breathing ozone results in short-term decreases in lung function and damages the lining the lungs. A warmer climate is projected to accelerate troposphere ozone formation and increase the frequency and duration of stagnant air masses that allow pollution to accumulate, which will exacerbate health symptoms. Under constant pollutant emissions, by the middle of this century, red-ozone-alert days (when the air is unhealthy for everyone) in the 50 largest cities in the eastern United States are projected to increase by 68% as a result of warming alone.[91]

Heavy downpours have increased in recent decades and are projected to increase further as the world continues to warm and the amount of water vapor increases in the atmosphere. This can lead to increased incidence of waterborne diseases due to pathogens such as Cryptosporidium and Giardia.[92] Downpours can trigger sewage overflows,

---

[89] A. Zanobetti and J. Schwartz, "Temperature and Mortality in Nine US Cities," *Epidemiology* 19, no. 4 (2008): 563–570.

[90] K. Hayhoe, D. Cayan, C. B. Field, et al., "Emissions Pathways, Climate Change, and Impacts on California," *Proceedings of the National Academy of Sciences* 101, no. 34 (2004): 12422–12427.

[91] M. L. Bell, R. Goldberg, C. Hogrefe, P. et al., "Climate Change, Ambient Ozone, and Health in 50 U.S. Cities," *Climatic Change* 82, no. 1–2 (2007): 61–76.

[92] K. L. Ebi, J. Balbus, P. L. Kinney, et al., "Effects of Global Change on Human Health."

contaminating drinking water and endangering beachgoers. The consequences will be particularly severe in the roughly 770 U.S. cities and towns, including New York, Chicago, Washington, Milwaukee, and Philadelphia, that have combined sewer systems, an older design that carries storm water and sewage in the same pipes. During heavy rains, these systems often cannot handle the volume, and raw sewage spills into lakes or waterways, including drinking-water supplies and places where people swim.

- Some diseases transmitted by food, water, and insects are likely to increase.
- Cases of food poisoning due to Salmonella and other bacteria peak within one to six weeks of the highest-reported ambient temperatures.
- Cases of waterborne Cryptosporidium and Giardia increase following heavy downpours. These parasites can be transmitted in drinking water and through recreational water use.
- Climate change affects the life cycle and distribution of the mosquitoes, ticks, and rodents that carry West Nile virus, equine encephalitis, Lyme disease, and Hantavirus.
- Heavy rain and flooding can contaminate certain food crops with feces from nearby livestock or wild animals, increasing the likelihood of foodborne disease associated with fresh produce.
- Vibrio species (shellfish poisoning) accounts for 20% of the illnesses and 95% of the deaths associated with eating infected shellfish. The U.S. infection rate increased 41% from 1996 to 2006 concurrent with rising temperatures.
- As temperatures rise, tick populations that carry Rocky Mountain spotted fever are projected to expand northward to new regions.
- The introduction of disease-causing agents from other regions of the world is a threat.

Communities have the capacity to adapt to climate change, but during extreme weather and climate events, actual practices have not always protected people and property. Vulnerability to extreme events is variable (Figure 6.10), with disadvantaged groups experiencing more disruption to their lives than other groups. Adaptation tends to be reactive, unevenly distributed, and focused on coping rather than on preventing problems.

CLIMATE IMPACTS TO GEOGRAPHIC REGIONS

At the end of the first decade of the 21st century, North America experienced significant cooling. This caused a decline in the public's appreciation for the reality of global warming, potentially influencing policy development aimed at managing greenhouse gas emissions. A joint study[93] by researchers in government and academic institutions concluded that the cooling was localized to the North American continent, it was short-term in nature, and the climate was not likely to be embarking upon prolonged cooling. Using model simulations the authors concluded that the anthropogenic impact in 2008 was to warm the region's temperatures but that it was overwhelmed by a particularly strong bout of naturally induced cooling resulting from the continent's sensitivity to widespread coolness of the tropical and northeastern Pacific sea surface temperatures. The implication is that the pace of North American warming was likely to resume in subsequent years. Indeed, shortly after this study was published, the U.K. Meteorological Office, NASA, NOAA, and the Japan Meteorological Agency all announced that 2010 had set a global record as the warmest year in recorded history (see Chapter 1) and 2011 and 2012 witnessed record-setting heat waves and drought throughout North America.

[93] J. Perlwitz, M. Hoerling, J. Eischeid, T. Xu, and A. Kumar, "A Strong Bout of Natural Cooling in 2008," *Geophysical Research Letters*, 36 (2009): L23706, doi:10.1029/2009GL041188.

Figure 6.10. U.S. locations with existing vulnerability to climate-sensitive health issues. **a,** Location of hurricane landfalls. **b,** Extreme heat events, defined as temperatures 5.5°C (10°F) or more above the average high temperature for the region and lasting for several weeks. **c,** Percentage of population older than 65 years; dark blue shows percentage that is greater than 17.5% and light blue shows percentage that is 14.4% to 17.5%. **d,** Locations of West Nile Virus cases reported in 2004.

Source: USGCRP, http://www.globalchange.gov/resources/gallery?func=viewcategory&catid=1 (accessed July 12, 2012).

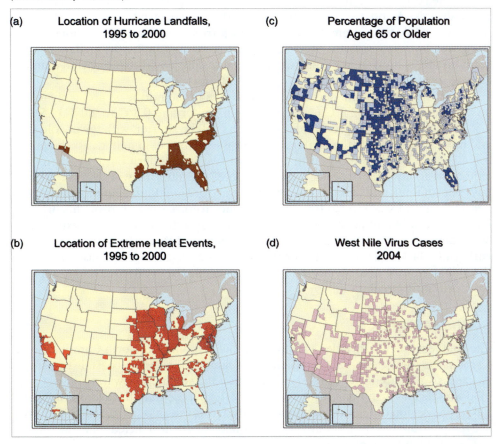

As global warming continues, the U.S. Global Change Research Program[94] provides American citizens and decision makers with periodic updates on the climate-related changes observed in the United States, its coastal waters, and globally. These climate changes include increases in heavy downpours, rising temperature and sea level, rapidly retreating glaciers, thawing permafrost, lengthened growing seasons, lengthened ice-free seasons in the ocean and on lakes and rivers, earlier snowmelt, expanding drought, and shifts in stream characteristics. Nearly all of these changes are projected to grow. An important contribution of USGCRP assessments is regional descriptions of climate-related impacts.

Northeast Region

The northeast region of the United States includes Maine, Vermont, New Hampshire, Massachusetts, Connecticut, New York, Pennsylvania, Rhode Island, New Jersey, Delaware, Maryland, West Virginia, and Washington, D.C. The climate in this area has changed in noticeable ways: more frequent days with temperatures above 32°C (90°F); a longer growing season; increased heavy precipitation; less winter precipitation falling as snow and more as rain; reduced snowpack; earlier breakup of winter

---

[94] T. R. Karl, J. M. Melillo, and T. C. Peterson (eds.), *Global Climate Change Impacts in the United States* (New York, Cambridge University Press, 2009).

ice on lakes and rivers; earlier spring snowmelt resulting in earlier peak river flows; and rising sea surface temperatures and sea level.[95] All of these measured changes are consistent with the rise of atmospheric temperature.

Since 1970, the yearly average temperature in the northeast has increased by 1°C (2°F). Winter temperatures have risen twice this much.[96] In one study,[97] researchers examined daily wintertime temperature extremes since 1948 and found that the warm extremes were much more severe and widespread than the cold extremes during the Northern Hemisphere winters of 2009–2010 (which featured an extreme snowfall episode on the East Coast dubbed "Snowmaggedon") and 2010–2011. Moreover, while the extreme cold was mostly attributable to a natural climate cycle, the extreme warmth was not. Overall, by late this century, under the higher emission scenario (IPCC A1FI), residents of New Hampshire could experience a summer climate similar to what occurs today in North Carolina.

Over the next several decades, temperatures in the northeast are likely to rise an additional 1.4°C to 2.2°C (2.5°F to 4°F) in winter and 0.8°C to 2.0°C (1.5°F to 3.5°F) in summer.[98] It is projected that winters will be shorter, with fewer cold days and more precipitation; the winter snow season will be cut in half across the northern states and will be reduced to a week or two in southern parts of the region; cities that today experience few days above 37.8°C (100°F) will average 20 such days per summer, and certain cities, such as Hartford and Philadelphia, will average nearly 30 days over 37.8°C (100°F); short, one- to three-month droughts are projected to occur as frequently as once each summer in the Catskill and Adirondack Mountains and across New England; hot summer conditions will arrive three weeks earlier and last three weeks longer into the fall; and sea level will rise more than the global average because of localized land subsidence in this area and changes in North Atlantic circulation.

Southeast Region

The southeast region includes Virginia, Kentucky, Tennessee, North Carolina, South Carolina, Georgia, Florida, Louisiana, Alabama, Mississippi, coastal Texas, and Arkansas. Compared with the rest of the nation the southeast is warm and wet, with mild winters and high humidity. Over most of the past century the average temperature of the region did not change significantly. However, since 1970, the annual average temperature has risen about 1.1°C (2°F), and the greatest increase in temperature has occurred in the winter months.

The number of freezing days has declined by 4 to 7 days per year since the mid-1970s, and the average autumn precipitation has increased by 30% over the 20th century[99] (Figure 6.11). Regions experiencing moderate to severe drought in the spring and summer have increased by over 10% since the 1970s. Even in the fall months, when precipitation tended to increase in most of the region, the extent of drought increased by 9%. Higher temperatures lead to more evaporation of

---

[95] UUSGCRP, "Global Climate Change Impacts in the United States, 2009," Available at: http://www.globalchange.gov/publications/reports/scientific-assessments/us-impacts (accessed July 12, 2012).

[96] K. Hayhoe, C. P. Wake, T. G. Huntington, et al., "Past and Future Changes in Climate and Hydrological Indicators in the U.S. Northeast," *Climate Dynamics* 28, no. 4 (2007): 381–407.

[97] K., Guirguis, A. Gershunov, R. Schwartz, and S. Bennett, "Recent Warm and Cold Daily Winter Temperature Extremes in the Northern Hemisphere," *Geophysical Research Letters* 38 (2011): L17701, doi:10.1029/2011GL048762.

[98] K. Hayhoe, C. P. Wake, T. G. Huntington, et al., "Past and Future Changes in Climate and Hydrological Indicators in the U.S. Northeast."

[99] T. R. Karl and R.W. Knight, "Secular Trends of Precipitation Amount, Frequency, and Intensity in the United States," *Bulletin of the American Metrological Society* 79, no. 2 (1998): 231–241. See also, B. D. Keim, "Preliminary Analysis of the Temporal Patterns of Heavy Rainfall across the Southeastern United States," *Professional Geographer* 49, no. 1 (1997): 94–104.

Figure 6.11. Average fall precipitation in the southeast region has increased by 30% over the 20th century. However, the percentage of the region experiencing drought has increased in recent decades, and summer and winter precipitation declined by nearly 10% in the eastern part of the region.

SOURCE: USGCRP, http://www.globalchange.gov/resources/gallery?func=viewcategory&catid=1 (accessed July 12, 2012).

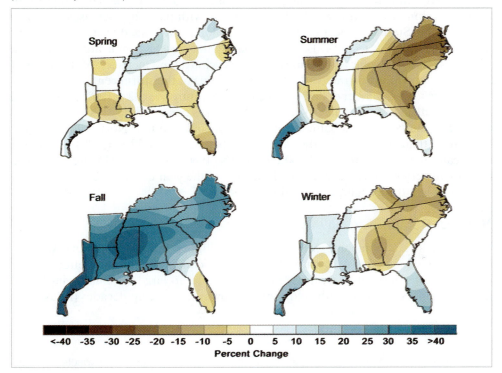

moisture from soils and water loss from plants; hence the frequency, duration, and intensity of droughts are likely to continue to increase.

Over the 21st century the number of very hot days is likely to rise at a greater rate than the average temperature. If greenhouse gas emissions are kept to a low level (IPCC B1), average temperatures could rise about 2.5°C (4.5°F) by the 2080s, but higher emissions (IPCC A1FI) could result in 5°C (9°F) of average warming; the increase in summer may be as much as 5.8°C (10.5°F).

The coastal area of the southeast is home to thousands of communities that have built in low-lying areas on the shores of the Atlantic Ocean and Gulf of Mexico. This region, more than any other in the United States, is prone to the deadly impacts of hurricanes, which pack winds capable of demolishing buildings and storm surges that rise as much as 4.5 to 6 m (15 to 20 ft) in the streets of coastal towns and cities. The destructive potential of Atlantic hurricanes has increased since 1970, correlated with an increase in sea-surface temperature. Notably, researchers have failed to establish a relationship between rising sea-surface temperature and the frequency of land-falling hurricanes.[100] However, in IPCC AR4, researchers conclude that the intensity of Atlantic hurricanes is likely to increase during this century, with higher

[100] See the following research on hurricanes: C. D. Hoyos, P. A. Agudelo, P. J. Webster, and J. A. Curry, "Deconvolution of the Factors Contributing to the Increase in Global Hurricane Intensity," *Science* 312, no. 577 (2006): 94–97; M. E. Mann and K. A. Emanuel, "Atlantic Hurricane Trends Linked to Climate Change," *Eos* 87, no. 24 (2006): 244; K. E. Trenberth and D. J. Shea, "Atlantic Hurricanes and Natural Variability in 2005," *Geophysical Research Letters* 33 (2006): L12704, doi:10.1029/2006GL026894; P. J. Webster, G. J. Holland, J. A. Curry, and H.-R. Chang, "Changes in Tropical Cyclone Number, Duration, and Intensity in a Warming Environment," *Science* 309, no. 5742 (2005): 1844–1846.

peak wind speeds, rainfall intensity, and storm surge height and strength.[101] Rising sea-surface temperatures are thought to be one reason for these increases. Even absent an increase in hurricane frequency, coastal inundation and shoreline erosion will increase as sea-level rise accelerates, which is one of the most certain and costly consequences of a warming climate.[102]

Studies[103] indicate that warming could cause the globally averaged intensity of tropical cyclones to shift toward stronger storms, with intensity increases of 2% to 11% by 2100. Studies project decreases in the globally averaged frequency of tropical cyclones by 6% to 34%. Balanced against this, research indicates substantial increases in the frequency of the most intense cyclones, and increases of the order of 20% in the precipitation rate within 100 km (62 mi) of the storm center.

Each year, the number of days with peak temperature over 32°C (90°F) is expected to rise significantly, especially under a higher emissions scenario (IPCC A1FI). By the end of the century, global circulation models indicate that North Florida could have more than 165 days (nearly six months) per year over 32°C (90°F), which is a significant increase from roughly 60 days in the 1960s and 1970s (Figure 6.12). The increase in very hot days to nearly half the days in the year could have consequences for human health, drought, and wildfires.

Midwest Region

The Midwest states include Michigan, Ohio, Illinois, Indiana, Missouri, Wisconsin, Minnesota, and Iowa. Located far from the climate-moderating effect of the ocean, the air temperature in the Midwest is subject to large seasonal swings. Hot, humid summers alternate with cold winters. However, in recent decades the average

Figure 6.12. The number of days each year that will exceed 32°C (90°F) is expected to rise significantly in the southeast region by the end of the century.
Source: USGCRP, http://www.globalchange.gov/resources/gallery?func=viewcategory&catid=1 (accessed July 12, 2012).

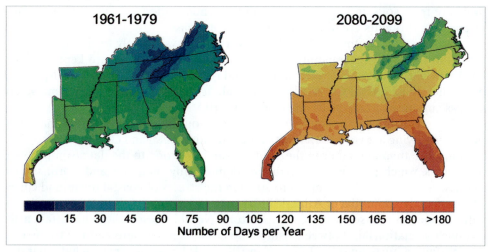

[101] G. A. Meehl, T. F. Stocker, W. D. Collins, et al., "Global Climate Projections." In T. R. Karl, G. A. Meehl, C. D. Miller, et al. (eds.), *Weather and Climate Extremes in a Changing Climate: Regions of Focus: North America, Hawaii, Caribbean, and U.S. Pacific Islands*. Synthesis and Assessment Product 3.3 (Washington, D.C., U.S. Climate Change Science Program, 2008), pp. 747–845.

[102] C. Tebaldi, B. Strauss, C. Zervas, "Modeling Sea Level Rise Impacts on Storm Surges along US Coasts," *Environmental Research Letters* 7 (2012): 014032, doi:10.1088/1748–9326/7/1/014032.

[103] T. R. Knutson, J. L. McBride, J. Chan, et al., "Tropical Cyclones and Climate Change," *Nature Geoscience* 3 (2009): 157–163.

annual temperature has increased, and the largest increase has been in wintertime.[104] Despite strong year-to-year variations, the length of the frost-free or growing season has extended by more than one week, mainly as a result of earlier dates for the last spring frost.

The major global warming issues for this region revolve around increases in both heat and flooding. Summer heat waves in cities can lead to health problems[105] as well as placing increased energy demand on public services. There may be reduced air quality, increases in insect and water-borne diseases, more heavy downpours, and increased evaporation in summer. This could produce more periods of both flooding and water deficits. A longer growing season provides the potential for increased crop yields in this important agricultural district, but growth in heat waves, floods, droughts, and insects and weeds migrating in from the south present mounting challenges to managing crops, livestock, and forests.

Scientists have also observed changes in rainfall in the Midwest; heavy downpours are now twice as frequent as they were a century ago, and both summer and winter precipitation have been above average for the last three decades, the wettest period in a century.[106] The Midwest has experienced both increasing extreme events and long-term trends: two record-breaking floods in the past 15 years, a decrease in lake ice (including on the Great Lakes), and increased frequency of large heat waves since the 1980s, which have been more frequent than any time in the past century, other than the Dust Bowl years of the 1930s.[107]

Models predict that Midwest summers could feel progressively more like summers currently experienced in states to the south and west (Figure 6.13).[108] By mid-century and toward the end of the century, Midwest states are projected to get considerably warmer and have less summer precipitation. Heat waves that are more frequent, more severe, and longer lasting are anticipated. The frequency of hot days and the length of the heat-wave season both may be more than twice as great under the higher-emissions scenario (IPCC A1FI) compared to the lower-emissions scenario (IPCC B1). In 1995 a heat wave hit the city of Chicago and resulted in more than 700 deaths. Events of this nature are expected to become more common. Under the B1 scenario, a heat wave equivalent to the 1995 event is projected to occur every other year in Chicago by the end of the century; under the A1FI scenario there would be approximately three such heat waves per year. Even more severe heat waves, such as the one that claimed tens of thousands of lives in Europe in 2003, are projected to become more frequent in a warmer world, occurring as often as every other year in the Midwest by the end of this century under the higher-emissions scenario.[109]

Great Plains Region

The Great Plains states include North and South Dakota, Nebraska, Kansas, Oklahoma, Texas, and portions of Colorado, Wyoming, and Montana. Major

[104] D. J. Wuebbles and K. Hayhoe, "Climate Change Projections for the United States Midwest," *Mitigation and Adaptation Strategies for Global Change* 9, no. 4 (2004): 335–363.

[105] S. C. Sheridan, A. J. Kalkstein, and L. S. Kalkstein, "Trends in Heat-Related Mortality in the United States, 1975–2004," *Natural Hazards* 50, no. 1 (2008): 145–160.

[106] NOAA National Climatic Data Center, "Climate of 2008: Midwestern U.S. Flood Overview," 2008, http://www.ncdc.noaa.gov/oa/climate/research/2008/flood08.html (accessed July 12, 2012).

[107] K. E. Kunkel, P.D. Bromirski, H.E. Brooks, et al., "Observed Changes in Weather and Climate Extremes."

[108] D. J. Wuebbles and K. Hayhoe, "Climate Change Projections for the United States Midwest."

[109] K. L. Ebi and G.A. Meehl, "The Heat Is On: Climate Change and Heat Waves in the Midwest." In *Regional Impacts of Climate Change: Four Case Studies in the United States* (Arlington, VA, Pew Center on Global Climate Change, 2007), 8-21. See: http://www. pewclimate.org/regional_impacts (accessed July 12, 2012).

Figure 6.13. Global circulation model projections of summer average temperature and precipitation changes in Illinois and Michigan by mid-century (2040–2059) and end of the century (2080–2099). These indicate that summers in the Midwest are expected to feel progressively more like summers currently experienced in states located to the south and west. Both states are projected to get considerably warmer and have less summer precipitation.

Source: USGCRP, http://www.globalchange.gov/resources/gallery?func=viewcategory&catid=1 (accessed July 12, 2012).

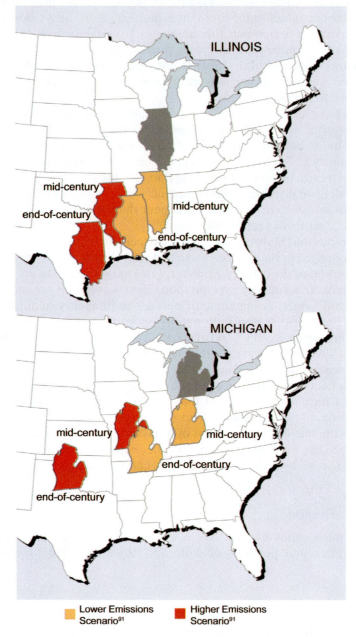

global warming issues in this area include increases in temperature (Figure 6.14), evaporation, extreme weather events, and drought frequency and magnitude. These trends are likely to lead to declining water resources with impacts on agriculture, ranching, and natural lands as well as on key habitats such as playa lakes, prairie potholes, and other wetland ecosystems. Human population shifts toward cities will lead to heat-related problems as well.[110]

---

[110] USGCRP, "Global Climate Change Impacts in the United States, 2009," http://www.globalchange.gov/publications/reports/scientific-assessments/us-impacts (accessed July 12, 2012).

Figure 6.14. Summer temperature change by end of the century. Temperatures in the Great Plains are expected to increase with global warming. By the end of the century the northern portion of the region is projected to experience the greatest temperature increase.
Source: USGCRP, http://www.globalchange.gov/resources/gallery?func=viewcategory&catid=1 (accessed July 12, 2012).

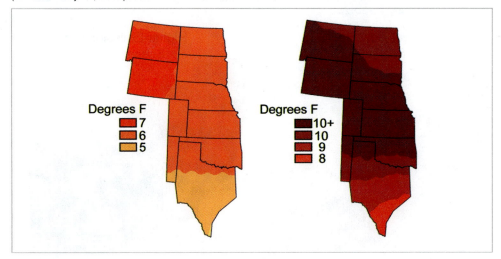

Climate has changed in the past few decades in the Great Plains. Average temperatures have increased the most in the northern states, and the largest increases have occurred in the winter. Relatively cold days are becoming less frequent and relatively hot days more frequent.[111] Temperatures are projected to continue to increase over the 21st century, with larger changes expected under scenarios of higher greenhouse gas emissions.

Summer changes are projected to be larger than those in winter in the southern and central Great Plains.[112] There has also been an increase in rainfall, with the greatest increases in states to the southeast. However, with continued global warming, conditions are anticipated to become wetter in the north and drier in the south. Changes in long-term climate will include more-frequent extreme events such as heat waves, droughts, and heavy rainfall. These will affect many aspects of life in the Great Plains including threats to water resources, essential agricultural and ranching activities, unique natural and protected areas, and the health and prosperity of inhabitants.[113]

Southwest Region

The Southwest states include California, Nevada, Utah, New Mexico, Arizona, and portions of Colorado and Texas. As global warming continues, the biggest problems that could develop in this region are related to water scarcity, drought, and heat. Studies indicate that much of the region is likely to have more than twice as many days per year above 32°C (90°F) by the end of the century.

The prospect of future droughts becoming more severe as a result of global warming is a significant concern, especially because the Southwest continues to

[111] A. T. DeGaetano and R. J. Allen, "Trends in Twentieth-Century Temperature Extremes across the United States," *Journal of Climate* 15, no. 22 (2002): 3188–3205.

[112] J. H. Christensen, B. Hewitson, A. Busuioc, et al., "Regional Climate Projections." In S. Solomon, D. Qin, M. Manning, et al. (eds.), *Climate Change 2007: The Physical Basis*. Contribution of Working Group I to the Fourth Assessment Report of the Intergovernmental Panel on Climate Change. Cambridge, U.K., Cambridge University Press, 2007), pp. 847–940.

[113] W. Parton, M. Gutmann, and D. Ojima, "Long-Term Trends in Population, Farm Income, and Crop Production in the Great Plains," *Bioscience* 57, no. 9 (2007): 737–747.

Figure 6.15. Future precipitation is likely to decrease dramatically in the Southwest. The figure shows the percentage change in precipitation for March, April, and May for 2080–2099 compared to 1961–1979 for a lower-emissions scenario (left) and a higher-emissions scenario (right). Confidence in the projected changes is highest in the hatched areas.

SOURCE: Figure from USGCRP, http://www.globalchange.gov/resources/gallery?func=viewcategory& catid=1 (accessed July 12, 2012).

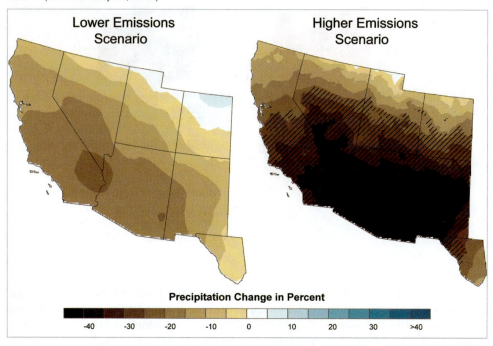

lead the nation in population growth. In an area that already wrestles with competing demands for scarce water resources, warming could force tradeoffs among rival water uses, potentially leading to conflict. Temperature increases throughout the century could amplify the frequency of drought and wildfire and could accentuate problems related to invasive species and shifts in agriculture.

The southwest region is one of the most rapidly warming in the United States, with some areas significantly exceeding the global average. Temperature increases are driving declines in spring snowpack, and consequently river discharge in the region is down as well.[114] Model projections (Figure 6.15) indicate that strong warming will continue under low-emissions scenarios (IPCC B1), with much-larger increases likely under higher scenarios (IPCC A1F1). There will almost certainly be serious water supply shortages in the future along with expanding urban heat island effects.[115]

Northwest Region

The Northwest region includes the states Washington, Oregon, Idaho, and western Montana. Global warming has caused the average annual temperature to rise 0.8°C (1.5°F) throughout the region. Some areas experienced an increase of 2.2°C (4°F) over the same period. Models project increases of an additional 1.7°C to 5.5°C (3°F to 10°F) this century, with the higher-emissions scenarios resulting in warming at the upper end of this range. Warming is likely to bring increased winter precipitation and decreased summer precipitation, with related changes to streamflow,

---

[114] T. P. Barnett, D. W. Pierce, H. G. Hidalgo, et al., "Human-Induced Changes in the Hydrology of the Western United States," *Science* 319, no. 5866 (2008): 1080–1083.

[115] S. A. Rauscher, J. S. Pal, N. S. Diffenbaugh, and M. M. Benedetti, "Future Changes in Snowmelt-Driven Runoff Timing over the Western United States," *Geophysical Research Letters* 35 (2008): L16703, doi:10.1029/2008GL034424; see also S. Guhathakurta and P. Gober, "The Impact of the Phoenix Urban Heat Island on Residential Water Use," *Journal of the American Planning Association* 73, no. 3 (2008): 317–329.

snowpack, forest ecosystems, wildfires, and other important aspects of life and ecology in the Northwest.[116]

The key issues related to global warming in the region include decreased spring snowpack (reducing summer stream flow and straining water resources), increased wildfires and insect outbreaks, shifting species composition in forest ecologies and impacts on the lumber industry, and stresses to salmon ecosystems with rising water temperatures and declining discharge. Sea-level rise and increased wave height along vulnerable coastlines could result in accelerated coastal erosion and land loss.[117]

Snow that collects throughout the winter feeds streams and groundwater. These sources of freshwater sustain human communities, aquatic ecosystems, and forest environments. Human demands for water in the Northwest are intense. Seasonal snow pack provides water to meet growing demand from municipal and industrial uses, agricultural irrigation, hydropower production, navigation, recreation, and fish industries. As global warming raises temperatures in the Northwest, more precipitation could fall as rain rather than snow and contribute to earlier snowmelt. The thickness and cover of April 1 snowpack, a key indicator of natural water storage available for the warm season, has declined substantially throughout the Northwest. For example, in the Cascade Mountains the average snowpack declined approximately 25% over the past 40 to 70 years (mostly due to a 1.4°C [2.5°F] increase in cool season temperatures).[118] It is likely that continued warming will contribute to further snowpack declines. The April 1 snowpack is projected to decline as much as 40% in the Cascades by mid-century.[119]

Coastal Regions

It has been estimated that approximately 3.9 million people in the United States[120] and more than 145 million people worldwide[121] live within 1 m (3.3 ft) of modern sea level and thus risk losing their land and property under most scenarios of global warming by the end of the century. The resulting disruption threatens the economy and social well-being of many more. This realization is driving some coastal communities to consider various ways to adapt to sea-level rise, including the development of guidance in the form of new government policies, engineering solutions, and other strategies to accommodate rising waters and its attendant problems. However, making the transition to an adapted community that has successfully reduced vulnerability to sea-level rise impacts is only beginning.

A study[122] of lands that are vulnerable to sea-level rise reveals that almost 60% of the land below 1 m (3.3 ft) along the U.S. Atlantic coast is expected to be developed and thus not able to accommodate the inland migration of wetlands, beaches, estuarine zones, and other tidal ecosystems. Less than 10% of the land below 1 m has

[116] USGCRP, "Global Climate Change Impacts in the United States, 2009," http://www.globalchange.gov/publications/reports/scientific-assessments/us-impacts (accessed July 12, 2012).

[117] A. W., Petersen, *Anticipating Sea Level Rise Response in Puget Sound*. M.M.A. thesis, School of Marine Affairs, University of Washington, Seattle, 2007.

[118] P.W., Mote, "Climate-Driven Variability and Trends in Mountain Snowpack in Western North America," *Journal of Climate* 19, no. 23 (2006): 6209–6220.

[119] J. T. Payne, A. W. Wood, A. F. Hamlet, R. N. Palmer, and D. P. Lettenmaier, "Mitigating the Effects of Climate Change on the Water Resources of the Columbia River Basin," *Climatic Change* 62, no. 1–3 (2004): 233–256.

[120] C. Tebaldi, Strauss, B., Zervas, "Modeling Sea Level Rise Impacts on Storm Surges along US Coasts," *Environmental Research Letters* 7 (2012): 014032, doi:10.1088/1748-9326/7/1/014032; and B. Strauss, R. Ziemlinski, J. Weiss, and J. Overpeck, "Tidally Adjusted Estimates of Topographic Vulnerability to Sea-Level Rise and Flooding for the Contiguous United States," *Environmental Research Letters* 7 (2012): 014033, doi:10.1088/1748-9326/7/1/014033.

[121] D. Anthoff, R. J. Nicholls, R. S. J. Tol, and A. Vafeidis, *Global and Regional Exposure to Large Rises in Sea-Level: A Sensitivity Analysis*. Working Paper 96 (Norwich, Tyndall Centre for Climate Change Research, (2006), http://www.tyndall.ac.uk/biblio/working-papers?biblio_year=2006 (accessed July 12, 2012).

[122] J. G. Titus, D. E. Hudgens, D. L. Trescott, et al., "State and Local Governments Plan for Development of Most Land Vulnerable to Rising Sea Level along the U.S. Atlantic Coast," *Environmental Research Letters* 4 (2009): 044008.

Figure 6.16. Surging population growth in the coastal zone exposes more people to the dangers of geologic hazards, such as storms, hurricanes, tsunamis, and others, than in any other geologic environment. The world's coasts are home to fragile ecosystems, beautiful vistas, pristine waters, and major growing cities, all coexisting in a narrow and constricted space. Expanding communities compete for more space at the expense of extraordinary wild lands. There are problems with coastal erosion, waste disposal, a dependency on imported food and water, and rising sea level.

IMAGE CREDIT: iStockphoto.

been set aside for conservation. Development not only threatens the migration path of tidal ecosystems but also entails population growth on the world's riskiest lands.

It has been estimated that about one third of all Americans live in counties that border the ocean coasts,[123] and coastal and ocean activities contribute more than $1 trillion to the nation's gross domestic product. The ecosystems of the coast and the 322 km (200 mi) wide Exclusive Economic Zone holds rich biodiversity and provides invaluable services.[124] However, over the past 50 years population growth in the coastal zone outpaced the ability of resource managers and community leaders to ensure the sustainability of coastal environments and resources (Figure 6.16). Fish stocks have been severely diminished by overfishing, large dead zones in coastal waters are depleted of oxygen because of excess nitrogen runoff, toxic algae blooms are growing in frequency and geographic diversity, seawall construction results in beach loss, and coral reefs are in decline in some areas from human causes. About half of the nation's coastal wetlands have been lost, and most of this loss has occurred during the past 50 years.[125]

[123] M. Crowell, S. Edelman, K. Coulton, and S. McAfee, "How Many People Live in Coastal Areas?" *Journal of Coastal Research* 23, no. 5 (2007): iii–vi.

[124] U.S. Commission on Ocean Policy, *An Ocean Blueprint for the 21st Century* (Washington, D.C., U.S. Commission on Ocean Policy, 2004) http://www.oceancommission.gov/documents/full_color_rpt/welcome.html (accessed July 12, 2012).

[125] USGCRP, "Global Climate Change Impacts in the United States," 2009, http://www.globalchange.gov/publications/reports/scientific-assessments/us-impacts (accessed July 12, 2012).

Global warming places new stresses on this situation. Rising sea level is eroding shorelines, drowning wetlands, and threatening communities on the coast.[126] The potential of Atlantic tropical storms and hurricanes to cause damage has grown since 1970 because more people have moved onto and built along the nation's coastlines and because rising Atlantic sea surface temperatures are fueling increased hurricane rainfall and wind speeds.[127] Studies[128] reveal that because of sea-level rise, the odds of flooding by catastrophic "100 year" floods (floods expected only once per century) will double for most coastal cities by 2030.

Over the past 50 years, coastal water temperatures have risen by about 1.1°C (2°F) in several regions, and the distribution of marine species has shifted.[129] Where rainfall has increased on land, greater river runoff pollutes coastal waters with nitrogen and phosphorous, sediments, and other contaminants that are carried from farm fields and polluted streets.

Among other stressors, coral reefs are affected by the mixture of atmospheric carbon dioxide with seawater, which lowers the pH of seawater, causing ocean acidification. This threatens corals, mollusks, plankton, and other marine organisms that form their shells and skeletons from calcium carbonate, which is not as stable in the new seawater chemistry. Ocean acidification threatens the ability of these organisms to secrete the calcium carbonate materials they need to live (see Chapter 1).[130] All of these forces converge and interact at the coasts, making these areas particularly sensitive to the impacts of climate change.

Alaska

Arctic temperatures have reached their warmest level of any decade in at least 2,000 years. To determine this, researchers[131] used geologic records and computer simulations that provide new evidence that the Arctic would be cooling if not for greenhouse gas emissions that are overpowering natural climate patterns. Part of this recent trend has been shown to originate with the positive climate feedback relating to the loss of arctic sea ice. Sea ice melting is changing the albedo (sunlight reflectivity) of the high north and causing dark ocean water to absorb solar radiation, whereas previously the white icy surface reflected the radiation back to space.[132]

Global warming is hitting Alaska in profound ways. As in many high-latitude locations, warming has exceeded the global average, and in Alaska the rate of warming has been more than twice the rate in the rest of North America. The primary impacts of global warming have already been seen. These include: an increase in wildfires and insect outbreaks; declining lakes and ponds resulting from drying; longer summers[133] (Figure 6.17) and higher temperatures causing drier conditions even in the absence of strong trends in precipitation; thawing permafrost that damages roads, pipelines, airports, water and sewer systems, and other infrastructure designed

[126] S. J. Williams, B. T. Gutierrez, J. G. Titus, et al., "Sea-Level Rise and its Effects on the Coast". In J. G. Titus, K. E. Anderson, D. R. Cahoon, et al., *Coastal Elevations and Sensitivity to Sea-level Rise: A Focus on the Mid-Atlantic Region*, Synthesis and Assessment Product 4.1 (Washington, D.C., U.S. Environmental Protection Agency, 2009), pp. 11–24.

[127] K. E. Kunkel, P. D. Bromirski, H. E. Brooks, et al., "Observed Changes in Weather and Climate Extremes."

[128] B. Strauss, R. Ziemlinski, J. Weiss, and J. Overpeck, "Tidally Adjusted Estimates of Topographic Vulnerability to Sea-Level Rise and Flooding for the Contiguous United States."

[129] W. J. Gutowski, G. C. Hegerl, G. J. Holland, et al., "Causes of Observed Changes in Extremes and Projections of Future Changes."

[130] J. C. Orr, V. J. Fabry, O. Aumont, et al., "Anthropogenic Ocean Acidification over the Twenty-First Century and Its Impact on Calcifying Organisms," *Nature* 437, no. 7059 (2005): 681–686.

[131] D. Kaufman, D. Schneider, N. McKay, et al., "Recent Warming Reverses Long-Term Arctic Cooling," *Science* 325 (2009): 1236–1239.

[132] J. Screen and I. Simmonds, "The Central Role of Diminishing Sea Ice in Recent Arctic Temperature Amplification," *Nature* 464 (2010): 1334–1337, doi:10.1038/nature09051.

[133] ACIA, *Impacts of a Warming Arctic: Arctic Climate Impact Assessment* (Cambridge, U.K., Cambridge University Press, 2004).

Figure 6.17. Over the last 100 years the length of the frost-free season in Fairbanks, Alaska, has increased by 50%. The trend toward a longer frost-free season will likely produce benefits in some sectors and detriments in others.

Source: USGCRP, http://www.globalchange.gov/resources/gallery?func=viewcategory&catid=1 (accessed July 12, 2012).

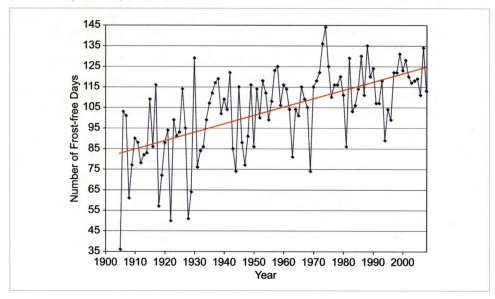

for colder conditions; coastal erosion that increases the risk to fishing villages, coastal towns; and growing storm vulnerability.

Alaska's annual average temperature has increased 1.9°C (3.4°F), and winters have warmed by 3.5°C (6.3°F). Warming is reducing sea ice, bringing an earlier spring snowmelt, melting permafrost, eroding coastlines,[134] and causing the retreat of glaciers throughout the state. These changes are consistent with model predictions that warming will exceed the pace of the rest of the nation, especially in winter. Sea ice reductions also alter the timing and location of plankton blooms, which is expected to drive important shifts in marine species such as pollock and other commercial fish stocks.[135]

Islands

Island communities in the Pacific and the Caribbean are isolated, trade-dependent, and ocean-oriented cultures that are especially vulnerable to climate change. In both the Caribbean[136] and Pacific,[137] air and ocean temperatures are rising, rainfall is decreasing in some areas and increasing in others, sea level is rising, and the ocean is acidifying. These trends signal decreased water resources, increased coastal erosion and marine inundation, and increasing economic expense. Rainfall on high Pacific islands is related to the orographic effect, a condensation process that takes place at high elevations on volcanic islands. The resulting water is the

---

[134] B. M. Jones, C. D. Arp, M. T. Jorgenson, K. M. Hinkel, J. A. Schmutz, and P. L. Flint, et al., "2009: Increase in the Rate and Uniformity of Coastline Erosion in Arctic Alaska," *Geophysical Research Letters*, 36 (2009): L03503, doi:10.1029/2008GL036205.

[135] J. M., Grebmeier, J. E. Overland, S. E. Moore, et al., "A Major Ecosystem Shift in the Northern Bering Sea," *Science* 311, no. 5766 (2006): 1461–1464.

[136] G. A. Meehl, T. F. Stocker, W. D. Collins, et al., "Global Climate Projections." In T. R. Karl, G. A. Meehl, C. D. Miller, et al. (eds.), *Weather and Climate Extremes in a Changing Climate: Regions of Focus: North America, Hawaii, Caribbean, and U.S. Pacific Islands.* Synthesis and Assessment Product 3.3 (Washington, D.C., U.S. Climate Change Science Program, 2008), pp. 747–845.

[137] C. H. Fletcher, *Hawai'i's Changing Climate* (Honolulu, University of Hawaii Sea Grant College Program, Center for Island Climate Adaptation and Policy, 2010) http://seagrant.soest.hawaii.edu/hawaiis-changing-climate-briefing-sheet-2010 (accessed July 12, 2012).

communities, and with rising air temperatures this precious resource is growing scarcer.

Marine and coastal ecosystems of the islands are particularly vulnerable to the impacts of climate change.[138] Sea-level rise, increasing water temperatures, rising coastal inundation and flooding from extreme events, beach erosion, ocean acidification, increased incidences of coral disease, and increased invasions by non-native species are among the threats that endanger the ecosystems that provide safety, sustenance, economic viability, and cultural and traditional values to island communities.

Reefs are under stress owing to rising water temperatures and acidification. Many fringing reefs are already stressed from a history of overfishing and polluted runoff from nearby watersheds. Changing ocean conditions further threaten these reefs, as does population growth along the desirable shorelines of islands, where development often leads to declining water quality related to sewage and other pollutants.

CONCLUSION

The USGCRP is scheduled to publish its next report to the U.S. Congress in 2013. The Intergovernmental Panel on Climate Change[139] is planning its next assessment report a year later in 2014. These are going to be important studies that set the tone for climate research over the following decade. Congress continues to investigate the science of global warming [140] and it is hoped that as public understanding of climate change grows and improves, the United States will eventually adopt a national carbon-control law that contributes to mitigating the problem of greenhouse gas emissions and thereby improves the prospects for avoiding the very worst aspects of long-term climate change.

ANIMATIONS AND VIDEOS

"White House Releases Landmark Climate Report," http://www.globalchange.gov/publications/reports/scientific-assessments/us-impacts/newsroom

"Climate Change and the Water Cycle," http://www.youtube.com/watch?v=BIbys6VQpVk&feature=related

"Research Meteorologists See More Severe Storms Ahead: The Culprit: Global Warming," http://www.sciencedaily.com/videos/2009/0109-global_warming_causes_severe_storms.htm

"End of Climate Change Skepticism," http://democrats.naturalresources.house.gov/hearings@id=0108.html

COMPREHENSION QUESTIONS

1. Describe the role of the U.S. Global Change Research Program.

2. How much has the U.S. average temperature risen over the past 50 years?

3. Describe the U.S. average temperature increase that might occur under the low-emissions scenario and under the high-emissions scenario by the end of the century.

4. What is the general rule among climatologists with regard to precipitation changes due to global warming?

5. Describe how global warming has changed precipitation in the United States.

6. Describe some ways climate change could affect the water resources and transportation sectors of the U.S. economy.

7. Describe some ways climate change could affect human health and energy supply and use in the United States.

8. Describe some ways climate change could affect ecosystems and agriculture in the United States.

9. How has the summer growing season changed in the Great Plains region?

10. Describe the impact of global warming where U.S. population growth has been the greatest in recent decades.

---

[138] R. G. Gillespie, E. M. Claridge, and G. K. Roderick, "Biodiversity Dynamics in Isolated Island Communities: Interaction between Natural and Human-Mediated Processes," *Molecular Ecology* 17, no. 1 (2008): 45–57.

[139] See the IPCC home page http://www.ipcc.ch/ (accessed July 12, 2012).

[140] See the video at the end of the chapter, "End of Climate Change Skepticism."

THINKING CRITICALLY

1. There are regions in the United States where annual precipitation has increased but there has been an increase in seasonal drought. Describe why this is a concern.

2. Rising air temperature is causing changes in the length of seasons. Describe how this can affect water resources.

3. Is global warming causing more or less extreme weather? Explain your answer and describe the type of extreme weather you are referring to.

4. Transportation operations are affected by global warming in several ways. Describe these.

5. How will changing air temperature affect demand for electricity? In what areas will this be most noticeable?

6. You are mayor of a town in New England. What effects can you expect from climate change and what should you do to prepare the town for these effects?

7. What special risks and vulnerabilities do coastal communities have in the face of climate change?

8. How is climate change affecting snowfall and what are the positive and negative impacts to local communities?

9. If the USGCRP has documented such wide-ranging impacts to climate change, why has the U.S. Congress failed to take significant action on the problem?

10. The U.S. Southwest and Southeast have rapidly growing populations. Describe the special risks they face from climate change.

CLASS ACTIVITIES (FACE TO FACE OR ONLINE)

ACTIVITIES

1. Visit the USGCRP website http://www.globalchange.gov/ and answer the following questions.

  a. What types of reports have they produced other than the 2009 national assessment?

  b. What is adaptation science? Why is it important?

  c. Visit the agencies that are coordinated under the USGCRP and describe the climate activities and concerns of six of them.

  d. Visit the "Related Federal Climate Efforts" page and describe some of the other types of federal climate work being performed.

2. Watch the video "Climate Denial Crock of the Week: Bad, Badder, BEST" http://www.youtube.com/watch?v=tciQts-8Cxo and answer the following questions.

  a. Describe the BEST study and what its conclusions were.

  b. What is the urban heat island effect and what has the BEST study concluded about the effect?

  c. Go to the Web and see if you can find any blogs of climate denialists and describe their reaction to the BEST study.

  d. Are climate denialists driven by facts or by some other motivation?

3. Watch the video "Lone Star State of Drought" http://www.youtube.com/watch?v=0VMpes8Eylw and answer the following questions.

  a. Describe the 2011 Texas drought and its effects.

  b. How does the information in this video compare to the description of climate change in the Great Plains states?

  c. Are extreme events expected in a warming climate? What types of events are likely?

  d. Describe the relationship between drought in Texas and the Pacific pattern known as ENSO.

  e. How will the ratio of record warm days to record cold days change by mid-century compared to the current pattern?

WHAT IS THE LATEST WORD ON CLIMATE CHANGE?

FIGURE 7.0. Global warming is making hot days hotter, rainfall and flooding heavier, hurricanes stronger, and droughts more severe. This supercell thunderstorm brings heavy rain and strong winds to a small farm on the plains of Oklahoma.
IMAGE CREDIT: Sean Waugh, NOAA/NSSL

CHAPTER SUMMARY

It is useful to review the latest evidence from the scientific realm confirming that global warming and climate change are still actively changing the planet we call home. This last chapter provides a review of some of the important climate issues we have touched on: climate change confirmed, a new record in global emissions, warming the high latitudes (Arctic and Antarctic), extreme weather, drought, dangerous climate, ecosystem impacts, and climate sensitivity.

In 2010 greenhouse emissions rose 5.9%, the largest annual increase on record. In 2011 global emissions rose another 3.2% above the 2010 level—and could reach 560 ppm CO_2, twice the natural level—in the second half of the century, producing 2°C to 4.7°C (3.6°F to 8.46°F) of warming. Researchers[1] find this would set Greenland on the path of unstoppable melting, resulting in several meters of sea-level rise affecting the lives of many millions of people. The National Oceanographic and Atmospheric Administration (NOAA) reports that the Arctic is shifting into a new state: warmer, greener, and less ice. Scientists have found that species are responding to climate change up to three times faster than previously estimated. Changes in habitat quality cause changes in the distribution of food sources and place wildlife populations under stress. Studies of the contributions of volcanism, the El Niño Southern Oscillation (ENSO), and solar activity to recent temperature trends verify that these natural processes are negligible contributors to the observed warming of the past several decades.

In this chapter you will learn that:

- At least four independent scientific organizations have confirmed the reliability of surface temperature data sets and the validity of global warming.
- The warming caused by human activities ranges from 0.170°C to 0.175°C per decade (0.31°F to 0.32°F per decade).
- 2010 had the largest annual increase in greenhouse gas emissions on record.
- The Arctic has settled into a new normal since 2006: It is warmer, greener, less icy, and suffering extreme ecosystem damage.
- NOAA has declared 2011 among the most extreme weather years in history. Day-to-day weather has grown increasingly erratic and extreme, with significant fluctuations in sunshine and rainfall affecting more than a third of the planet.

[1]A. Robinson, R. Calov, and A. Ganopolski, "Multistability and Critical Thresholds of the Greenland Ice Sheet," *Nature Climate Change* 2 (2012): 429–432, doi: 10.1038/NCLIMATE1449.

Learning Objectives

As the atmosphere continues to warm, researchers continue to document the consequences; the polar regions are melting, the oceans are heating and acidifying, extreme weather is increasing, drought is spreading, ecosystems are altering, and the world is poised to change in ever more dangerous ways.

- 2012 saw the worst heat wave on record and the most severe drought conditions to hit the United States since the Dust Bowl era.

- If global warming persists as expected, it is estimated that almost a third of all plant and animal species worldwide could become extinct.

- Sea surface temperature has increased by an average of 0.6°C (1°F) in the past 100 years, and the acidity of the ocean surface has increased tenfold.

- The oceans have absorbed about one third of the carbon dioxide emitted by humans over the past two centuries.

- An increase of 3.1°C (5.6°F) in global average surface temperatures seems most likely as a result of doubling the CO_2 concentration above preindustrial levels.

Tracking the science of climate change can be an alarming yet fascinating enterprise. There are essentially two types of climate information: scientific literature, including peer-reviewed articles and government-sponsored updates and reports (discussed in Chapter 2) and the media dedicated to delivering scientific climate news, which ranges from reliable to unreliable.[2] Climate news ranges from reports of the rising intensity and frequency of weather events around the world to economic debate on the relative merits of various steps to mitigate use of carbon energy. The wave of climate information available to the informed and aware listener is delivered non-stop, and disturbing climate news, it sadly turns out, is a weekly event.

But tracking the latest word on climate change is not a spectator sport. You will find yourself getting involved. Once you begin gaining climate awareness, it is hard not to consider it in your daily life. Conversations with family and friends present an opportunity to educate the less aware. Attending political events becomes an opportunity to question candidates on their level of knowledge about climate change and their willingness to acknowledge the preponderance of scientific evidence—acts that could take some political courage on both of your parts.

In becoming climate aware, you will find how surrounded we are by subtle expressions of misinformation. One of my neighbors, of whom I think very highly, insists on telling me that he reads "both sides of the debate." He is confused when I ask him "What is the debate?" and falls silent as I let him know that among scientists there is no debate about the existence and causes of global warming. The only debate is about the details of climate variability: What will climate change look like regionally and locally? How fast will climate change happen, and what will the impacts be as the atmosphere continues to warm?

Because it is always useful to review the latest evidence from the scientific realm confirming that global warming and climate change are still actively changing the planet we call home, this last chapter provides a review of some of the important climate issues we have touched on: confirmation of global climate change, global emissions, arctic amplification, extreme weather, drought, dangerous climate, ecosystem impacts, and climate sensitivity.

[2]See the video "Intent to Intimidate" at the end of the chapter.

CLIMATE CHANGE CONFIRMED . . . AGAIN

In December 2011 the highly respected journal Science published a special section on "Data Replication and Reproducibility." Articles reviewed data related to various scientific fields including one article that reviewed the history of data collection describing atmospheric temperature change.[3] The article describes the extraordinary efforts by teams of scientists in England and the United States to recover surface temperature data from stations located around the world and to build a time series of consistent, reliable, and reproducible measurements now over a century in length. Counting the NASA–Goddard Institute of Space Studies (GISS), the National Climatic Data Center (NCDC), the Berkeley-based BEST team (discussed below), and the University of East Anglia Climatic Research Unit (joined with the U.K. Meteorological Office Hadley Centre [HadCRU]), at least four independent bodies have confirmed the reliability of surface temperature data sets and the validity of global warming.

The second half of the Science article discussed an alleged cooling of the lower troposphere since 1979, based on flawed analysis of satellite microwave data. Published[4] in 2003 by researchers at the University of Alabama, the analysis cast doubt on the reality of global warming. The claims of a cooling atmosphere did not withstand rigorous testing, however. Scientists at the commercial firm Remote Sensing Systems (RSS) identified two serious errors in the Alabama work that, once corrected, showed a net warming of about 0.5°C (0.9°F) over the life of the satellite record.[5] Correction of these errors has produced greater consistency between ground- and satellite-based estimates of warming trends, and independent research[6] continues to identify widespread warming as a reality of modern climate.

BEST Skeptic

At several points in this text we have referred to climate skepticism: the fact that portions of the U.S. public doubt that climate change is real and doubt that humans are the primary cause. For instance, in a recent poll[7] 66% of Americans believe global warming is happening (a 3% increase since November 2011), but the proportion who said global warming is caused by human activities decreased from 50% to 46%.

Until recently, a prominent skeptic was Dr. Richard Muller, a prize-winning physicist on the faculty at the University of California at Berkeley. He and a group of scientists, calling themselves the Berkeley Earth Surface Temperature (BEST) team, studied land-based climate station data going back as far as the early 19th century and found "reliable evidence of a rise in the average world land temperature of approximately 1°C since the mid-1950s."[8] What is notable about this finding (Figure 7.1) is that Muller had been a well-known skeptic, and because of his strong scientific credentials, his skepticism lent credence to the larger community of climate change deniers.

Among the BEST conclusions was acknowledgement that temperature analyses conducted by the government teams at NCDC, GISS, and HadCRU were "done carefully and the potential biases identified by climate change skeptics did not seriously affect their conclusions."

[3]B. D. Santer, T. M. L. Wigley, and K. E. Taylor, "The Reproducibility of Observational Estimates of Surface and Atmospheric Temperature Change," *Science* 334, no. (2011): 1232–1233. (2011).

[4]J. R. Christy, R. W. Spencer, W. B. Norris, W. D. Braswell, and D. E. Parker, "Error Estimates of Version 5.0 of MSU-AMSU Bulk Atmospheric Temperatures," *Journal of Atmospheric and Oceanic Technology* 20 (2003): 613–629.

[5]B. D. Santer, C. Mears, C. Doutriaux, et al., "Separating Signal and Noise in Atmospheric Temperature Changes: The Importance of Timescale," *Journal of Geophysical Research* 116 (2011): D22105, doi: 10.1029/2011JD016263.

[6]R. S. Vose, S. Applequist, M. J. Menne, C. N. Williams Jr., and P. Thorne, "An Intercomparison of Temperature Trends in the U.S. Historical Climatology Network and Recent Atmospheric Reanalyses," *Geophysical Research Letters* 39 (2012): L10703, doi: 10.1029/2012GL051387.

[7]See Yale Project on Climate Change Communication, http://environment.yale.edu/climate/news/Climate-Beliefs-March-2012/ (accessed July 12, 2012).

[8]See the Berkeley Earth Surface Temperature (BEST, 2011) project press release at the site http://berkeleyearth.org/pdf/berkeley-earth-summary-20-october-2011.pdf (accessed July 12, 2012).

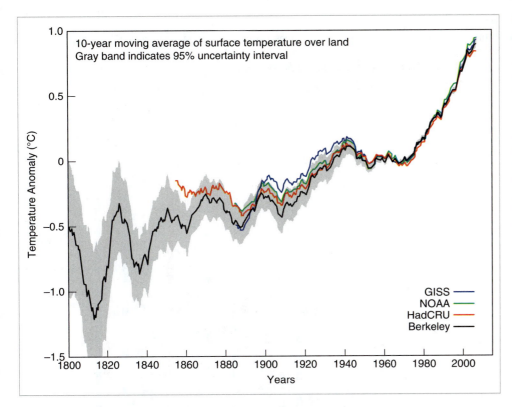

Figure 7.1. The decadal land-surface average temperature using a 10-year moving average of surface temperatures over land. Anomalies are relative to the January 1950 to December 1979 mean. The gray band indicates 95% statistical and spatial uncertainty of temperature data. GISS, NASA Goddard Institute for Space Studies; NOAA, National Oceanic and Atmospheric Administration; HadCRU, U.K. Met Office Hadley Center and the Climatic Research Unit of East Anglia University.

Source: Surface Temperature (BEST, 2011); http://www.berkeleyearth.org/ (accessed July 12, 2012).

The BEST study has concluded that despite issues raised by climate-change deniers, "Global warming is real." The study combined 1.6 billion daily and monthly temperature records from 10 data archives to identify 39,390 unique station records, more than five times the number of stations used by other climate studies. They found that the urban heat-island effect can be locally large and real but does not contribute significantly to the average land-temperature rise of about 1°C since the mid-1950s, and about one third of temperature sites around the world show net cooling and two thirds show net warming. The overall land-surface temperature record synthesized by BEST and the records produced by other climate teams are shown in Figure 7.1.

Continued Warming . . . . of a Sort

Earth's changing climate is expressed in several studies that reveal the signature of global warming in various datasets. Global temperatures in 2011 were the tenth highest on record and higher than any previous year characterized by La Niña (Figure 7.2). (La Niña events have a relative cooling influence on the years in which they occur, and 2011 was the warmest La Niña year on record.)

Two studies in 2011–2012 confirm the human component of global warming. In the first, Huber and Knutti[9] estimate that approximately 100% of the observed surface warming since the 1950s has been caused by human effects. This estimate corresponds to a total warming of approximately 0.55°C (0.99°F), most of which has occurred since mid-1970. The study is based on a global circulation model analysis of Earth's total heat-content increase since 1850. The study calculates how much of the increase results from various natural and anthropogenic factors that influence atmospheric temperature. The authors conclude that there is 95% certainty that external (human) forcing of global temperature is responsible for between 74% and 122% of the observed warming since 1950, with a most likely value of close to 100%.[10]

---

[9]M. Huber and R. Knutti, "Anthropogenic and Natural Warming Inferred from Changes in Earth's Energy Balance," *Nature Geoscience* 5 (2012): 31–36.

[10]See SkepticalScience.com for discussion of this paper, http://www.skepticalscience.com/huber-and-knutti-quantify-man-made-global-warming.html (accessed July 12, 2012).

Figure 7.2. 2011 was the warmest La Niña year on record.

Source: World Meteorological Organization, http://www.wmo.int/pages/mediacentre/press_releases/gcs_2011_en.html (accessed July 12, 2012).

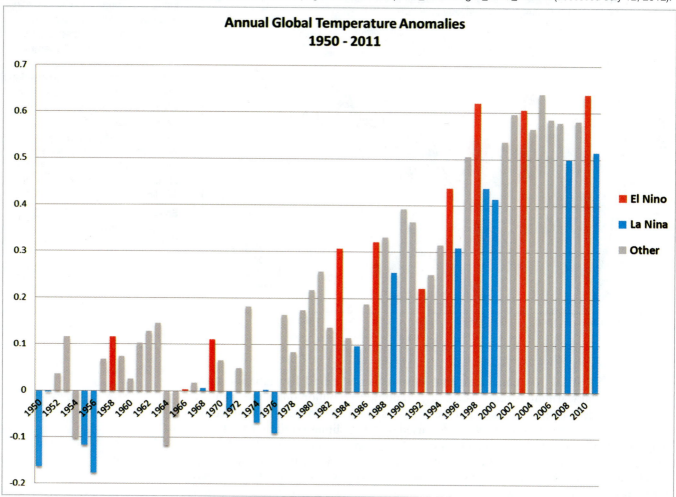

In another study, Foster and Rahmstorf[11] analyzed the surface and lower-troposphere temperature data from 1979 to 2010 after filtering out the effects of solar activity, volcanic emissions, and the El Niño Southern Oscillation (ENSO). They tested various published data sets describing each of these natural factors and found their results were not changed by which data set was used. For instance, solar activity was tested using both sunspot activity and solar irradiance. They used statistical methods to estimate the influence of each of these factors on five temperature datasets from GISS, NCDC, HadCRU, RSS, and University of Alabama. The study also analyzed lag effects, because these factors can have a delayed effect on temperatures.

Foster and Rahmstorf found that warming of surface temperatures as a result of human activities ranges from 0.170°C to 0.175°C per decade (0.31°F to 0.32°F per decade). Temperatures in the lower troposphere, as measured by satellite microwave sensors, have warmed from 0.141 to 0.157°C per decade (0.25 to 0.28°F per decade) due to human activities. The filtering of natural factors revealed their average delayed effect on the global mean surface temperature. ENSO has a delayed effect of 2 to 4 months, volcanic aerosols have a delayed effect of 5 to 7 months, and changes in solar activity have an average delayed effect of 1 month. An important conclusion of this work is that these natural factors account for many variations in the global temperature data over the past 32 years. Thus, there has not been any slowing of

[11]G. Foster and S. Rahmstorf, "Global Temperature Evolution 1979–2010," *Environmental Research Letters* 6 (2011): 044022, doi: 10.1088/1748-9326/6/4/044022

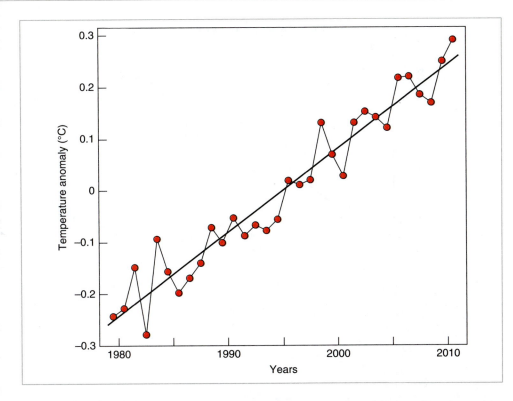

Figure 7.3. Average temperature of five datasets—GISS, NCDC, HadCRU, RSS, and University of Alabama—adjusted for the effect of natural factors (solar activity, volcanism, and ENSO). The rate of global warming has been remarkably steady during the 32 years from 1979 through 2010 and shows no indication of any slowdown or acceleration.

Source: G. Foster and S. Rahmstorf, "Global Temperature Evolution 1979–2010," *Environmental Research Letters* 6 (2011): 044022, doi: 10.1088/1748-9326/6/4/044022. Copyright IOP Publishing, Inc.

global warming, and any deviations from a linear warming trend are explained by the influence of ENSO, volcanoes, and solar variability. In all five adjusted datasets, 2009 and 2010 are the two hottest years on record.

Foster and Rahmstorf adjusted all five temperature datasets for these natural factors and averaged them together to produce a single "composite record of planetary warming showing the true global warming signal" as isolated from the natural agents that also affect climate (Figure 7.3).

NO END OF EMISSIONS

At the end of 2011 new data emerged[12] showing that global carbon dioxide emissions rose a total of 49% since 1990, and 5.9% in 2010 alone. Over the course of 2011, carbon dioxide emissions[13] rose another 1.0 Gt (gigatons) above the level of 2010, or 3.2%. By the end of 2011, these emissions followed slightly below the IPCC-AR4 A1FI economic scenario and slightly above the A2 economic scenario (Figure 7.4). Coal accounted for 45% of total energy-related CO_2 emissions in 2011, followed by oil (35%) and natural gas (20%). As the largest annual increase on record, the 2010 rise showed that an earlier decrease over the period 2008–2009 (due to the global economic recession) had come to an end, and carbon dioxide production was reaching new all-time highs. On average, global carbon dioxide (CO_2) emissions rose 3.1% each year between 2000 and 2010—three times the rate of increase in the previous decade—and emissions were projected to continue to increase by 3.1% in 2011, a prediction that has been more than fulfilled.

Carbon dioxide levels swing up and down in natural seasonal cycles, and roughly half of the emissions are stored in the atmosphere. The other half are stored in equal parts in ocean and land reservoirs (such as forests). However,

---

[12]G. P. Peters, G. Marland, C. Le Quéré, et al., "Rapid Growth in CO_2 Emissions after the 2008–2009 Global Financial Crisis," *Nature Climate Change* 2, no. 2–4 (2011) doi: 10.1038/nclimate1332.

[13]International Energy Agency, "Global Carbon-Dioxide Emissions Increase by 1.0 Gt in 2011 to Record High," 2012, http://www.iea.org/newsroomandevents/news/2012/may/name,27216,en.html (accessed July 12, 2012).

Figure 7.4. a, Global emissions of carbon dioxide follow the high end of the IPCC AR4 economic scenarios. **b,** Emissions still have the opportunity to track any of the IPCC AR5 representative concentration pathways (RCPs).

SOURCE: Skeptical Science.com; http://www.skepticalscience.com/iea-co2-emissions-update-2011.html (accessed July 12, 2012).

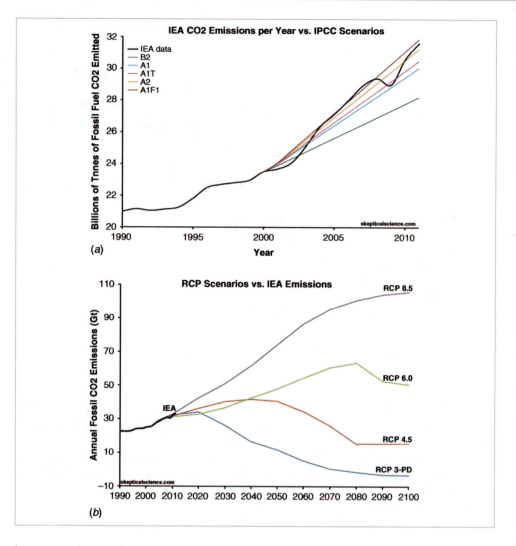

human activities (primarily the burning of coal, oil, and gas for transportation and power) have consistently driven up emission concentrations. Before the widespread burning of coal and oil associated with the Industrial Revolution of the 19th century, carbon dioxide concentration in the atmosphere was about 280 parts per million (ppm). By mid-2012 the global average was 395 ppm,[14] and in the Arctic it had reached the disturbing benchmark of 400 ppm.[15,16]

Are these high CO_2 levels unusual? According to ice-core data[17] the answer is "yes." Climate proxies from Antarctica (Figure 7.5) show that atmospheric carbon dioxide levels have been rising at steadily increasing rates since 1850 and represent the highest levels in the past 1,000 years; according to additional research,[18] they are the highest of the past 15 million years.

---

[14] You can track the ever-rising global carbon dioxide level at the NOAA Earth System Research Laboratory, http://www.esrl.noaa.gov/gmd/ccgg/trends/global.html (accessed July 12, 2012).

[15] See "NOAA: Carbon Dioxide Levels Reach Milestone at Arctic Sites," http://researchmatters.noaa.gov/news/Pages/arcticCO2.aspx (accessed July 12, 2012).

[16] See the video "Time History of Atmospheric Carbon Dioxide from 800,000 Years Ago until January, 2009" at the end of the chapter.

[17] J. Ahn, E. J. Brook, L. Mitchell, et al., "Atmospheric CO_2 over the Last 1000 Years: A High-Resolution Record from the West Antarctic Ice Sheet (WAIS) Divide Ice Core," *Global Biogeochemical Cycles* 26 (2012): GB2027, doi: 10.1029/2011GB004247.

[18] A. K. Tripati, D. R. Roberts, and R. A. Eagle, "Coupling of CO_2 and Ice Sheet Stability over Major Climate Transitions of the Last 20 Million Years," *Science* 326, no. 5958 (2009): 1394–1397, http://www.sciencemag.org/cgi/content/abstract/1178296 (accessed July 12, 2012).

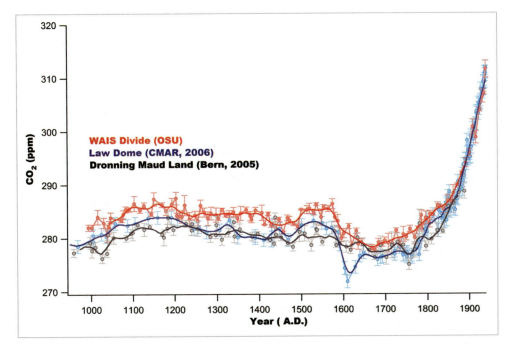

Figure 7.5. Atmospheric CO_2 for the last 1,000 years. Data from ice cores collected at three Antarctic localities (Law Dome, blue; West Antarctic Ice Sheet Divide, red; and Dronning Maud Land, black) are compared. Atmospheric CO_2 variations over the time period A.D. 1000–1800 are statistically correlated with Northern Hemisphere climate and tropical Indo-Pacific sea-surface temperature.

SOURCE: J. Ahn, E. J. Brook, L. Mitchell, et al., "Atmospheric CO_2 over the Last 1000 Years: A High-Resolution Record from the West Antarctic Ice Sheet (WAIS) Divide Ice Core," *Global Biogeochemical Cycles* 26 (2012): GB2027, doi: 10.1029/2011GB004247.

According to the National Oceanographic and Atmospheric Administration (NOAA) Global Monitoring Division,[19] methane (CH_4) levels rose in 2010 to 1,799 parts per billion (ppb). After remaining nearly constant from 1996 to 2006, they have since risen for five consecutive years. Methane measured 1,714 ppb in 1990 and 1,794 ppb in 2009. Pound for pound, methane is 25 times more potent as a greenhouse gas than carbon dioxide, but there's less of it in the atmosphere, and once it is introduced to the atmosphere, methane only resides for a decade or so compared to the millennia-long residence time of carbon dioxide.

Levels of nitrous oxide (N_2O), a greenhouse gas emitted from natural sources and as a byproduct of agricultural fertilization, livestock manure, sewage treatment, and some industrial processes, rose steadily in 2010. CFC11 and CFC12 molecules continued their long-term decline. Greenhouse gases that belong to the class of human-made chemicals called chlorofluorocarbons, CFC11 and CFC12 cause the depletion of Earth's ozone layer. They have been dropping at a rate of about 1% per year since the late 1990s because of an international agreement, the Montreal Protocol, to protect Earth's ozone layer.

The largest carbon dioxide contributors are (in order) China, the United States, India, Russia, and the European Union. The high level of CO_2 production in 2010 and 2011 track the worst-case projections of the IPCC AR4, which, according to models, could produce warming in excess of 2°C by 2100. According to the World Meteorological Organization[20] (WMO), this level of warming could trigger far-reaching and irreversible changes in the environment. The WMO estimated the global combined sea-surface and land-surface air temperature for 2011 (January through October) at 0.41±0.11°C (0.74±0.20°F) above the 1961–1990 annual average of 14.00°C (57.2°F), making 2011 the tenth warmest year since the start of recordkeeping in 1850. In March of 2012, the WMO issued a report[21] that concluded the rate of global temperature increase (Figure 7.6) since 1971 has been "remarkable." Over the past four decades, global temperature has increased at an average rate of 0.166°C (0.3°F) per decade

[19]See "NOAA Greenhouse Gas Index Continues to Climb," http://researchmatters.noaa.gov/news/Pages/aggi2011.aspx (accessed July 12, 2012).

[20]WMO Press Release No. 935, Nov. 29, 2011: http://www.wmo.int/pages/mediacentre/press_releases/pr_935_en.html (accessed July 12, 2012).

[21]See the 2012 World Meteorological Organization annual statement on the status of the global climate: http://www.wmo.int/pages/prog/wcp/wcdmp/documents/1085_en.pdf (accessed July 12, 2012). See the Press Release here: http://www.wmo.int/pages/mediacentre/press_releases/pr_943_en.html (accessed July 12, 2012).

Figure 7.6. Since 1971, the global temperature has increased at a rate (red line) of 0.166°C (0.3°F) per decade compared to the average rate (blue line) of 0.06°C (0.11°F) per decade computed over the full period (1881–2010). Global land- and sea-surface temperatures were estimated to be 0.46°C (0.83°F) above the long-term average (1961–1990) of 14.0°C (25.2°F). It was the warmest decade ever recorded for the global land surface and sea surface and for every continent.

SOURCE: Figure from World Meteorological Organization 2012 Annual Report: http://www.wmo.int/pages/mediacentre/press_releases/documents/943_en-figure1-3.pdf (accessed July 12, 2012).

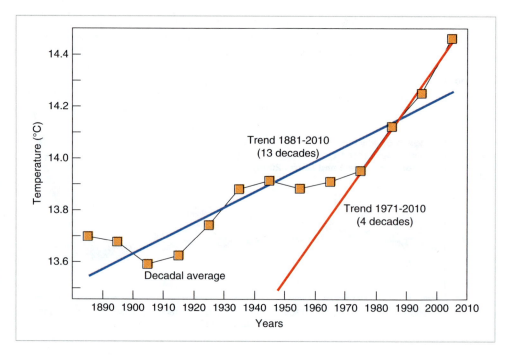

compared to the average rate of 0.06°C (0.11°F) per decade computed over the full period of recordkeeping, 1881 to 2010.

A NEW NORMAL

The Arctic continued to experience warming in 2012 and set a new record low in sea ice extent. In fact, NOAA reports[22] that since 2006 the Arctic has settled into a "new normal," characterized by persistent decline in thickness and extent of sea ice and a warmer, fresher upper ocean. Relative to lower latitudes, air temperatures in the Arctic deviate from historical averages by a factor of two or more. This phenomenon is known as Arctic amplification, a term that embodies the fact that atmospheric carbon dioxide concentration in the Arctic is higher than the global average[23] and warming seawater is releasing methane[24] that bubbles through the sea ice into the air.

According to the 2012 National Snow and Ice Data Center (NSIDC),[25] Arctic amplification is caused primarily by increased summer sea-ice loss (Figure 7.7) and northward transport of heat by the atmosphere and ocean. The NSIDC reported that the seasonal Arctic sea-ice area minimum, reached on September 16, 2012, was 50% below the 1979–2000 average. Sea-ice volume was even further below average and set a new record low that surpassed the record set in 2010.

NOAA reports that the Arctic is shifting into a new state: warmer, greener, and with less ice. NOAA[26] and others report the following:

- **Atmosphere:** In 2011, the average annual near-surface air temperatures over much of the Arctic Ocean were approximately 1.5°C (2.5°F) warmer than the 1981–2010 baseline period.

[22]NOAA Press Release: "Arctic Settles into New Phase—Warmer, Greener, and Less Ice," http://www.noaanews.noaa.gov/stories2011/20111201_arcticreportcard.html (accessed July 12, 2012).

[23]See "NOAA: Carbon Dioxide Levels Reach Milestone at Arctic Sites," http://researchmatters.noaa.gov/news/Pages/arcticCO2.aspx (accessed July 12, 2012).

[24]E. A. Kort, S. C. Wofsy, B. C. Daube, et al., "Atmospheric Observations of Arctic Ocean Methane Emissions up to 82° North," *Nature Geoscience* 5 (2012): 318–321. See the article http://earthobservatory.nasa.gov/IOTD/view.php?id=77868&src=eorss-iotd (accessed July 12, 2012).

[25]See the National Snow and Ice Data Center press release, September 19, 2012, http://nsidc.org/news/press/2012_seaiceminimum.html (last accessed October 25, 2012).

[26]NOAA Press Release, "Arctic Settles into New Phase—Warmer, Greener, and Less Ice," http://www.noaanews.noaa.gov/stories2011/20111201_arcticreportcard.html (accessed July 12, 2012).

Figure 7.7. The volume of Arctic sea ice plotted as an anomaly relative to the 1979–2012 average. The trend for the period 1979 to the present is shown in blue. Shaded areas show one and two standard deviations from the trend. Error bars indicate the uncertainty of the monthly anomaly plotted once per year. Monthly averaged ice volume for September 2012 was 3,400 km³. This value is 72% lower than the mean over this period, 80% lower than the maximum in 1979, and 2.0 standard deviations below the 1979–2012 trend.[27]

Source: Fig.1 Arctic sea ice volume anomaly from PIOMAS updated once a month. "Arctic Sea Ice Volume Anomaly, version 2, Polar Science Center. httppsc.apl.washington.eduwordpressresearchprojectsarctic-sea-ice-volume-anomaly (Retrieved June 19, 2012).

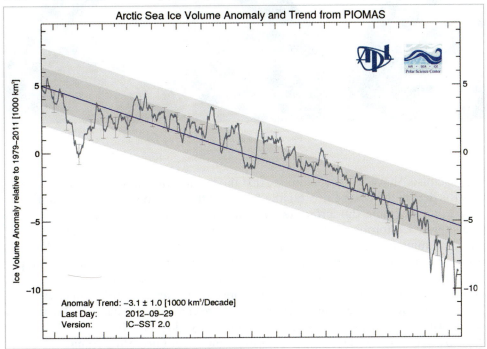

- **Sea ice:** Minimum Arctic sea-ice area in September 2012 was the second lowest recorded by satellite since 1979. Sea ice loss has been tied to impacts from global warming.[28]

- **Ocean:** Arctic Ocean temperature and salinity may be stabilizing after a period of warming and freshening. Acidification of sea water (ocean acidification) as a result of carbon dioxide absorption has also been documented in the Beaufort and Chukchi seas.

  o An anomalous pool of freshwater (27 m [88.5 ft] thick in places) fed by inland meltwater has been found floating atop Arctic seawater.[29]

- **Land:** Arctic tundra vegetation continues to increase and is associated with higher air temperatures over most of the Arctic land mass.

  o Thawing permafrost and warming seawater will release huge amounts of carbon and accelerate climate change.[30]

  o Arctic coastlines are eroding on average by 0.5 m [1.6 ft] per year.[31]

[27]See the video "Witness a Glacier's Staggering Seven Year Retreat" at the end of the chapter.

[28]D. Notz and J. Marotzke, "Observations Reveal External Driver for Arctic Sea-Ice Retreat," *Geophysical Research Letters* 39 (2012): L08502, doi: 10.1029/2012GL051094.

[29]B. Rabe, "An Assessment of Arctic Ocean Freshwater Content Changes from the 1990s to the 2006–2008 Period," *Deep Sea Research I* 58 (2011): 173, doi: 10.1016/j.dsr.2010.12.002.

[30]E. G. Schuur and B. Abbott; Permafrost Carbon Network, "High Risk of Permafrost Thaw," *Nature* 480 (2011): 32–33.

[31]D.L. Forbes (ed.), *State of the Arctic Coast (2010) Scientific Review and Outlook*, a report of the International Arctic Sciences Committee, Land-Ocean Interactions in the Coastal Zone, Arctic Monitoring and Assessment Programme and International Permafrost Association (Geesthacht, Germany, Helmholtz-Zentrum, 2011).

Figure 7.8. A NASA study revealed that the oldest and thickest Arctic sea ice (**a**, 1980) is disappearing at a faster rate than the younger and thinner ice at the edges of the Arctic Ocean's floating ice cap. Multi-year ice extent—which includes all areas of the Arctic Ocean where multi-year ice covers at least 1% of the ocean surface—is diminishing at a rate of 15.1% per decade (**b**, 2012).[32]

Source: NASA/Goddard Scientific Visualization Studio.

(a)

(b)

Scientists[33] have found (Figure 7.8) that the average thickness of the Arctic sea ice cover is declining because it is rapidly losing its thickest component, the multi-year ice. At the same time, the surface temperature in the Arctic is going up, which results in a shorter ice-forming season. The continued loss of sea ice is causing profound and continuing changes to Arctic marine ecosystems. For example, because the open water season is lasting longer and the area of open water is significantly larger, primary production by phytoplankton in the Arctic Ocean increased approximately 20% between 1998 and 2009. In addition, the composition, range, and total biomass of sea-floor communities in the shallow Arctic Ocean are changing dramatically. Polar bears and walrus are also experiencing negative effects from loss of habitat, and whales now have greater access to the Northwest Passage and other northern feeding areas.

Greenland

Greenland and neighboring ice caps (e.g., Penny Ice Cap on Baffin Island[34]) have continued their meltdown. The area and duration of melting at the surface of the ice sheet in summer 2011 were the third highest since 1979.[35] Total ice sheet mass loss in 2011 was 70% larger than the 2003–2009 average annual loss rate of 250 Gt per year and, according to satellite gravity data obtained since 2002, ice-sheet mass loss is accelerating.

Combinations of satellite data (gravity, altimetry, and radar) are being used to document the distribution of Greenland melting, calving, and rainfall and snowfall rates.[36] Researchers found that Greenland's ice loss through melting and iceberg calving during the last 10 years is unusually high compared to the last 50 years. They show that the Greenland ice sheet continues to lose mass and thus contributes at about 0.7 mm/yr (0.03 in/yr) to the currently observed sea level change of about

[32]NASA finds the thickest parts of the Arctic ice cap melting faster: http://www.nasa.gov/topics/earth/features/thick-melt.html (accessed July 12, 2012).

[33]J. Comiso, "Large Decadal Decline of the Arctic Multiyear Ice Cover," *Journal of Climate* 25 (2012): 1176–1193. doi: http://dx.doi.org/10.1175/JCLI-D-11-00113.1.

[34]C. Zdanowicz, A. Smetny-Sowa, D. Fisher, et al, "Summer Melt Rates on Penny Ice Cap, Baffin Island: Past and Recent Trends and Implications for Regional Climate," *Journal of Geophysical Research*, 117 (2012): F02006, doi: 10.1029/2011JF002248.

[35]J. Richter-Menge, M. O. Jeffries, and J. E. Overland (eds.), "Arctic Report Card 2011." http://www.arctic.noaa.gov/reportcard (accessed July 12, 2012).

[36]I. Sasgen, M. van den Broeke, J. Bamber, "Timing and Origin of Recent Regional Ice-Mass Loss in Greenland," *Earth and Planetary Science Letters* (2012): 333–334, 293 doi: 10.1016/j.epsl.2012.03.033.

3 mm/yr. This trend increases each year by a further 0.07 mm/yr (0.003 in/yr). The pattern and timing of loss is complex, with largest losses occurring in southwest and northwest Greenland; the respective contributions of melting, iceberg calving, and fluctuations in snow accumulation differ considerably.

Remote-sensing imagery revealed the impact of the dramatic summer-melt season of the past decade on the continental ice sheet covering Greenland. Each summer, melting snow along the coast reveals an ever-widening swath of newly exposed rock, meltwater runs in streams across the ice surface and disappears through deep vertical shafts in the ice (called moulins), and ice-dammed lakes collapse under their own weight and cascade into the ocean. Satellite gravity measurements[37] show that the mass loss from the entire Greenland ice sheet during 2010–2011 was the largest annual loss in the satellite record beginning in 2002, and it contributed just over 1 mm (0.04 in) to global sea-level rise.

Antarctica

Antarctica may be losing mass—exactly how much is not clear.[38] East Antarctica is a high, dry ice sheet averaging a little over 2 km (1.2 mi) thick and covering two thirds of the continent; it is remote and difficult to observe. Historically, measurements have indicated that little, if any, surface warming is occurring in this region, and mass loss at the edges of East Antarctica (where ice flows into the sea) is offset by accumulation of snow in the interior. This traditional view has, however, been challenged by researchers at the NASA/German Aerospace Center's Gravity Recovery and Climate Experiment (GRACE), who suggest that there has been more ice loss from East Antarctica than previously thought. Data[39] suggest that East Antarctica as a whole is losing mass, mostly in coastal regions, at a rate of -57 ± 52 Gt per year, apparently because of increased ice loss since 2006. Research[40] suggests, however, that meltwater production is not the primary process causing this loss, and some other process is likely responsible.

East Antarctica holds approximately 61% of all fresh water on Earth: a massive amount of ice, the equivalent of 230 ft (70 m) of global sea level. It is roughly the size of Australia, and making measurements of its annual ice loss (by glacier flow into the sea) and gain (through snowfall) is a highly complex business[41] that depends on a number of uncertainties and assumptions. The difficulty of calculating the balance between annual Antarctic-wide snowfall and ice loss emerged in the scientific literature in the summer of 2011, with a paper by NASA scientists Jay Zwally and Mario Giovinetto that challenged previous estimates.[42] Their paper reassesses the uncertainties in the various measurement techniques used by previous workers, and they offer revised estimates of net change in Antarctic ice, ranging from $+27$ to -40 billion tons per year. For the period from 1992 to 2001, they estimate a loss of only 31 billion tons per year. Although these sound like huge numbers, they represent only a net gain or loss in the range $+1.1\%$ to -1.7% of the 2400 billion tons of snow that fall in Antarctica each year.

West Antarctica, losing mass faster than the rest of the continent, consists of several ice streams (concentrated ice flows within a larger continental glacier) that have increased their rate of flow over the past decade. The largest, Pine Island Glacier,

---

[37]J. Richter-Menge, M. O. Jeffries, and J. E. Overland (eds.), "Arctic Report Card 2011."

[38]R. A. Kerr, "Antarctic Ice's Future Still Mired in Its Murky Past," *Science* 333, no. 6041 (2011): 401, doi: 10.1126/science.333.6041.401.

[39]J. L. Chen, C. R. Wilson, D. Blankenship and B. D. Tapley, "Accelerated Antarctic Ice Loss from Satellite Gravity Measurements," *Nature Geoscience* 2 (2010): 859–862.

[40]P. Kuipers Munneke, G. Picard, M. van den Broeke, J. Lenaerts and E. van Meijgaard, "Insignificant Change in Antarctic Snowmelt Volume since 1979," *Geophysical Research Letters* 39 (2012): L01501, doi: 10.1029/2011GL050207.

[41]R. A. Kerr, "Antarctic Ice's Future Still Mired in Its Murky Past."

[42]H. J. Zwally and M. B. Giovinetto, "Overview and Assessment of Antarctic Ice-Sheet Mass Balance Estimates: 1992–2009," *Surveys in Geophysics* 32, nos. 4–5 (2011): 351–376, doi 10.1007/s10712-011-9123-5.

has quadrupled its rate of flow between 1995 and 2006.[43] An important aspect of ice stream acceleration in coastal settings is the collapse of ice shelves that lie at the foot of these glaciers. Ice shelves that are grounded on the seafloor are thought to buttress ice streams, preventing them from dramatically accelerating into the ocean. Where ice shelves have collapsed, the adjoining ice streams have accelerated significantly. The Larsen B, for example, was a 12,000-year-old ice shelf the size of Rhode Island that in 2002 disintegrated in only three weeks.[44]

Oceanographer Eric Rignot hypothesized in 1998[45] that given the landward retreat of the grounding line (the boundary between the floating ice shelf and the portion of the glacier that sits on land), acceleration of Pine Island Glacier was the result of contact with warm ocean water that promotes shelf collapse. This hypothesis has been supported by subsequent studies,[46] fundamentally shifting the traditional view of glaciers as slow giants to behaving in some cases as raceways of streaming ice. (Figure 7.9)[47]

Figure 7.9. This image is the first complete map of the speed and direction of ice flow in Antarctica. The thick black lines delineate major ice divides. Subglacial lakes in Antarctica's interior are also outlined in black.

Source: NASA Jet Propulsion Laboratory, http://www.jpl.nasa.gov/spaceimages/search_grid.php?q=antarctic ice fl ow&img_search_submit.x=66&img_search_submit.y=8 (accessed July 12, 2012).

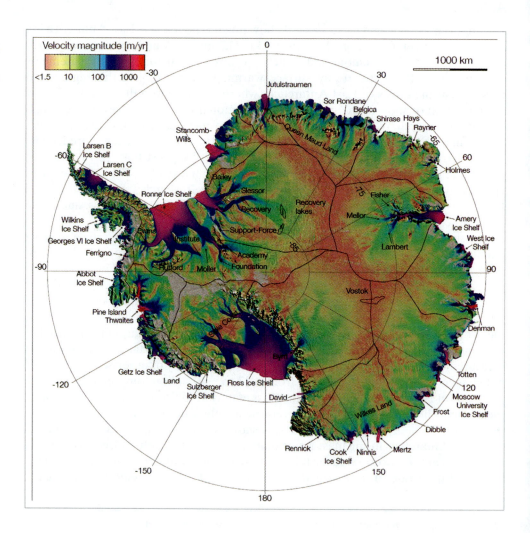

[43]D. J. Wingham, D. W. Wallis and A. Shepherd, "Spatial and Temporal Evolution of Pine Island Glacier Thinning, 1995–2006," *Geophysical Research Letters* 36 (2009): L17501.

[44]See "Ice Shelves" at the National Snow and Ice Data Center, http://nsidc.org/cryosphere/sotc/iceshelves.html (accessed July 12, 2012).

[45]E. J. Rignot, "Fast Recession of a West Antarctic Glacier," *Science* 281 (1998): 549–551.

[46]P. A. Mayewski, M. P. Meredith, C. P., Summerhayes, et al., "State of the Antarctic and Southern Ocean Climate System," *Reviews of Geophysics* 47 (2009): 1–38.

[47]See the NASA animation "Flow of Ice Across Antarctica" at the end of the chapter.

Measurements[48] by GRACE satellites show that the Antarctic ice sheet is not only losing mass, but it is also losing mass at an accelerating rate; that is, each year the amount of ice lost increases over the previous year. What makes this particularly significant is that it isn't just GRACE data that show accelerating loss, satellite radar data do as well. A related study[49] of the rate of ice loss from both Greenland and Antarctica shows that if current ice-sheet melting rates continue for the next four decades, by 2050 their total loss could raise global mean sea level 15 cm (5.9 in). When combined with the projected contribution from glacial ice caps (8 cm [3.1 in]) and oceanic thermal expansion (9 cm [3.5 in]), the total amount of global mean sea-level rise could reach 32 cm (over a foot) by mid-century.

EXTREME WEATHER

In the spring and summer of 2012, one month after another set records for high temperatures as the Northern Hemisphere was enveloped in a heat wave that refused to end. The hottest March in U.S. history turned into an April that set a record for average global land temperature. Global temperatures in May 2012 were the second warmest since recordkeeping began in 1880. In the United States, June was 1.1°C (2°F) above the twentieth century average,[50] and temperatures late in the month broke or tied over 170 all-time records across North America. June also culminated the warmest 6-month and 12-month periods in national history. July 2012 was the hottest July on record, and the period June–August 2012 became the hottest 3 month period since record keeping began.

The spring and summer heat brought drought to the nation's agriculture from northern Florida to eastern Washington state. Conditions ranging from "abnormally dry" to "exceptional drought" prompted the U.S. Department of Agriculture to declare[51] more than 1,000 counties in 26 states natural disaster areas. This nationwide emergency established the largest natural-disaster area in U.S. history.

In 2011 the United States was pummeled by 14 extreme weather events, each of which caused more than $1 billion in damage; in several states the months of January to October were the wettest ever recorded, and March 2012 set over 1,000 record-high temperatures[52] across the nation. According to NOAA[53] scientists, 2011 was a record-breaking year for climate extremes, as much of the United States faced historic levels of heat, precipitation, flooding, and severe weather. Japan also registered record rainfalls, and the Yangtze River basin in China suffered a record drought. Similar record-breaking events occurred also in previous years. In 2010, Western Russia experienced the hottest summer in centuries, and in Pakistan and Australia record-breaking amounts of rain fell. Europe had its hottest summer in at least half a millennium in 2003[54]; in 2002, Germany measured more rain in one day than ever

---

[48]I. Velicogna, "Increasing Rates of Ice Mass Loss from the Greenland and Antarctic Ice Sheets Revealed by GRACE," *Geophysical Research Letters* 36 (2009): L19503.

[49]E. Rignot, I. Velicogna, M. R. van den Broeke, A. Monaghan and J. Lenaerts, "Acceleration of the Contribution of the Greenland and Antarctic Ice Sheets to Sea Level Rise," *Geophysical Research Letters* 38 (2011): L05503, doi: 10.1029/2011GL046583.

[50]See "June 2012 Brings More Record-Breaking Warmth to U.S.," http://www.climatewatch.noaa.gov/image/2012/june-2012-brings-more-record-breaking-warmth-to-u-s (accessed July 19, 2012).

[51]See "USDA Announces Streamlined Disaster Designation Process with Lower Emergency Loan Rates and Greater CRP Flexibility in Disaster Areas," http://www.usda.gov/wps/portal/usda/usdahome?contentid=2012/07/0228.xml&navid=NEWS_RELEASE&navtype=RT&parentnav=LATEST_RELEASES&edeployment_action=retrievecontent (accessed July 19, 2012).

[52]NASA, Earth Observatory, "Historic Heat in North American Turns Winter to Summer," http://earthobservatory.nasa.gov/IOTD/view.php?id=77465&src=eoa-iotd (accessed July 12, 2012).

[53]NOAA, "2011 a Year of Climate Extremes in the United States," http://www.noaanews.noaa.gov/stories2012/20120119_global_stats.html (accessed July 12, 2012).

[54]"Extreme Weather of Last Decade Part of Larger Pattern Linked to Global Warming," http://www.sciencedaily.com/releases/2012/03/120325173206.htm (accessed July 12, 2012).

before, followed by the worst flooding of the Elbe River for centuries. Are these climate extremes coincidental, or are they the product of global warming? Scientists investigating[55] this question have concluded that a clear link connects extreme rainfall and heat waves to human-caused global warming, a link that is supported by elementary physical principles, statistical trends, and computer simulations.

Attribution

For years, most scientists have been hesitant to connect single weather events, such as powerful rain storms and hot nights, to climate change. When queried about the connection between global warming and extreme weather, scientists usually begged off with statements such as, "It's too early to say for sure that weather is changing," or "Climate and weather are different and you cannot predict one with the other." That attitude is changing, however, as new studies emerge that link daily weather to climate change.

For example, a study[56] by the Australian government's Pacific Climate Change Science Program reported that future weather and climate in the region will be characterized by more-intense tropical cyclones, more-frequent deluges, and a greater proportion of hot days and warm nights. Already, people living in Pacific Islands are experiencing changes in their climate, such as higher temperatures, shifts in rainfall patterns, changing frequencies of extreme events, and rising sea levels. These changes are affecting peoples' lives and livelihoods as well as important industries like agriculture and tourism. Extreme rainfall events that currently occur once every 20 years on average are projected to occur four times per year on average by 2055 and seven times per year on average by 2090 under a high emissions scenario. By 2030, the projected regional warming for the south Pacific is around +0.5 to +1.0°C (+0.9 to +1.8°F), regardless of the emissions scenario.

In another study, researchers[57] examining daily wintertime temperature extremes since 1948 found that the warmest days were much more severe and widespread than the coldest days during the Northern Hemisphere winters of 2009–2010. Furthermore, whereas the extreme cold was mostly attributable to a natural climate cycle, the extreme warmth was not. In another study,[58] researchers found that the number of unusually warm nights increased during the second half of the 20th century, the rate of increase was greatest in the most recent period, the increase could not be explained by natural climate variability alone, and at least part of the change was attributable to global warming. Unusually warm nights and daily winter temperature extremes: These are just some of the weather phenomena that are attributable to climate change.

In their 2012 Annual Statement, the World Meteorological Organization[59] concluded that the decade 2002–2011 was probably the warmest globally for at least a millennium and that 2011 was the eleventh hottest on record. Researchers are now concluding[60] that it is very likely that the record of extreme weather events would not have occurred without global warming resulting from industrial greenhouse gas emissions.

Extreme weather events are devastating in their impacts and affect nearly all regions of the globe. Events include severe floods and heat waves. Nearly twice as many record hot days as record cold days occur in the United States and Australia. The length of summer heat waves in Western Europe has almost doubled and the

---

[55]D. Coumou and S. Rahmstorf, "A Decade of Weather Extremes," *Nature Climate Change* 2011, doi: 10.1038/NCLIMATE1452.

[56]The Pacific Climate Change Science Program, "Climate Change in the Pacific: Scientific Assessment and New Research," 2011, www.cawcr.gov.au/projects/PCCSP/publications.html (accessed July 12, 2012).

[57]K. Guirguis, A. Gershunov, R. Schwartz, and S. Bennett, "Recent Warm and Cold Daily Winter Temperature Extremes in the Northern Hemisphere," *Geophysical Research Letters* 38 (2011): L17701, doi: 10.1029/2011GL048762.

[58]S. Morak, G. C. Hegerl, and J. Kenyon, "Detectable Regional Changes in the Number of Warm Nights," *Geophysical Research Letters* 38 (2011): L17703, doi: 10.1029/2011GL048531.

[59]World Meteorological Organization, *Annual Statement,* http://www.wmo.int/pages/mediacentre/press_releases/pr_943_en.html (accessed July 12, 2012).

[60]D. Coumou and S. Rahmstorf, "A Decade of Weather Extremes."

frequency of hot days has almost tripled. Extremely hot summers are now observed in over 10% of the global land area, compared with only about 0.1% to 0.2% for the period 1951 to 1980. A record number of tropical storms and hurricanes in the Atlantic occurred in 2005. In 2010, Russia had the hottest summer since 1500 and Pakistan had the worst flooding in its history. In 2011 alone, the United States suffered 14 weather events that caused losses of more than $1 billion each, and during March 13–19, 2012, historical heat records were exceeded in more than 1,000 places in North America. This high number of extremes is not normal. Although single weather extremes have always occurred and are related to localized processes, they are now unfolding against a background of a warmer atmosphere and amplified water cycle that can turn extreme weather into a record-breaking event.

NOAA has declared[61] 2011 among the most extreme weather years in history. Jane Lubchenco, NOAA's chief, declared that the 2011 U.S. record of more than a dozen billion-dollar weather disasters in one year is "a harbinger of things to come" and "not an aberration." NOAA scientists found that 2011 was the latest and worst year in an annually increasing trend of natural disasters (Figure 7.10). According to Lubchenco, "at least some of the ongoing increase in natural disasters appears to be driven by climate change."[62]

Studies are now beginning to support the opinion among climatologists that climate change is leading to more extreme weather. For instance, one study[63] identified a link between the increase in atmospheric water-holding capacity and increases in heavy precipitation. As the atmosphere warms, its ability to hold water (its water vapor content) is expected to increase exponentially with temperature. Thus far, atmospheric water content is increasing in agreement with this theoretical expectation,[64] leading researchers to suggest that global warming may be partly responsible for increases in

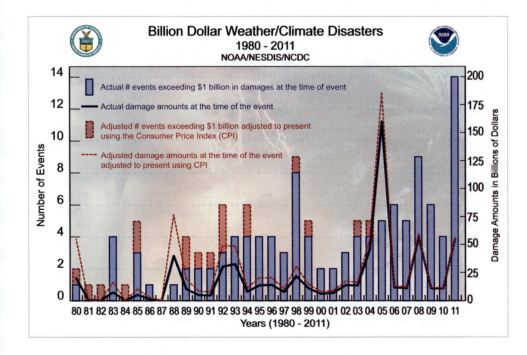

Figure 7.10. In 2011 the United States set a record with 14 separate billion-dollar weather and climate disasters, with an aggregate damage total of approximately $52 billion. 2011 broke the previous record of $9 billion weather and climate disasters in one year, which occurred in 2008.

Source: Billion Dollar U.S. Weather/ Climate Disasters, http://www.ncdc. noaa.gov/oa/reports/billionz.html).

[61]See "NOAA Makes It Official: 2011 Among Most Extreme Weather Years in History," *Scientific American*, http://www.scientificamerican.com/article.cfm?id=noaa-makes-2011-most-extreme-weather-year (accessed July 12, 2012).

[62]See the Climate Progress column at http://thinkprogress.org/romm/2011/12/07/384524/noaa-us-sets-record-with-a-dozen-billion-dollar-weather-disasters-in-one-year/ (accessed July 12, 2012).

[63]S.-K. Min, X. Zhang, F. W. Zwiers, and G. C. Hegerl, "Human Contribution to More Intense Precipitation Extremes," *Nature* 470 (2011): 378–381, doi: 10.1038/nature09763.

[64]B. D. Santer, C. Mears, F. J. Wentz, et al., "Identification of Human-Induced Changes in Atmospheric Moisture Content," *Proceedings of the National Academy of Sciences* 104 (2007): 15248–15253.

heavy precipitation.[65] Because of the limited availability of daily observations, however, previous studies have examined only model results.[66] In early 2011, a study[67] emerged that connected observations of extreme precipitation events with human-induced climate change. The study used observational data of precipitation events over two thirds of Northern Hemisphere land areas, and showed that increases in extreme precipitation were consistent with the increase in greenhouse gases and that models[68] may be underestimating observed increases in heavy precipitation with warming.

Overall, NOAA experts report that extreme weather events have grown more frequent in the United States since 1980. According to Tom Karl, Director of the National Climatic Data Center, part of that shift is due to climate change. "Extremes of precipitation are generally increasing because the planet is actually warming and more water is evaporating from the oceans," he said. "This extra water vapor in the atmosphere then enables rain and snow events to become more extensive and intense than they might otherwise be."[69]

Increasingly, published research[70] is drawing connections between weather and climate change. October and November 2000 marked the wettest autumn in England and Wales since recordkeeping began in 1766. Flooding damaged nearly 10,000 properties, severely disrupted services, and caused billions of dollars in losses. To assess the role of atmospheric warming in these events, researchers used thousands of climate-model simulations of autumn 2000 weather under current conditions and under conditions as they might have been had global warming not occurred. Their data show that in nine out of 10 cases, 20th-century anthropogenic greenhouse gas emissions increased the risk of floods occurring in England and Wales in autumn 2000 by more than 20%, and in two out of three cases by more than 90%.

The U.S. Climate Change Science Program (CCSP) studied the issue of extreme weather under global warming.[71] They found that extreme events have significant impacts on our economy and environment and are among the most serious challenges associated with climate change. Over the past three decades North America has experienced an increase in unusually hot days and nights, a decrease in unusually cold days and nights, and a reduction in frost days.[72]

According to the CCSP study, heavy rain events and severe droughts have become more frequent and more intense.[73] Hurricanes have increased in power and frequency, although the annual number of North American land-falling storms has not shown any trend.[74] Mid-latitude storm tracks are shifting northward, and the strongest storms are becoming even stronger. Throughout these patterns the CCSP finds extreme weather

[65]K. E. Trenberth, A. Dai, R. M. Rasmussen, and D. B. Parsons, "The Changing Character of Precipitation," *Bulletin of the American Meteorological Society* 84 (2003): 1205–1217.

[66]S.-K. Min, X. B. Zhang, F. W. Zwiers, P. Friederichs, and A. Hense, "Signal Detectability in Extreme Precipitation Changes Assessed from Twentieth Century Climate Simulations," *Climate Dynamics* 32 (2009): 95–111.

[67]S.-K. Min, X. Zhang, F. W. Zwiers, and G. C. Hegerl, "Human Contribution to More Intense Precipitation Extremes."

[68]R. P. Allan and B. J. Soden, "Atmospheric Warming and the Amplification of Precipitation Extremes," *Science* 321 (2008): 1481–1484.

[69]See "NOAA Makes It Official: 2011 Among Most Extreme Weather Years in History."

[70]P. Pall, A. Tolu, D. A. Stone, et al., "Anthropogenic Greenhouse Gas Contribution to Flood Risk in England and Wales in Autumn 2000," *Nature* 470, no. 7334, (2011): 382, doi: 10.1038/nature09762.

[71]U.S. Climate Change Science Program, *Weather and Climate Extremes in a Changing Climate. Regions of Focus: North America, Hawaii, Caribbean, and U.S. Pacific Islands*. A Report by the U.S. Climate Change Science Program and the Subcommittee on Global Change Research. (Washington, D.C., Department of Commerce, NOAA's National Climatic Data Center, 2008).

[72]T.C. Peterson, X. Zhang, M. Brunet-India, and J.L. Vázquez-Aguirre, "Changes in North American Extremes Derived from Daily Weather Data," *Journal of Geophysical Research* 113 (2008): D07113, doi: 10.1029/2007JD009453.

[73]D. R. Easterling, T. Wallis, J. Lawrimore, and R. Heim, "The Effects of Temperature and Precipitation Trends on U.S. Drought," *Geophysical Research Letters* 34 (2007): L20709, doi: 10.1029/2007GL031541.

[74]T.R. Knutson and R.E. Tuleya, "Tropical Cyclones and Climate Change: Revisiting Recent Studies at GFDL." In H. Diaz and R. Murnane (eds.), *Climate Extremes and Society*, (Cambridge, U.K., Cambridge University Press, 2008), pp. 120–144.

events can be attributed to anthropogenic climate change. For instance, increased atmospheric water vapor due to warming is theoretically and statistically associated with the increase in heavy precipitation events. Although no studies have formally attributed changes in drought severity in North America to climate change,[75] early snowmelt due to short winters has been shown to be responsible for extended summer drought.

The CCSP found several potential effects of global warming. Continued global warming will lead to future increases in the frequency and intensity of heat waves and heavy downpours. Droughts of greater severity and frequency are likely to occur across substantial areas of North America. There will be future increases in hurricane wind speeds, rainfall intensity, and storm surge levels. During winter, the strongest storms are likely to be more frequent and have stronger winds and more extreme wave heights.

One study[76] found that day-to-day weather has grown increasingly erratic and extreme, with significant fluctuations in sunshine and rainfall affecting more than a third of the planet. Researchers reported that extremely sunny or cloudy days are more common today than they were in the early 1980s. Analysis of daily weather data revealed that swings from thunderstorms to dry days rose considerably since the late 1990s. These swings can have consequences for ecosystem stability and the control of pests and diseases as well as for industries such as agriculture and solar-energy production, all of which are vulnerable to inconsistent and extreme weather.

A special report[77] by the Intergovernmental Panel on Climate Change (IPCC) provides an analysis of extreme weather events, their attribution to climate change, and probabilities of future changes. Extreme weather events vary from year to year and place to place, but overall the number of events and the economic losses they cause have increased over time. The probability that the frequency of heavy precipitation will increase in the 21st century over many regions is 66% to 100%, and it is virtually certain (99% to 100% probability) that increases in the frequency of warm daily temperature extremes and decreases in cold daily extremes will occur on a global scale throughout the 21st century. It is also very likely (90% to 100% probability) that heat waves will increase in length, frequency, and/or intensity over most land areas. Projected precipitation and temperature changes imply changes in floods; however, because of limited evidence and because the causes of regional climate changes are complex, there is low confidence overall at the global scale regarding climate-driven changes in magnitude or frequency of river-related flooding.

It is likely (66% to 100% probability) that the average maximum wind speed of tropical cyclones will increase throughout the coming century, although possibly not in every ocean basin. It is also likely that overall there will be either a decrease or essentially no change in the number of tropical cyclones. It is very likely (90% to 100% probability) that average sea-level rise will contribute to upward trends in extreme coastal (high) water levels.

Overall, the IPCC report finds that human activity related to global warming has driven increases in some extreme weather and climate events around the world in recent decades. Those events and other weather extremes will worsen in coming decades as greenhouse gases build. As a result, the report recommends that society take "low-regret measures" to manage the problem. Such measures might include development of early-warning systems, land-use planning, ecosystem management, and improvements to water supplies, irrigation, and drainage systems. Such measures would benefit society in dealing with the current climate as well as with almost any range of possible future climates.

Extreme weather is consistent with what we know is occurring as a result of climate change. For instance, on average, the United States is 2°F (1.1°C) warmer than

[75]P. Y. Groisman and R.W. Knight, "Prolonged Dry Episodes over North America: New Tendencies Emerged During the Last 40 Years," *Advances in Earth Science* 22, no. 11 (2007): 1191–1207.

[76]D. Medvigy and C. Beaulieu, "Trends in Daily Solar Radiation and Precipitation Coefficients of Variation since 1984," *Journal of Climate* 25 (2011): 1330–1339, doi: 10.1175/2011JCLI4115.1.

[77]Intergovernmental Panel on Climate Change (IPCC), "Summary for Policymakers." In C. B. Field, V. Barros, T.F, Stocker, et al. (eds.), *Intergovernmental Panel on Climate Change Special Report on Managing the Risks of Extreme Events and Disasters to Advance Climate Change Adaptation* (Cambridge, U.K., Cambridge University Press, 2011).

it was 40 years ago. Warmer air increases the odds of extreme precipitation[78] because the air holds more moisture and can release more of it during rainstorms and snow-storms. Heavy precipitation, both rain and snow, is happening more often[79] than it used to. Heat-related extreme events are also on the rise around the globe, and global warming has significantly increased the odds of some specific events, including the killer European heat wave of 2003,[80] the Russian heat wave of 2010, and the intense U.S. drought of 2012.[81] Even small increases in average temperatures raise the risk of heat waves, droughts, and wildfires. Twice as many record highs have been set in the past decade as record lows in the United States.[82] By 2050, record highs could outpace record lows by 20 to one in the United States. By the end of the century, the ratio could jump to 100 to one if greenhouse-gas emissions continue unabated.

In the United States, setting climate records is becoming commonplace. One climate science website[83] calculated that 20 major U.S. cities had their wettest year on record during 2011, smashing the previous record from 1996 of 10 cities with a wettest year. Despite this fact, precipitation across the United States was near-average during 2011 (the 45th driest year in the 117-year record) because heavy rains in some places were balanced out by dry conditions across much of the southern United States (Texas had its driest year on record). The year 2011 ranked as the 23rd warmest in U.S. history, but the summer ranked as the hottest in 75 years, exceeded only by the Dust Bowl summer of 1936.

DROUGHT

As we learned in Chapter 1, the region around 30° latitude is characterized by dry, sinking air associated with the Hadley Cell (part of global atmospheric circulation). Known as the subtropics, this great belt around the globe is characterized by deserts, few clouds, and little precipitation. Climate studies[84] indicate that as global warming continues to force the tropics around the equator to expand,[85] the subtropics will expand as well, and precipitation patterns will change.

In Assessment Report 4, the IPCC projected that expanding drought will be associated with expansion of the subtropical belt, with high-latitude areas getting more precipitation. In their more-recent Special Report on Managing the Risks of Extreme Events and Disasters to Advance Climate Change Adaptation, the IPCC calculated a large drying trend over many Northern Hemisphere land areas since mid-1950 and an opposite trend in eastern North and South America.[86] They report that one study found that very dry land areas across the globe have more than doubled in extent since the 1970s, initially as a result of a short-term El Niño event and subsequently due to surface heating (air warming).[87]

[78]S.-K. Min, X. Zhang, F. W. Zwiers, and G. C. Hegerl, "Human Contribution to More Intense Precipitation Extremes."

[79]L.V. Alexander, X. Zhang, and T.C. Peterson, "Global Observed Changes in Daily Climate Extremes of Temperature and Precipitation," *Journal of Geophysical Research – Atmospheres*, 111.D5 (2006): D05109, doi: 10.1029/2005JD006290.

[80]N. Christidis, P.A. Stott, and S. Brown, "The Role of Human Activity In the Recent Warming of Extremely Warm Daytime Temperatures," *Journal of Climate*, (2011): doi: 10.1175/2011JCLI4150.1.

[81]S. Rahmstorf and D. Coumou, "Increase in Extreme Events in a Warming World," *Proceedings of the National Academy of Sciences* 108, no. 44 (2011): 17905–17909, doi 10.1073/pnas.1101766108.

[82]G. A. Meehl, C. Tebaldi, and G. Walton, "Relative Increase of Record High Maximum Temperatures Compared to Record Low Minimum Temperatures in the U.S.," *Geophysical Research Letters* 36 (2009): L23701, doi: 10.1029/2009GL040736.

[83]http://www.wunderground.com/blog/JeffMasters/comment.html?entrynum=2012 (accessed July 12, 2012).

[84]A. Dai, "Drought under Global Warming: A Review," *Climate Change* 2 (2011): 45–65. doi: 10.1002/wcc.81.

[85]J. Lu, C. Deser, and T. Reichler, "Cause of the Widening of the Tropical Belt since 1958," *Geophysical Research Letters* 36 (2009): L03803, doi: 10.1029/2008GL036076.

[86]Intergovernmental Panel on Climate Change (IPCC), "Summary for Policymakers."

[87]See CO2NOW.org: http://co2now.org/Know-the-Changing-Climate/Climate-Changes/ipcc-faq-changes-in-extreme-events.html (accessed July 12, 2012).

Drought can also take on seasonal patterns. For instance, as climate warms more precipitation will take the form of rain and less as snow, and snow that does accumulate during winter will melt faster and earlier in the spring. Planners and managers have long worried that the consequence of this pattern will be growing flood risk early in the year and decreasing discharge in the mid- and late summer, producing drought in the summer and fall growing season. Researchers[88] have now found this pattern verified in model studies that show faster and earlier snowpack melting due to rising air temperature, which in turn produces an increase in catastrophic events such as flooding and summer droughts. That this pattern is present today and is historically unusual has been confirmed by a reconstruction[89] of 800 years of snowpack size for the watersheds feeding the Colorado, Columbia, and Missouri rivers. Results show that snowpack in the northern Rocky Mountains has shrunk at an unusually rapid pace during the past 30 years. The research documents that recent declines are nearly unprecedented, owing to a combination of natural variability and human-induced atmospheric warming.

Drought has many impacts. For instance, in the U.S. Rocky Mountain region, a steady decline in the winter snowfall[90] over the past few decades has produced some important effects. As the snowpack at high elevations decreases, elk browse on plants that were previously inaccessible during the snow season. As a result, deciduous trees and associated songbirds in mountainous Arizona have decreased over the past two decades.[91] The increased browsing results in a trickle-down effect, such as lowering the quality of habitat for mountain songbirds.

Dry periods are not unusual in history, and they have occurred many times over the past thousand years. North America, West Africa, and East Asia have experienced megadroughts triggered by irregular tropical sea-surface temperatures. La-Niña-like sea-surface temperature conditions lead to drought in North America, and El-Niño-like sea-surface temperatures affect the famous wet season known as the monsoon, causing drought in East Asia and westward.

Models project increased aridity in the 21st century over most of Africa, southern Europe, the Middle East, most of the Americas, Australia, and Southeast Asia. Research indicates this process is already under way and that drought has expanded and deepened over the 20th century.[92] Studies found that the percentage of Earth's land area afflicted by serious drought more than doubled from the 1970s to the early 2000s, and as a result, some of the world's major rivers are losing water,[93] threatening drinking water and crop irrigation in previously stable areas.

Droughts are events associated with reduced precipitation, dry soils leading to crop failure, and imperiled drinking-water supplies. Drought is measured by the Palmer Drought Severity Index (PDSI), which tracks precipitation and evaporation and compares them to historical patterns. Model studies[94] indicate that by the 2030s some regions could experience particularly severe drought, including much of the central and western United States; lands bordering the Mediterranean, Central America, and the Caribbean region; and portions of Europe and Asia. By the end of the century, drought could intensify and spread with continued warming. Many populated areas, including the United States, could reach unprecedented levels of drought severity (Figure 7.11).

[88]A., Molini, G. Katul, and A. Porporato, "Maximum Discharge from Snowmelt in a Changing Climate," *Geophysical Research Letters* 38 (2011): L05402, doi: 10.1029/2010GL046477.

[89]G. Pederson, S. Gray, C. Woodhouse, et al., "The Unusual Nature of Recent Snowpack Declines in the North American Cordillera," *Science* 333 (2011): 332–335.

[90]D. W. Pierce, T. P. Barnett, H. G. Hidalgo, et al., "Attribution of Declining Western US Snowpack to Human Effects," *Journal of Climatology* 21 (2008): 6425–6444.

[91]T. E. Martin and J. L. Maron, "Climate Impacts on Bird and Plant Communities from Altered Animal-Plant Interactions," *Nature Climate Change* 2 (2012): 195–200, doi: 10.1038/nclimate1348.

[92]A. Dai, K.E. Trenberth, and T. Qian, "A Global Dataset of Palmer Drought Severity Index for 1870–2002: Relationship with Soil Moisture and Effects of Surface Warming," *Journal of Hydrometeorol* 5 (2004): 1117–1130.

[93]See "Climate Change: Drought May Threaten Much of Globe Within Decades," http://www2.ucar.edu/news/2904/climate-change-drought-may-threaten-much-globe-within-decades (accessed July 12, 2012).

[94]A. Dai, "Drought under Global Warming: A Review," *Climate Change* 2 (2011): 45–65. doi: 10.1002/wcc.81.

Figure 7.11. These four maps illustrate the potential for future drought worldwide, based on current projections of future greenhouse-gas emissions. The maps use the Palmer Drought Severity Index, which assigns positive numbers when conditions are unusually wet for a particular region and negative numbers when conditions are unusually dry. A reading of −4 or below is considered extreme drought. Regions that are blue or green will likely be at lower risk of drought, and those in red and purple could face more unusually extreme drought conditions.

IMAGE CREDIT: "Climate Change: Drought may Threaten Much of Globe within Decades," http://www2.ucar.edu/news/2904/climate-change-drought-may-threaten-much-globe-within-decades (accessed July 12, 2012).

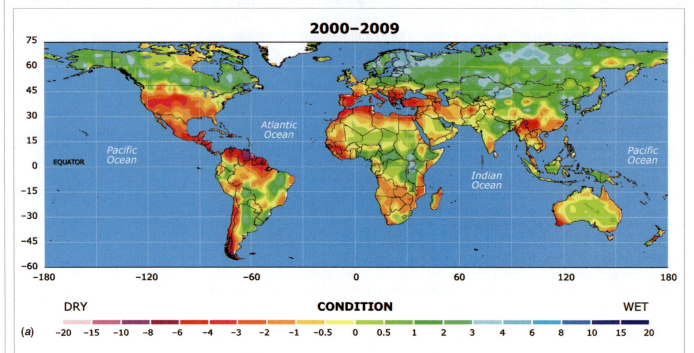

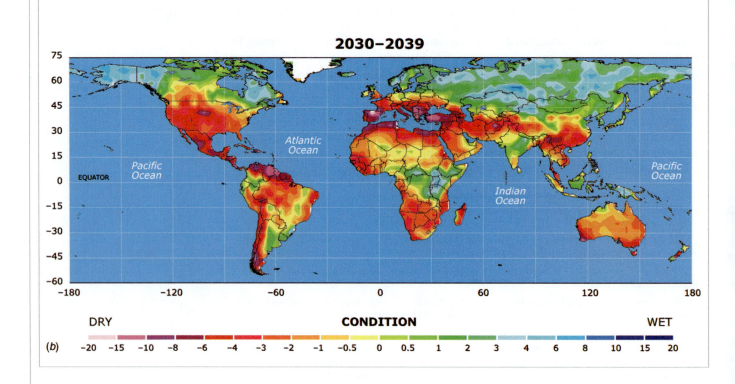

2060–2069

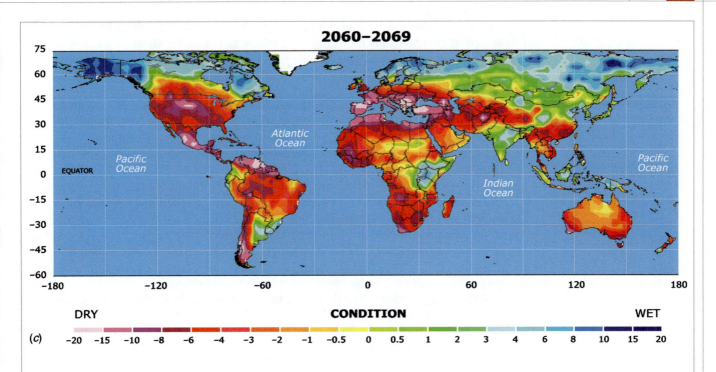

DRY   **CONDITION**   WET

(c) -20 -15 -10 -8 -6 -4 -3 -2 -1 -0.5 0 0.5 1 2 3 4 6 8 10 15 20

2090–2099

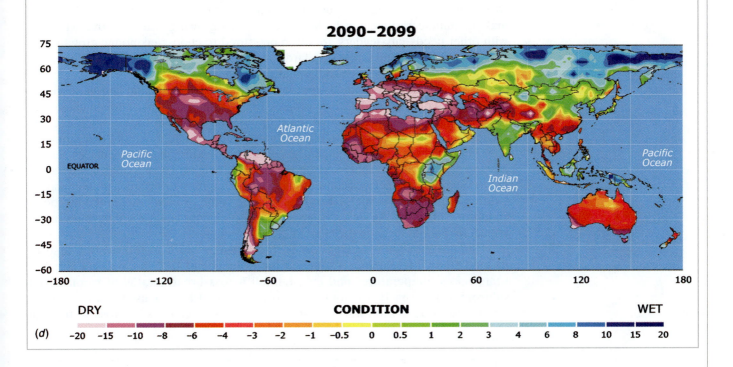

DRY   **CONDITION**   WET

(d) -20 -15 -10 -8 -6 -4 -3 -2 -1 -0.5 0 0.5 1 2 3 4 6 8 10 15 20

DANGEROUS CLIMATE

Drought, extreme weather, heat waves, and other aspects of global warming are considered dangerous by many planners and scientists because they threaten human safety. How likely we are to avoid the most dangerous aspects of warming depends on how fast and how far humans can cut greenhouse gas emissions. One study[95] characterized current goals for cutting emissions as falling far short of what is needed. By merging estimates of greenhouse-gas emissions in this century with the potential climate response, researchers calculated that only three out of 193 model simulations in the peer-reviewed literature are both plausible and likely to succeed. All three plausible scenarios require that world emissions peak this decade, start dropping immediately, and be far less than half of current levels by 2050. These scenarios would also require intense efforts to remove greenhouse gases from the atmosphere (such as carbon sequestration, the process of capturing and removing carbon dioxide from the atmosphere for long-term storage).

An examination[96] of urban policies at the National Center for Atmospheric Research (NCAR) found that even though billions of urban dwellers are vulnerable to heat waves, sea-level rise, and other changes associated with warming temperatures, cities worldwide are failing to take the necessary steps to protect residents from the likely impacts of climate change. Not only are most cities failing to reduce emissions of carbon dioxide and other greenhouse gases, they are falling short in preparing residents for the likely impacts of climate change.

The study further noted that more than half the world's population lives in cities, where construction patterns are often dense, housing substandard, and access to reliable drinking water, roads, and basic services poor—all conditions that magnify the potential for humanitarian disaster. Potential threats associated with climate include storm surges, which can inundate coastal areas; development of steep hillsides and floodplains; and prolonged hot weather, which can heat heavily paved cities more than surrounding areas, exacerbate existing levels of air pollution, and cause widespread health problems. The study also identified factors that keep city leaders from making climate resilience a higher priority: Fast-growing cities are overwhelmed with other needs, city leaders are often pressured to choose economic growth over the need for health and safety standards, and climate projections are rarely fine-scale enough to predict impacts on individual cities.

Another study[97] combined climate change data with a global census of nearly 97% of the world's population to project human vulnerability to climate change by mid-century. The study concluded that populations in low-latitude tropical regions, such as central South America, the Arabian Peninsula, and much of Africa, may be most vulnerable to climate change. Those communities already experience extremely hot and arid conditions that make agriculture challenging. Even a small temperature increase would have serious consequences on their ability to sustain a growing population. Communities in high-latitude temperate zones are already limited by cooler conditions, however. As such, researchers expect climate change will have less of an impact on people living in these areas.

Climate change in stressed low-latitude nations can lead to war. In a first-of-its-kind study,[98] researchers examined the influence of El Niño, which every few years raises temperatures and cuts rainfall across broad swaths of tropical and subtropical regions. It was found that the onset of El Niño, used in the study as a proxy for longer-scale warming, doubles the risk of civil wars across 90 tropical

[95]J. Rogeli, W. Hare, J. Lowe, et al., "Emission Pathways Consistent with a 2°C Global Temperature Limit," *Nature Climate Science* (2011): 413–418; doi: 10.1038/nclimate1258.

[96]J. Hardoy and P.R. Lankao, "Latin American Cities and Climate Change: Challenges and Options to Mitigation and Adaptation Responses," *Current Opinion in Environmental Sustainability* (2011), doi: 10.1016/j.cosust.2011.01.004.

[97]J. Samson, D. Berteaux, B.J. McGill, and M.M. Humphries, "Geographic Disparities and Moral Hazards in the Predicted Impacts of Climate Change on Human Populations," *Global Ecology and Biogeography* (2011): doi: 10.1111/j.1466-8238.2010.00632.x.

[98]S. M. Hsiang, K. C. Meng, and M. A. Cane, "Civil Conflicts Are Associated with the Global Climate," *Nature*; 476, no. 7361 (2011): 438, doi: 10.1038/nature10311.

countries and might account for a fifth of worldwide conflicts during the past half century. Study authors did not investigate why climate feeds conflict; however, they point out that a community characterized by poverty has underlying tensions, and it may be that warming delivers the final blow to peaceful solutions to persistent problems related to basic survival. For instance, when crops fail, or water dries up, or other fundamental resources grow scarce, people may take up a gun simply to make a living. In fact, social scientists have shown in the past that individuals can become more aggressive when temperatures rise, but whether this behavior applies to whole societies is still speculative.

Climate change might already be affecting some fundamental resources, such as crops, on a global scale. A study[99] found that global wheat production since 1980 was 5.5% lower than it would have been had climate remained stable and that global corn production was lower by almost 4%. In the United States, Canada, and Northern Mexico, a very slight cooling trend over the study period resulted in no significant production impacts. Outside of North America, most major agricultural countries experienced some decline in wheat and corn yields related to the rise in global temperature. Although crop yields in most countries are still going up because of improvements in technology, fertilization, and other factors, they are not rising as fast as they would be without warming. Russia, India, and France experienced the greatest drop in wheat production, and China and Brazil experienced the largest losses in corn production.

Most evaporation and precipitation takes place over the oceans, and as the atmosphere warms the rate of these processes accelerates. A study in the spring of 2012 revealed just how much the water cycle has sped up as a result of global warming.[100] Using 50 years of ocean surface salinity data (1.7 million measurements), scientists documented how the salinity of the ocean surface has changed as a result of changes in evaporation and precipitation. A map of their results (Figure 7.12) reveals that, as expected, wet areas are getting wetter and dry areas are getting drier; high-latitude and equatorial parts of the oceans, where there is greater precipitation than average, became less salty; and mid-latitude areas (the central regions of ocean basins), where evaporation dominates, became saltier. The results indicate that the water cycle had sped up roughly 4% while the surface warmed 0.5°C, roughly twice as fast as predicted by most climate models. The study authors conclude that if the world warms 2°C to 3°C by the end of the century, the water cycle will accelerate 16% to 24%. An amplified water cycle such as this would fuel violent storms in wet areas from tornadoes to tropical cyclones and produce severe and frequent flooding, and in dry areas it could mean long and intense droughts.

There are other considerations when it comes to dangerous climate change. Global drought[101] is pushing governments in semi-arid regions to consider the possibility that national water supplies may be insufficient to meet demand. Flooding by the sea and during extreme rainfall events[102] can lead to stagnant water that breeds pathogen-carrying insects, cholera bacteria, and other causes of disease. Sea-level rise, decreases in freshwater, increases in climate hazards, and the spread of disease—all of these trends could result in a growth of the number of environmental

[99]D.B. Lobell, W. Schlenker, and J. Costa-Roberts, "Climate Trends and Global Crop Production since 1980," *Science* (2011), doi: 10.1126/science.1204531.

[100]P. J. Durack, S. E. Wijffels, and R. J. Matear, "Ocean Salinities Reveal Strong Global Water Cycle Intensification during 1950 to 2000," *Science* 336, no. 6080 (2012): 455, doi: 10.1126/science.1212222.

[101]J. Samson, D. Berteaux, B.J. McGill, and M.M. Humphries, "Geographic Disparities and Moral Hazards in the Predicted Impacts of Climate Change on Human Populations."

[102]See "The Impact of Climate Change on Water, Sanitation, and Diarrheal Diseases in Latin America and the Caribbean," http://www.prb.org/Articles/2007/ClimateChangeinLatinAmerica.aspx (accessed July 12, 2012).

Figure 7.12. Absolute surface salinity change over the period 1950–2000. Rainfall and evaporation changes are making the oceans less salty in vast regions (blue) and more salty elsewhere (red). Research shows that while the surface warmed 0.5°C, the water cycle has sped up roughly 4%, twice as fast as predicted by most climate models. These results also indicate that in general, wet areas got wetter and dry areas got drier.[103]

Source: Ocean Change: Salinity. http://www.cmar.csiro.au/oceanchange/salinity.php (accessed July 12, 2012). Durack, P and Wijffels, SE, Fifty-Year Trends in Global Ocean Salinities and Their Relationship to Broad-Scale Warming, Journal of Climate, 23, (16) pp. 4342–4362. ISSN 0894-8755 (2010).

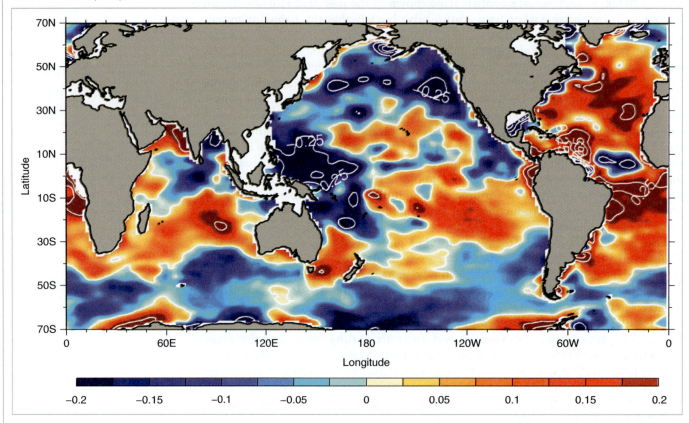

refugees[104] displaced from traditional homelands. Behind these trends of human vulnerability are two unmitigated factors: global warming and human population growth.[105] Until both these global issues are effectively managed, the trend of climate change leading to dangerous impacts on human communities is likely to continue.

ECOSYSTEM IMPACTS

In a study[106] based on decades of observations, researchers have documented how a broad group of plant species living in open conditions (rather than a controlled laboratory) have responded to rising temperatures. Data from historical records of 1,558 species of wild plants on four continents show that leafing and flowering advances, on average, five to six days per degree Celsius of warming. The power of this finding is the global distribution of the database and the fact that it records plant behavior under real-life conditions of seasons, weather, predator–prey

[103]P. J. Durack, S. E. Wijffels, and R. J. Matear, "Ocean Salinities Reveal Strong Global Water Cycle Intensification during 1950 to 2000."

[104]A. de Sherbinin, M. Castro, F. Gemenne, et al., "Preparing for Resettlement Associated with Climate Change," *Science* 334, no. 6055 (2011): 456–457, doi: 10.1126/science.1208821.

[105]Population, Special Section, *Science* 333 (2011): 540–546.

[106]E. M. Wolkovich, B. I. Cook, J. M. Allen, et al., "Warming Experiments Underpredict Plant Phenological Responses to Climate Change," *Nature* (2012), doi: 10.1038/nature11014.

relationships, and other natural wild conditions. The results are statistically consistent across species and geographic datasets. When compared to the usual method of understanding how plants react to warming (small-scale experiments of a few plants under laboratory conditions), these data show that previous estimates of plant response to global warming grossly underpredict advances in flowering by eight and a half times and advances in leafing by four times. These results suggest that the way global warming experiments on plant health are currently conducted needs to be re-evaluated, especially because data of this type are used to parameterize global climate models when predicting responses to global warming and changes in the carbon cycle.

Ecosystems are sensitive to the balance of multiple stressors, both natural and human-related. Studies[107] show that natural decreases in biodiversity are as potentially damaging as the negative impacts resulting from climate change, pollution, and other major forms of environmental stress. Because natural stressors are ever present, the growth of negative impacts related to climate change and human population growth could cause increasing damage to ecosystems that are already stressed as a natural condition. Researchers combined data from published accounts of how environmental factors affect two important ecosystem processes: plant growth and decomposition of dead plants by bacteria and fungi. They found that species losses of 1% to 20% have negligible effects on ecosystem plant growth; losses of 21% to 40% reduce plant growth by 5% to 10%, which is comparable to the impact of global warming and increased ultraviolet radiation due to stratospheric ozone loss; and losses of 41% to 60% equate with the effects of major damage such as ozone pollution, acid deposition on forests, and nutrient pollution. This research suggests that natural stressors to global biodiversity will be amplified by the growth of climate change.

The ways humans use land, the ocean, and other natural resources affect the distribution and quality of plant and animal habitats. The area of undeveloped space for wildlife is continually declining under the pressure of a growing human population. Essential freshwater systems are affected by pollution, damming, and diversion of water for human use. No area of the ocean is untouched by human pollution in some form.[108] Climate change is driving aquatic and forest ecosystems toward the heads of their watersheds at the highest elevations, with little recourse thereafter as warming continues. One group of researchers[109] have concluded that if global warming persists as expected, almost a third of all flora and fauna species worldwide could become extinct and that by 2080 more than 80% of genetic diversity within species could disappear in certain groups of organisms.

Climate change causes many terrestrial species to shift to higher elevations and higher latitudes in order to maintain the same climate conditions that are optimal to their survival. But researchers[110] have found that species are responding to climate change up to three times faster than previously appreciated. Species have moved toward the poles at three times the rate previously accepted in the scientific literature, and they have moved to cooler, higher altitudes at twice the rate previously realized. On average, species have moved to higher elevations at 12.2 m (40 ft) per decade and, more dramatically, to higher latitudes at 17.6 km (11 mi) per decade. Scientists estimate that these changes are equivalent to animals and plants shifting away from the equator at around 20 cm (8 in) per hour, every hour of the day, and

[107]D. U. Hooper, E. C. Adair, B. J. Cardinale, et al., "A Global Synthesis Reveals Biodiversity Loss as a Major Driver of Ecosystem Change," *Nature* (2012), doi: 10.1038/nature11118.

[108]See "A Global Map of Human Impacts to Marine Ecosystems," http://www.nceas.ucsb.edu/globalmarine (accessed July 12, 2012).

[109]M. Bálint, S. Domisch, C. H. M. Engelhardt, et al., "Cryptic Biodiversity Loss Linked to Global Climate Change," *Nature Climate Change* (2011), doi: 10.1038/NCLIMATE1191.

[110]I.-C. Chen, J. K. Hill, R. Ohlemuller, D. B. Roy, and C. D. Thomas, "Rapid Range Shifts of Species Associated with High Levels of Climate Warming," *Science* 333, no. 6045 (2011): 1024, doi: 10.1126/science.1206432.

every day of the year. They estimate that this trend has been going on for the last 40 years and that it will continue for at least the rest of this century.

Global warming is even changing the routine of America's home gardeners. On the back of seed packets bought by 80 million U.S. gardeners each year is a color-coded map[111] of plant hardiness zones. The map provides guidance on where various species of flowers, vegetables, and ornamental plants will have an optimal growing climate. For the first time since 1990, the U.S. Department of Agriculture revised this official guide and shifted about half the continental United States approximately a half zone to the north. The new map reflects the fact that climate (and growing) zones have shifted strongly to the north as a consequence of changing climate. Nearly entire states, including Ohio, Nebraska, and Texas, have been updated to warmer zones.

Changes in habitat quality cause changes in the distribution of food sources and place wildlife populations under stress. Species extinction and the degradation of ecosystems are proceeding rapidly, and the pace is accelerating. Biodiversity is declining throughout the world, and the challenges of conserving the world's species are made even larger in light of the negative effects of global climate change. The world is losing species at a rate that is 100 to 1000 times faster than the natural extinction rate.[112]

Climate change can affect species in relation to their role in an ecosystem. Scientists hypothesize that species in rich, biodiverse ecosystems are exposed to heightened threats by the consequences of global warming, specifically extreme weather events. High winds, torrential downpours, and droughts have become more frequent; this increases the risk for species extinction in diverse ecosystems such as coral reefs and tropical rainforests. In a rainforest or on a coral reef there are a wide variety of species of primary producers. Primary producers are organisms (such as green plants and algae) that produce biomass from inorganic compounds and thus provide a foundation to the food web. Because they are competitors, relatively few individuals of the same species exist, exposing them to a greater risk of extinction should environmental conditions change, such as during and after an extreme weather event. This could result in a depletion of food sources for species (such as herbivores) that rely on primary producers. This extinction, in turn, affects a predator at the top of the food web. Biologists call this transformation a cascading extinction. Using models of this process, researchers[113] found that flora and fauna in these conditions are 100 to 1,000 times more likely to become extinct than normal.

Researchers have found that water temperatures in many streams and rivers throughout the United States are increasing. Analysis[114] of historical records from 20 major U.S. streams and rivers reveals that annual mean water temperatures increased by 0.009°C to 0.077°C per year (0.02°F to 0.14°F per year). Long-term increases in stream water temperatures were correlated with increases in air temperatures, and rates of warming were most rapid in urbanized areas. Warming water can affect basic ecological processes, aquatic biodiversity, biological productivity, and the cycling of contaminants through the ecosystem.

As global warming continues, many plant and animal species face increasing competition for survival as well as significant species turnover as some species invade

---

[111]See the map here: http://planthardiness.ars.usda.gov/PHZMWeb/; and find a media article about the change here: http://wwwp.dailyclimate.org/tdc-newsroom/usda/climate-change-comes-to-your-backyard (accessed July 12, 2012).

[112]See the website, Biodiversity Crisis Is Worse the Climate Change Experts Say, http://www.sciencedaily.com/releases/2012/01/120120010357.htm (accessed July 12, 2012).

[113]L. Kaneryd, C. Borrvall, S. Berg, et al., "Species-Rich Ecosystems Are Vulnerable to Cascading Extinctions in an Increasingly Variable World," *Ecology and Evolution* (2012) 29 March, doi: 10.1002/ece3.218.

[114]S. S. Kaushal, G. E. Likens, N. A. Jaworski, et al., "Rising Stream and River Temperatures in the United States," *Frontiers in Ecology and the Environment* 8 (2010): 461–466, http://dx.doi.org/10.1890/090037 (accessed July 12, 2012).

areas occupied by other species. NASA scientists have investigated[115] the influence of doubled CO_2 on ecological sensitivity, and their results show that changes accompanying higher carbon dioxide levels lead to increasing ecological change and stress in Earth's biosphere. Most of Earth's land that is not covered by ice or desert is projected to undergo at least a 30% change in plant cover—a change that will require humans and animals to adapt and often relocate. Other studies[116] have confirmed these results, finding that as species migrate at different rates to new ecosystems, conflict grows owing to competition for space and resources. The collision course that results leads to new ecosystems for which there is no historical analogue, and it increases stress and extinctions among the affected plant and animal community.

Marine Ecosystems

The impact of shifting climate on marine ecosystems has also been measured.[117] When temperatures rise, plants and animals that need a cooler environment move to new regions. Land warms about three times faster than the ocean, but species do not necessarily move three times faster on land. If the land temperature becomes too hot, some species can move to higher elevations, where temperatures are cooler. That's not an easy option, however, for marine species that live at the surface of the ocean. When the temperature of seawater rises, species such as fish will be able to move into deeper water to find the cooler environments they prefer. However, deeper water has reduced light levels, potentially changing aspects of metabolism and predator–prey relationships. Other species, such as marine plants or corals, are tied to specific characteristics of shallow water including light levels, water circulation, and oxygen content. These species have to move horizontally to find suitable habitats, and they could become trapped if there are no cooler places for them to go. Rising temperatures could leave some marine species with nowhere to go.

Sea surface temperature has increased by an average of 0.6°C in the past 100 years,[118] and the acidity of the ocean surface has increased 10-fold. Corals cannot tolerate severely warming waters, however, and temperature stress causes a phenomenon known as bleaching, whereby corals expel the symbiotic algae that live in their tissues. In 1997 and 1998 an unusually strong El Niño event caused high sea-surface temperatures, which led to coral bleaching that was observed in almost all of the world's reefs during that record-setting year. An estimated 16% of the world's corals died in that strong bleaching event, an unprecedented occurrence. Bleaching occurred again in 2005, 2009, and 2010. The frequency and intensity of bleaching may be growing, and scientists wonder how much more coral populations can withstand.[119]

In response to warming temperatures, apparently, reef-forming coral species along the coast of Japan have been shifting their range into cooler waters to the north since the 1930s at rates as high as 14 km (8.7 mi) per year.[120] Many coral reefs also have the unfortunate circumstance of being located immediately adjacent to

[115]J. Bergengren, D. Waliser, and Y. Yung, "Ecological Sensitivity: A Biospheric View of Climate Change," *Climatic Change* 107, nos 3–4, (2011), doi: 10.1007/s10584-011-0065-1.

[116]M. C. Urban, J. J. Tewksbury, and K. S. Sheldon, "On a Collision Course: Competition and Dispersal Differences Create No-Analogue Communities and Cause Extinctions during Climate Change," *Proceedings of the Royal Society B: Biological Sciences* (2012), doi: 10.1098/rspb.2011.2367.

[117]M. T. Burrows, D. S. Schoeman, L. B. Buckley, et al., "The Pace of Shifting Climate in Marine and Terrestrial Ecosystems," *Science* 334, no. 6056 (2011): 652, doi: 10.1126/science.1210288.

[118]IPCC, Working Group I: The Scientific Basis: http://www.ipcc.ch/ipccreports/tar/wg1/005.htm (accessed July 12, 2012).

[119]C. M. Eakin, J. A. Morgan, S. F. Heron, et al., "Caribbean Corals in Crisis: Record Thermal Stress, Bleaching, and Mortality in 2005," *PLoS ONE* 5, no, 11 (2010): e13969, doi:10.1371/journal.pone.0013969.

[120]H. Yamano, K. Sugihara, and K. Nomura, "Rapid Poleward Range Expansion of Tropical Reef Corals in Response to Rising Sea Surface Temperatures," *Geophysical Research Letters* 38 (2011): L04601, doi: 10.1029/2010GL046474.

urbanized watersheds. Fifty-eight percent of the world's coral reefs are potentially threatened by human activity,[121] ranging from coastal development and destructive fishing practices to overexploitation of resources, marine pollution, and polluted runoff from inland deforestation and farming. The list of reef stressors is long: eroded silt in muddy runoff, pollutants of various types that cause coral disease, overfishing of species that are important in cropping back invasive algae that compete with corals for seafloor space, excessive levels of nitrogen, phosphorus and other nutrients, direct human impact with anchors, explosive fishing methods, and other human impacts all add to the stress that threatens coral reefs around the world.

Coral reefs are also experiencing the effects of sea-level rise. In locations where turbid runoff from exposed watersheds has delivered mud to the coastal zone and muddy shorelines line the landward edges of coral reefs, scientists[122] fear that a rising sea will permit additional wave action across the reef flat to erode the muddy coast. Muddy coastal waters, formed by higher wave energy, could stab reef ecosystems in the back, killing coral communities on the seaward reef edges that currently enjoy open ocean conditions of clean and appropriately cool water. But not all reefs are experiencing these threats. Researchers[123] have also documented increases in coral reef growth under rising seas in more than one locality.[124] The reason is simple: Over the past millennium or so, many reefs have grown upward to the limit of the water column. Wave energy, hot summer temperatures, and shallow water do not allow any further upward growth in many reef flats of the world. By raising sea level, global warming offers the possibility of additional upward growth in some locations, thus stimulating new coral growth in waters that were previously too shallow.

The oceans have absorbed about one third[125] of the carbon dioxide emitted by humans over the past two centuries. Increasing ocean acidification, brought on by dissolved carbon dioxide that mixes with seawater to form carbonic acid, makes it difficult for calcifying organisms (corals, mollusks, and many types of plankton[126]) to secrete the calcium carbonate they need for their skeletal components (a process called calcification). Scientists have found[127] that carbon dioxide emissions in the last 100 to 200 years have already raised ocean acidity far beyond the range of natural variations. In some regions, the rate of change in ocean acidity since the Industrial Revolution is 100 times greater than the natural rate of change between the Last Glacial Maximum and preindustrial times.

When Earth started to warm 17,000 years ago, terminating the last glacial period, atmospheric CO_2 levels rose from 190 ppm to 280 ppm over 6,000 years, giving marine ecosystems ample time to adjust. Now, for a similar rise in CO_2 concentration to the present level near 400 ppm, the adjustment time is reduced to only 100 to 200 years and might have decreased the overall calcification rates by 15%. On a global scale, pH conditions that support coral reefs are currently found in about 50% of the

[121]D. Bryant, L. Burke, J. McManus, and M. Spalding, *Reefs at Risk: A Map-Based Indicator of Threats to the World's Coral Reefs* (Washington, D.C., World Resources Institute, 1998).

[122]M. E. Field, A. S. Ogston, and C. D. Storlazzi, "Rising Sea Level May Cause Decline of Fringing Coral Reefs," *Eos* 92 (2011): 273–280.

[123]B. Brown, R. Dunne, N. Phongsuwan, and P. Somerfield, "Increased Sea Level Promotes Coral Cover on Shallow Reef Flats in the Andaman Sea, Eastern Indian Ocean," *Coral Reefs* 30 (2011): 867–878.

[124]J. Scopelitis, A. Andrefouet, S. Phinn, T. Done, and P. Chabanet, "Coral Colonization of a Shallow Reef Flat in Response to Rising Sea Level: Quantification from 35 Years of Remote Sensing Data at Heron Island, Australia," *Coral Reefs* 30 (2011): 951–965.

[125]See http://www.sciencedaily.com/releases/2011/08/110803133517.htm (accessed July 12, 2012).

[126]L. Beaufort, I. Probert, T. de Garidel-Thoron, et al., "Sensitivity of Coccolithophores to Carbonate Chemistry and Ocean Acidification," *Nature* 476, no. 7358 (2011): 80, doi: 10.1038/nature10295.

[127]T. Friedrich, A. Timmermann, and A. Abe-Ouchi, et al., "Detecting Regional Anthropogenic Trends in Ocean Acidification against Natural Variability," *Nature Climate Change* (2012), doi: 10.1038/NCLIMATE1372.

Figure 7.13. The upper panels show modeled surface seawater aragonite ($CaCO_3$) saturation for the years 1800, 2012, and 2100, respectively. Aragonite is a form of calcium carbonate that corals and other organisms use to build skeletons. As seawater becomes less saturated with aragonite (reddish colors) it becomes more difficult for corals and other organisms to secrete their skeletal components; below zero, aragonite dissolves. White dots indicate present-day main coral reef locations. The lower panel shows atmospheric CO_2 concentration in parts per million, simulated for the years 1750 to 2100.

Source: University of Hawaii, International Pacific Research Center.

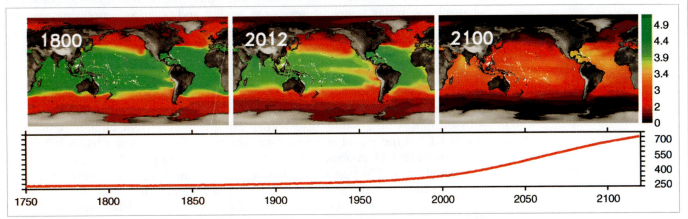

ocean, mostly in the tropics. By the end of the 21st century (Figure 7.13), this fraction is projected to be less than 5%. The Hawaiian Islands, which sit just on the northern edge of the tropics, will be one of the first to feel the impact.

Ocean acidification has other impacts. Acidification of seawater decreases the absorption of sound by up to 50% in the frequency range that is important to whales and other acoustic organisms.[128] Ship traffic, seismic testing, and industrial activities that were previously muted in the world's oceans will become more acute and potentially affect marine species. High levels of low-frequency sound have a number of behavioral and biological effects on marine life, including tissue damage, mass stranding of cetaceans, and temporary loss of hearing in dolphins.

Ocean acidification has also damaged a $273 million per year oyster farming industry in the Pacific Northwest. A study[129] found that increased dissolved carbon dioxide levels in seawater resulted in more-corrosive ocean water and inhibited larval oysters from developing their shells. Because of this, larvae grew at a pace that prohibited cost-effective commercial production and contributed to a collapse of the oyster farming industry.

Ocean acidification is one of the consequences of CO_2 buildup that could have a great impact on the world's ocean ecology, which depends on the secretion of calcium carbonate by thousands of different species. As carbon dioxide emissions increase, it is anticipated that 450 ppm CO_2 will be reached before 2050. At that point, corals may be on a path to extinction within a matter of decades.[130] By 2050, the remaining coral reefs could fall victim to ocean acidification. Such a catastrophe would not be confined to reefs but could be the start of a domino-like sequence of the fall of other marine ecosystems.

The loss of healthy coral reefs affects all the species that dwell there (such as turtles, mollusks, crabs, and fish) as well as the animals that depend on reef habitats

[128]T. Ilyina, R.E. Zeebe, and P.G. Brewer, "Future Ocean Increasingly Transparent to Low-Frequency Sound Owing to Carbon Dioxide Emissions," *Nature Geoscience* 3 (2009): 18–22.

[129]A. Barton, B. Hales, G. Waldbusser, C. Langdon, and R. Feely, "The Pacific Oyster, *Crassostrea gigas*, Shows Negative Correlation to Naturally Elevated Carbon Dioxide Levels: Implications for Near-Term Ocean Acidification Effects," *Limnology and Oceanography* 57, no. 3 (2012): 698, doi: 10.4319/lo.2012.57.3.0698.

[130]Zoological Society of London "Coral Reefs Exposed to Imminent Destruction From Climate Change," 2009. See http://www.sciencedaily.com/releases/2009/07/090706141006.htm (accessed July 12, 2012).

as a food source (including sea birds, mammals, and humans). One quarter of all sea animals spend time in coral reef environments during their life cycle. There are economic impacts as well. Tourism and commercial fisheries generate billions of dollars in revenue annually. Biodiversity, food supplies, and economics could thus all be affected by global climate change. Reef loss is a complex issue, however. Reefs can suffer from coastal pollution, overfishing, and other types of human stresses. Exactly what roles warming temperatures, ocean acidity, and other anthropogenic impacts will play in global reef health has yet to be fully defined by researchers.

CLIMATE SENSITIVITY

How sensitive is climate to high levels of carbon dioxide? This issue is explored with estimates of a value researchers call the equilibrium climate sensitivity (ECS). ECS is the global mean near-surface temperature when it has equilibrated to atmospheric CO_2 concentrations that are double the preindustrial level of CO_2 (estimated to be 280 ppm). Another way of putting it is, "How warm will it be when the CO_2 concentration reaches 560 ppm?" Recall that today the CO_2 level is at or near 400 ppm and rising at about 2 ppm per year.[131] When will the level of CO_2 reach 560 ppm? This depends on the level of continued greenhouse-gas emissions, a major subject of IPCC research. Continued emissions at present rates, the "business as usual" scenario, would lead to doubled CO_2 levels toward the end of the century.

The IPCC[132] Fourth Assessment Report (AR4) concludes that ECS is "likely to be in the range of 2 to 4.5°C (3.6 to 8.1°F), with a best estimate of about 3°C (5.4°F), and is very unlikely to be less than 1.5°C (2.7°F). Values substantially higher than 4.5°C (8.1°F) cannot be excluded, but agreement of models with observations is not as good for those values." The problem with this wide range of estimates is that identifying the true effect of limiting fossil-fuel burning becomes highly uncertain, and it is difficult to rule out large temperature increases as a result of greenhouse-gas emissions. Improved estimates of ECS would encourage governments to set emission targets with better-understood consequences.

Researchers use various methods to estimate ECS, and they all have advantages and disadvantages.[133] The typical methods include reconstructing past temperature changes[134] that accompanied shifts in CO_2 concentration, which are estimated from geologic information (climate proxies; see Chapter 3); using global climate model simulations to estimate[135] ECS; and calculating ECS from measurements of modern climate change.[136]

The problem with paleoclimate reconstructions is twofold: Today's rapid climate changes and complex feedbacks might represent completely unique conditions that will not be accurately represented by paleoclimate history. Another problem is the

[131]See NOAA, Earth System Research Laboratory, Global Monitoring Division, http://www.esrl.noaa.gov/gmd/ccgg/trends/global.html#global_data (accessed July 12, 2012).

[132]IPCC, "Climate Sensitivity and Feedbacks." In R. K. Pachauri and A. Reisinger, (eds.), *Climate Change 2007: Synthesis Report*. Contribution of Working Groups I, II and III to the Fourth Assessment Report of the Intergovernmental Panel on Climate Change (Geneva, Intergovernmental Panel on Climate Change, 2007). http://www.ipcc.ch/publications_and_data/ar4/syr/en/mains2-3.html. Retrieved 2010-07-03 (accessed July 12, 2012).

[133]For a detailed discussion of climate sensitivity, see: http://www.skepticalscience.com/detailed-look-at-climate-sensitivity.html (accessed July 12, 2012).

[134]R.E. Zeebe, J.C. Zachos, and G.R. Dickens, "Carbon Dioxide Forcing Alone Insufficient to Explain Palaeocene-Eocene Thermal Maximum Warming," *Nature Geoscience* 2 (2009): 576–580.

[135]M. R. Allen, D. J. Frame, C. Huntingford, et al., "Warming Caused by Cumulative Carbon Emissions Towards the Trillionth Tonne," *Nature* (2009), doi: 10.1038/nature08019.

[136]P.M. Forster and J.M. Gregory, "The Climate Sensitivity and Its Components Diagnosed from Earth Radiation Budget Data," *Journal of Climate* 19, no. 1 (2006): 39–52, doi: 10.1175/JCLI3611.1.

high degree of uncertainty that accompanies the use of climate proxies (dating inaccuracies, chemical changes in proxies when they are buried in the crust, and low precision in characterizing climate). Nevertheless, despite these criticisms, the period over which climate shifted from the last glacial maximum 23,000 to 19,000 years ago to warmer, preindustrial conditions (pre-19th century) has been used[137] to estimate ECS. This period is a potentially valuable climate episode for characterizing ECS for several reasons: Earth's climate changed relatively rapidly (in a geologic sense) from a glacial state to modern interglacial conditions, it was relatively recent in geologic history (better-resolved proxies), and the last glacial maximum can be robustly characterized by a number of independent, globally distributed climate proxies.

Investigators[138] used detailed paleoclimate proxy data to reconstruct the climate of the last glacial maximum and ran a series of global climate model simulations over the same period, with each simulation using a different ECS value. By comparing the modeling results to the paleoclimate reconstruction, it was possible to identify the ECS value that most closely predicted the true paleoclimate equilibrium climate sensitivity. The study found that an increase of 3.1°C (5.6°F) in global average surface temperatures seems most likely as a result of doubling the CO_2 concentration above preindustrial levels. The range of most-probable temperatures varies from 2°C to 4.7°C (3.6°F to 8.46°F). Furthermore, the model simulations suggest that a 4.7°C (8.46°F) rise will be difficult to exceed as a result of carbon dioxide levels doubling.

This range of ECS—2°C to 4.7°C (3.6°F to 8.46°F)—centered on 3.1°C (5.6°F) agrees well with a comprehensive review of research on this topic that was published in 2008.[139] Figure 7.14 summarizes estimates of ECS using various approaches to the problem, all of which analyze how sensitive the mean near-surface temperature has been under various conditions: the modern instrumental period, the mean-climate state, global-climate modeling, paleoclimate patterns, perturbations by volcanic eruptions, expert judgment, and combinations of evidence. These various methodologies generally indicate that ECS falls within the range of 2°C to 4.5°C, but they leave open the possibility of lower and higher values.

IN CLOSING

The body of research that defines climate sensitivity is thorough and creative, and it generally provides a robust estimate that global near-surface temperature can reach or exceed 3°C (5.4°F) when atmospheric carbon dioxide concentration rises to 580 ppm sometime in the second half of this century. This level of warming could lead to a number of negative effects on human societies and the ecosystem, including drought, dangerous weather, accelerated sea-level rise, water stress, and widespread environmental damage. Studies show that the world is already committed to further warming, even if all emissions were to stop now.[140]

To provide some context to the question of how high global mean temperatures could rise this century and what the consequences may be, researchers at NASA[141] compared the present climate to the paleoclimate record of the Eemian. The most recent period of time marked by interglacial conditions similar to our own

[137]A. Schmittner, N. Urban, J. Shakun, et al., "Climate Sensitivity Estimated from Temperature Reconstructions of the Last Glacial Maximum," *Science* 334, no. 6061 (2011): 1385–1388, doi: 10.1126/science.1203513.

[138]A. Schmittner, N. Urban, J. Shakun, et al., "Climate Sensitivity Estimated from Temperature Reconstructions of the Last Glacial Maximum."

[139]R. Knuttie and G.C. Hegerl, "The Equilibrium Sensitivity of the Earth's Temperature to Radiation Changes," *Nature Geoscience* 1 (2008): 735–743, doi: 10.1038/ngeo337.

[140]K. Armour and G. Roe, "Climate Commitment in an Uncertain World," *Geophysical Research Letters* 38, no. 1 (2011), doi: 10.1029/2010GL045850.

[141]See "Secrets from the Past Point to Rapid Climate Change in the Future," http://climate.nasa.gov/news/index.cfm?FuseAction=ShowNews&NewsID=649 (accessed July 12, 2012).

Figure 7.14. Estimates of equilibrium climate sensitivity (ECS) from various published studies summarized in Knuttie and Hegerl (2008). The graph shows most likely values (circles), likely ranges (boxes, more than 66% probability), and very likely ranges (lines, more than 90% probability). The IPCC (AR4) likely range (vertical blue band) and most likely value (vertical black line) are shown. These temperatures are typically authors' best estimates of ECS, based on various lines of analysis. Individual values are typically uncertain by 0.5°C. Dashed lines indicate upper estimates that lack strong evidence.

SOURCE: Originally from R. Knuttie and G.C. Hegerl, "The Equilibrium Sensitivity of the Earth's Temperature to Radiation Changes," *Nature Geoscience* 1 (2008): 735–743; design follows SkepticalScience.com.

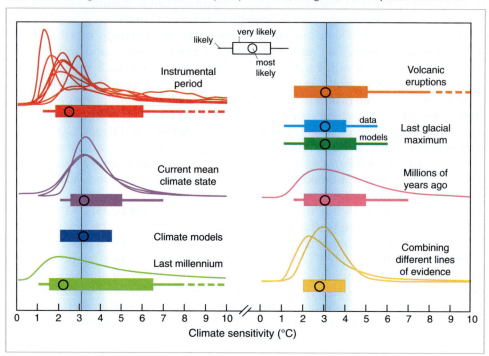

(see Chapters 3 and 5), the Eemian began about 130,000 years ago and lasted for 15,000 years. Eemian temperatures were less than 1°C (1.8°F) warmer than today. Currently, Earth's global-mean temperature is warming at 0.170°C to 0.175°C (0.31°F to 0.32°F) per decade,[142] and studies of the equilibrium climate sensitivity suggest additional warming could reach or exceed 3°C (5.4°F) when atmospheric carbon dioxide concentration rises to 580 ppm sometime before 2100. Thus, global mean temperature is on track to far exceed Eemian conditions in this century. During the Eemian, sea levels were 6 to 9 m (20 to 30 ft) higher than present—a prescription for disaster should the ocean respond the same way today. In the words of Jim Hansen of NASA, "We don't have a substantial cushion between today's climate and dangerous warming. Earth is poised to experience strong amplifying feedbacks in response to moderate additional global warming."[143]

Under the auspices of the United Nations Framework Convention on Climate Change[144] there are ongoing efforts within and between governments to come to an

---

[142]G. Foster and S. Rahmstorf, "Global Temperature Evolution 1979–2010," *Environmental Research Letters* 6 (2011): 044022, doi: 10.1088/1748-9326/6/4/044022.

[143]J. Hansen, M. Sato, P. Kharecha, et al., "Target Atmospheric CO₂: Where Should Humanity Aim?" *Open Atmospheric Science Journal*, 2 (2008): 217–231, doi: 10.2174/1874282300802010217.

[144]See the United Nations Framework Convention on Climate Change, http://unfccc.int/2860.php (accessed July 12, 2012).

agreement with regard to lowering future greenhouse-gas emissions in order to limit the amount of damage to the planet. Climate-change negotiations occur every year or so. These are events where international talks focus on defining global production of carbon dioxide and other heat-trapping gases. Often, there is disagreement between nations on the levels of sacrifice that should be apportioned among various economies. Some developing countries argue that most of the world's greenhouse-gas emissions are produced by a few industrial nations whose quality of life generally has benefited from their rapid growth fueled by fossil energy, and they ask why developing countries should not aspire to the same benefits. Under guidance by the U.N., countries are slowly rectifying these differences of opinion; however, even if significant emission thresholds are set in the near term, greenhouse-gas production cannot stop on a dime. Thus, even the strongest practical steps at limiting warming are going to take some time to be implemented and additional time before a response from the climate system is witnessed.

Figure 7.15 surfaced following the 2011 Durban, South Africa, climate negotiations. It shows that even if a new, binding climate treaty comes into effect by 2020, the pledged emissions cuts would still put the world on course to 3.5°C (6.3°F) warming, dramatic sea-level rise, and the spread of drought.

It is already too late to stop significant warming from occurring, but as the world watches ongoing climate negotiations, the question in many scientists' minds is, "Can we act in time to avoid the most dangerous aspects of climate change?"[145]

Figure 7.15. Emissions cuts voluntarily pledged by nations that are implemented by 2020 will commit the world to 3.5°C (6.3°F) warming, dramatic sea-level rise, and the spread of drought.
Source: J. Tollefson, "Durban Maps Path to Climate Treaty," *Nature* 480 (2011): 299–300.

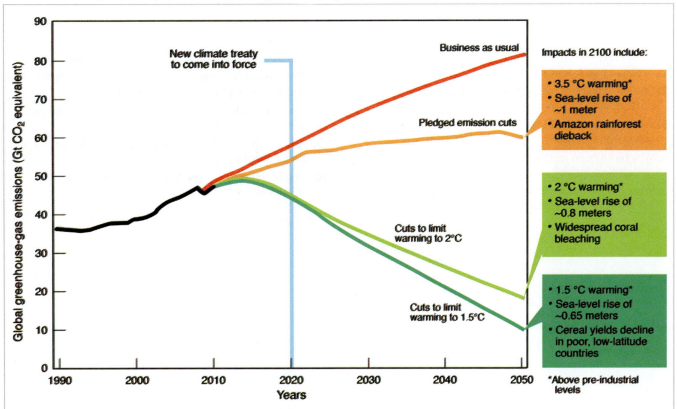

---

[145]See the talk "Global Warming: The Full Picture" at the end of this chapter.

Evangelical Christian
midate," http://www.skepti-
much-of-this-is-intended-

n dioxide from 800,000
://www.esrl.noaa.gov/
marvelous animation by
centration of CO_2 in the
in measurement stations

NASA, "Flow of Ice Across Antarctica," http://www.jpl.nasa.
gov/video/index.cfm?id=1015

"Witness a Glacier's Staggering Seven-Year Retreat," http://
io9.com/5905656/witness-a-glaciers-staggering-seven+year-
retreat

"Global Warming: The Full Picture," http://www.skepti-
calscience.com/Public-talk-Global-Warming-The-Full-
Picture.html

COMPREHENSION QUESTIONS

1. Describe the Berkeley-based BEST study and explain why it is significant.

2. What are the primary natural processes that contributed to global temperature rise in the past few decades?

3. Describe the trend in greenhouse gas emissions in recent years.

4. Which nations are the largest carbon dioxide contributors?

5. What effect is warming having on the Arctic?

6. Are global warming and the weather related? How?

7. What is drought and why is global warming causing it to change?

8. List the ways urban areas are vulnerable to climate change.

9. How is climate change affecting the world's ecosystems?

10. Describe equilibrium climate sensitivity.

THINKING CRITICALLY

1. Where is up-to-date information on climate change available that is reliable?

2. Satellite studies of global temperature have been controversial. Why?

3. What role does the El Niño Southern Oscillation play in the year-to-year climate?

4. Why has it been so difficult to decrease the production of greenhouse gases?

5. What are the causes of Arctic amplification? Describe why it is a global concern and not just a regional issue.

6. Describe the stability of the Antarctic ice sheet.

7. Extreme weather has increased in frequency. Why? How is global warming tied to extreme weather?

8. Describe the types of weather changes caused by global warming.

9. How are the world's ecosystems changing as a result of global warming?

10. Why is improving understanding of climate sensitivity relevant to controlling future climate change?

CLASS ACTIVITIES (FACE TO FACE OR ONLINE)

ACTIVITIES

1. Visit the Intergovernmental Panel on Climate change website http://www.ipcc.ch/ and answer the following questions.

  a. How is the IPCC organized and what is its purpose?

  b. Describe the Fifth Assessment Report and what its purpose is.

  c. What are the major conclusions of AR5?

2. Research climate change in your state.

  a. What agency is in charge of tracking climate change?

  b. What conclusions have they reached about threats related to climate change in your state?

  c. What is being done about these threats?

3. View the animation "Time History of Atmospheric Carbon Dioxide from 800,000 years ago until January 2009" at http://www.esrl.noaa.gov/gmd/ccgg/trends/history.html and answer the following questions.

  a. Describe what you learned as you watched this animation.

  b. Watch it a second time; what did you learn this time?

  c. Has it been valuable to spend taxpayers dollars to track global carbon dioxide concentration? Why?

  d. What have we learned that is valuable from this effort?

INDEX